THE WRITING MACHINE

The Edison Mimeograph Typewriter, Model One. (AC)

THE WRITING MACHINE

BY

MICHAEL H. ADLER

London

GEORGE ALLEN & UNWIN LTD

RUSKIN HOUSE   MUSEUM STREET

First published in 1973

Filmset and printed in Great Britain by
BAS Printers Limited, Wallop, Hampshire

*To my children
David and Sharon*

Acknowledgements

The long list of people to whom I am indebted for their assistance in the preparation of this book must begin with three who proved truly indispensable. In Paris, Barbara Spadaccini uncomplainingly devoted hundreds of hours of her time to research in stuffy dusty offices and I am as grateful to her for the information she unearthed (often literally) as I am for her patience in dealing with that peculiarly French breed of stuffy, dusty bureaucrat found in such offices. In London, there was Melvin Harris to whose encyclopedic knowledge I continually referred. He helped locate esoteric facts during the formative stages and eventually read the completed manuscript, making many valuable suggestions and corrections. He also ferreted out some of the machines in my collection, and likes to be remembered above all as the rescuer of the Kamm Zerograph. And in New York, Paul Lippman played an equally vital role. He read and corrected the manuscript, having previously supplied many details about American machines, from his own superb collection as well as from public and other private sources. He probably knows as much as any man alive about old typewriters—he, and his compatriot Dave Golden, who has assembled what is undoubtedly the world's largest private collection, including many unique pieces, and who gave freely of his time and knowledge during the years when the book was being compiled. Both these men supplied photographs and Dave was kind enough to let me use the Densmore ledger in his possession.

Mr W. E. Church of the Science Museum, London, was the most helpful of the many museum officials with whom I dealt, and I am grateful to him for his assistance on the many occasions I was obliged to bother him. I am similarly indebted to Mr K. O. B. Jorgensen of the Danish Technical Museum who helped supply the facts about his illustrious compatriot Malling Hansen, and to Mrs Kirsten Boideff who moved half a continent single-handed to get me the previously unpublished photographs of Malling Hansen's Writing Ball; to Mr C. W. Garland for assistance with machines for the blind; to Don Sutherland who snapped his mighty shutter on some unforgetable images (including that of the all-but-extinct Horton); to Dott. Gino Badini who helped with the Fantoni-Turri correspondence in the State Archives, Reggio Emilia; to Fr Ferrerius for details of machines in his museum; to Ursula Schulz, Leon Smith, Ed Quiring, Roberto Cabot, Léonie Heuer, Claude Fleury, Wilf Beeching, Robin Wyatt, Anne Smith, Julie Peverett, and to the many more whose response to cries for help was immediate and spontaneous. And especially to my good friend, the artist Maurizio Cruciani, who supplied all the drawings.

And last but, of course, not least, to my wife Linda for her tangible help in correcting and re-typing successive drafts of the manuscript, and for her intangible but more important contribution of patience, tolerance, encouragement—and coffee.

9

Contents

Illustrations

ILLUSTRATIONS

Introduction

Years ago, while rummaging through the local flea-market, I came across a peculiar little machine with the name FROLIO on it. I took it home, cleaned it up, and was amazed to find that it worked quite well. But neither I nor anyone to whom I appealed knew exactly what it was—the general consensus was that it was an old toy, and not even all that old, but nothing more concrete could be located. The further I went in trying to get information on it, the more intrigued I became. And, as fate would have it, I was no nearer an answer when a LAMBERT showed up . . . and I was hooked!

Pretty soon, I was to make many more discoveries, and not all of them were mechanical. First of all, I came to realise that ignorance about one of the world's most common appliances was ubiquitous. And then it slowly dawned on me that I had become the source of ridicule to all those to whom I was foolish enough to confess that I had begun collecting old typewriters. The image immediately conjured up by the confession was that of a wild-eyed beast surrounded by heaps of twisted, rusty uprights of Second World War vintage. No one had any idea what old typewriters are really like.

Things have changed quite a lot since those days, and typewriter collectors seem to have sprung up all over the place. But although the passage of time made it all the more respectable, knowledge of the subject did not become any more general. I could walk into a bookshop and buy a learnéd tome on just about anything from buttons to walking-sticks, matchboxes to brass rubbings. But to find any information on typewriters required days spent in museums, libraries, patent offices, state archives, second-hand bookshops selling old magazines, and so on.

Gradually, my accumulation of data grew, only to reveal that even the few people who had bothered to write about the subject years ago were hopelessly inaccurate.

And out of this came my eventual determination to write a comprehensive and authentic book on the history of the writing machine, once and for all! In one respect I was fortunate, for I happen to be fluent in almost all the languages in which early material was written: French, Italian, Spanish, German and, of course, English. I needed some help only with Danish and Latin. I was also fortunate in having friends and fellow-collectors around the world without whose assistance I could not possibly have compiled the necessary facts. For the proliferation of conflicting information in the relatively short bibliography on this subject is quite staggering. It is the product of ignorance, incomplete research, careless reporting and perverse national and commercial interests. Everything had to be checked—where the original source could still be located, that is.

Collectors made two worthy suggestions, both of which I tried. The first was to include every scrap of conflicting information from whatever source and attribute it to its origins. I abandoned this approach because of the sheer volume of useless data I was accumulating, for once I had succeeded in unearthing the correct fact there seemed little point in listing the incorrect alternatives. Having established, for instance, that the Frenchman was definitely called Progin made it unnecessary to include Porgin, Progrin, Pogrin or Projean, with a little footnote for each. The second suggestion was that I merely asterisk every disputed piece of information, and I started doing this, too, but abandoned it after finding that each page of manuscript had more stars than the Milky Way.

The more logical approach, and the one I eventually adopted, was to list the authentic facts and attribute to their source only those which I had not been able to confirm. Index numbers in the body of the text refer to the bibliographical origins of information of this type.

In the chapters dealing with historical development, I have opted for a strictly chronological order in preference to one which deals with inventions according to the functions for which they were designed. From the point of view of mechanical development as a whole, it is immaterial whether an early instrument was designed primarily as a shorthand machine, or as one for the blind, or as a printing telegraph. The design and its historical contribution are the important factors, for they influenced later inventors of all kinds of machines. Furthermore, the distinction between the various classifications is difficult to determine in the case of many early patents, because of the common practice among inventors of striving to protect as wide an area as possible. Where, for example, would one classify the application to telegraphy of a shorthand machine for the blind?

The other more or less arbitrary decision I made when planning this book was to limit its scope solely to 'unconventional' typewriters. There are literally hundreds and hundreds of makes and models of 'conventional' front-stroke, type-bar machines with four-row keyboards. Frankly, they are of little interest, and despite the extravagant claims of rival manufacturers, they are all virtually the same. It is the unconventional ones that are interesting; one day, someone will write a history of standardization and mass production in the twentieth century and list the others.

With unconventional machines, however, dating the various makes and models was a prolonged nightmare. Some manufacturers no longer exist; what few records were kept by others are usually inaccurate or incomplete, and the machines themselves rarely reveal these secrets. Sometimes, of course, they bear plates or transfers with patent information but, more often than not, these details are concealed beneath such blanket statements as 'Patents Pending' or 'Patents Applied For'. Furthermore, patent dates do not necessarily correspond to those of production or marketing, nor is a patent necessarily filed and granted in the same year. Despite all this, I have resorted to patents wherever possible, since these contain the only consistently reliable facts available. In those instances in which patents cannot be located, dates sometimes have to be supplied according to advertising information, exhibition awards or other such devious means. In these cases, the best that one can sometimes hope for is to quote the date 'give or take a year'.

Early typewriter historians invariably tried to make the machine sound like the most important invention of all, as if they felt obliged to justify the time they had spent chronicling its growth. Of course it was not the most important, nor anywhere near it; compared to some of its illustrious siblings—telegraph, telephone, phonograph, wireless, electric light, flight—to name but a few—it occupies a very humble place indeed. It has been credited with bringing about the emancipation of women, but this is not so, for the so-called Industrial Revolution was responsible for that phenomenon and the typewriter was merely one of the countless facets of that revolution. It also caused profound changes in business pace and methods, we are told, but in fact it was the transformation of the means of production which caused these changes, not the typewriter itself. What is undeniably true, however, is that the development of the typewriter, that ubiquitous gadget, provides us with an unparalleled insight into the inventiveness and ingenuity of man, for no other of his inventions can boast such a versatile and fascinating mechanical history.

In 1973, many Americans will be celebrating the centenary of the invention of the typewriter. That is, of course, their privilege. After all, the Italians celebrated the same event in 1955, the Austrians in 1964, and so on. I am all for having a party, myself, so I thoroughly approve of them all. It is not the centenary of the typewriter they are commemorating, of course, but what does it matter? I hope they have a good time, regardless.

Bibliography

BOOKS
1. Aliprandi, G. *Giuseppe Ravizza, Inventore della Macchina da Scrivere*. Novara, 1931
2. Aliprandi, G. *Cenni Storici della Macchina da Scrivere*. Padova, 1938
3. Aliprandi, G. *Giuseppe Ravizza Attraverso le Pagine del Suo Diario*. Novara, 1942
4. Baculo, L. *Manuale di Stenographie*. Napoli, 1849
5. Bliven, B. *The Wonderful Writing Machine*. New York, 1954
6. Budan, E. *Le Macchine da Scrivere dal 1714 al 1900*. Milano, 1902
7. Caizzi, B. *Gli Olivetti*. Torino, 1962
8. Chapuis, A. and Droz, E. *Les Automates*. Neuchatel, 1949
9. Chapuis, A. and Droz, E. *Les Automates des Jaquet-Droz*. Neuchatel, 1951
10. Chapuis, A. and Gelis, E. *Le Monde des Automates*. Neuchatel, 1928
11. Cooke, Thomas Fothergill. *Authorship of the Practical Electric Telegraph of Great Britain*. London, 1868
12. Current, R. *The Typewriter and the Men who made it*. Illinois, 1954
13. Denman, R. P. Q. *Electrical Communication*. London, 1926
14. Dupont, H. and Canet, L. F. *Les Machines à Ecrire*. Paris, 1901
15. Dupont, H. and Sénéchal, G. *Les Machines à Ecrire . . . Leur Evolution*. Limoges-Paris, 1906
16. Dupont, H. and Sénéchal, G. *Les Machines à Sténographier*. Limoges-Paris, 1907
17. Dyer, F. L. and Martin, T. C. *Edison, his Life and Inventions*. New York, 1910
18. Finlaison, J. *Some Remarkable Applications of the Electric Fluid . . . etc.* London, 1843
19. Freebody, J. W. *Telegraphy*. London, 1959
20. Granichstaedten-Czerva, R. *Peter Mitterhofer, Erfinder der Schreibmaschine*. Vienna, 1923
21. Herkimer County Historical Society. *The Story of the Typewriter, 1873–1923*. New York, 1923
22. Herrl, G. *The Carl P. Dietz Collection of Typewriters*. Milwaukee, 1965
23. Jones, C. LeRoy. *Typewriters Unlimited. History of the Typewriter*. Missouri, 1956

24. Kahn, D. *The Codebreakers*. New York, 1967
25. Karrass, Th. *Telegraphen- und Fernsprech-Technik*. Berlin, 1909
26. Krcal, R. *1864–1964 Peter Mitterhofer und Seine Schreibmaschine*. Aachen, 1964
27. Mares, G. *The History of the Typewriter*. London, 1909
28. Martin, E. *Die Schreibmaschine und ihre Entwicklungsgeschichte*. Pappenheim, 1949
29. Morelli, D. *La Storia della Macchina per Scrivere*. Brescia, 1956
30. Müller, F. *Schreibmaschinen*. Berlin, 1900
31. Oden, C. V. *Evolution of the Typewriter*. 1917
32. Quaife, M. *Henry Roby's Story of the Invention of the Typewriter*. Wisconsin, 1925
33. Rees, A. *Cyclopaedia*. London, 1819–20
34. Roblin, J. *Louis Braille*. London
35. Rochefort-Luçay, M. D. *Les Machines à Ecrire*. Paris, 1896
36. Rousset, J. *Les Machines à Ecrire*. Paris, 1911
37. Sénéchal, G. *Description des Machines à Ecrire Françaises*. Paris, 1903
38. Shaffner, T. P. *The Telegraph Manual*. New York, 1859
39. Spencer, H. *Autobiography*. London, 1904
40. Tilghman Richards, G. *The History and Development of Typewriters*. London, 1964
41. Turnbull, L. *The Electro-Magnetic Telegraph*. Philadelphia, 1853
42. Typewriter Topics. *History of the Typewriter*. Reprinted by Metropolitan Typewriter Co., Michigan, 1923
43. Weller, C. *The Early History of the Typewriter*. Indiana, 1921
44. Zetzsche, K. E. *Geschichte der Elektrischen Telegraphie*. Berlin, 1877

PERIODICALS AND OTHER PUBLICATIONS
45. *Berlingske Tidende*. 1878
46. *Bollettino della Accademia Italiana di Stenografia*. 1926, 1930, 1931, 1936
47. British Association for the Advancement of Science, York Meeting, 1844
48. *Bulletin de la Société d'Encouragement pour L'Industrie etc.* 1831, 1843, 1850
49. *Design and Work*. 1877
50. *Dinglers Polytechnisches Journal*. 1822, 1872, 1881, 1883, 1887, 1888
51. *English Mechanic and Mirror of Science*. 1866
52. Exhibition of the Works of Industry of All Nations. Official Catalogue and Reports by the Juries. London, 1851
53. *Frankfurter Ober-Postamts-Zeitung*. 1831
54. Hubert, P. G. *The Typewriter, its Growth and Uses*. 1888
55. *Illustrirte Zeitung*. 1872
56. *Inland Printer*. 1886
57. International Congress on Technology and Blindness, Proceedings of New York, 1963

58. International Exhibition 1862, London. Reports by the Juries
59. Jenkins, H. C. Cantor Lectures on Typewriting Machines. *Journal of the Society of Arts*, 1894
60. *La Lettura*. 1908. Antiche Lettere Scritte a Macchina, Umberto Dallari
61. *Magazin Aller Neuen Erfindungen. 1808*
62. Mathews, W. S. B. The Writing Machine. *Northwestern Christian Advocate*, 1891
63. *Mechanics Magazine*. 1832
64. Milwaukee County Historical Society Dedication Pamphlet. 1956
65. Museo Nazionale della Scienza e della Tecnica. Commemorative Exhibition Publication. Milano, 1955
66. Reports of the Paris Exhibition, 1867
67. *Scientific American*. 1867, 1887, 1903

In addition, advertising publications by IBM, Olivetti, Remington, Royal and Underwood were consulted. None of them offered new information of sufficient authenticity to warrant separate inclusion in the bibliography.

Chapter 1

On Writing Machines in General

By and large, today's typewriter client is faced with the difficult task of choosing between a vast number of virtually identical machines. All of them have keyboards, with four rows of keys, and all of them sport the same inefficient QWERTY letter order on the keyboard. With one sole exception, all have type-bars which strike at a common visible printing point, the type-bars lying horizontally or obliquely in front of the platen. Virtually the only decision to be made is whether to buy an electric or a manual machine.

The exception mentioned above is the 'revolutionary' IBM type-wheel electric, the one on which the little ball rotates and moves along electrically while the paper stays still. Thomas Edison patented a type-wheel machine with similar specifications exactly a hundred years ago, a piece of information which the present manufacturer appears reluctant to publicize. Edison was not the first to apply electricity to typewriters, however, for this was done as far back as 1854,

1. A 'newcomer' to the class of machines that used a type-wheel was the PORTABLE EXTRA (Helios Klimax), introduced in 1914.   (AC)

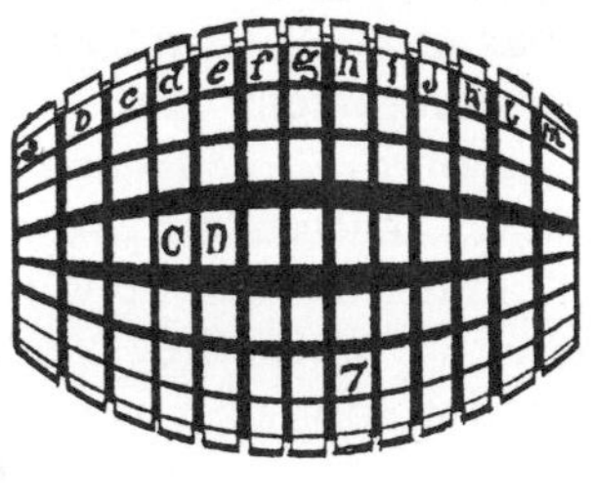

2. The 'golf ball' was precocious indeed, having acquired a modern profile as early as 1884, as H. B. Richardson's patent reveals.   (USP)

and even earlier, while the type-wheel design has been used on innumerable occasions from the very earliest days of typewriter development. And the first machine with moving type font and stationary paper appeared in 1823.

Our great grandfathers, then, were faced with the same alternatives of electric versus manual machines and type-bars versus type-wheels as we are today. In addition to this, they were confronted with a stunning variety of machines and mechanical principles which history, for better or for worse, eventually discarded. There were literally hundreds of these, and claims and counter-claims were hurled at the prospective buyer in bewildering profusion. In the absence of standardization of any kind, the client would have a difficult time understanding even the jargon, and if he ordered a typewriter for his office he might easily open the door to a liberated young lady toting an enormous box. *She* was the 'typewriter' and the onus was on her to provide her own 'type writer'. Employers very often married their typewriters, according to *How to Get Married* (1895) and one of the better of many later vaudeville jokes went: 'I saw you and your queen walking down Main Street last night.' 'Oh, that was not my queen. That was my typewriter.'[5] Of course, in those days the instrument was eulogized as the single most important factor in the emancipation of women. Today, our women's lib movement considers the machine the worst instrument of torture ever devised by man for the perpetual enslavement of woman, and anyone who believes otherwise is dismissed as a male chauvinist pig! One generation's meat is another generation's poison.

Confusion enveloped every aspect of the writing machine's early history. Once it got off the ground, it developed too quickly. There was no time to chronicle its growth. One early writer[27] complained in the preface that his manuscript was already obsolete by the time it reached the printer: he was obliged to postpone publication for two years (till 1909) to handle the flood of new machines on the market. A contemporary of his[15] estimated that no less than 750 typewriter patents had been granted in the world during the twelve years from 1880–92, to say nothing of before or after. There were probably more: by the author's count, the number granted in the year 1891 in the United States alone was 149, plus others for components and accessories. By 1905, the total issued in the United States is reported to have swollen to 2,678.[42] Many machines protected by these patents never materialized; others appeared in a flash of publicity which burnt itself out almost immediately, leaving barely a trace. Records were often not kept and, where they were, they were later lost or destroyed. The few who wrote of the machine's development were often inaccurate and their information was incomplete; their errors were blindly repeated in later generations by those who studied the typewriter's early days. Few people took the trouble to authenticate the facts that appear in those secondary sources. After all, why bother? Who cares? Who can tell the difference anyway? Or, as a museum official confessed when confronted with a basic error his predecessor had made: 'There just wasn't time to check, I suppose.' *Wasn't time to check . . .*

All of which led to a kind of happy historical anarchy in which general ignorance was interpreted as licence for perverting the facts to suit any purpose. The history of the machine was exploited at will and a kind of inverse alchemy has long ruled the subject: base conclusions are regularly drawn from otherwise perfectly valid information, like turning gold into lead, or, a little closer to home, like the British patent granted to a man called Holman in 1688 'for a new art or invenčon of makeing a certaine powder which, being put into faire water, beer, ale or wine, doth immediately turne the same into very good black writing ink.' One suspects he might have been better remembered by history had he invented a 'certaine powder' for turning black writing ink into very good ale.

The heritage handed down to us is about as confusing, then, as the array of machines our grandfathers were obliged to consider. Take the basic question of just who invented the typewriter, for instance: surely it sets a world record for rival claims!

According to Dr Richard Current,[64] it was the typewriter and not beer that made his town famous. 'The typewriter originated in Milwaukee', he stated categorically. It is clearly unnecessary to mention that the man was not only an American but also a keen student of Wisconsin history—as unnecessary as to spell out Dr Giuseppe Aliprandi's nationality. 'The Italian invention emigrated and then returned to Italy "Made in U.S.A."' he fumed.[1] To Giuseppe Ravizza goes the honour of 'absolute priority in the invention of the typewriter'.

I beg your pardon, say the English, but 'all typewriter history commences with a British patent granted to a Henry Mill.'[40]

Ach so? Well, Dr Granichstaedten-Czerva has some news for you.[20] 'Mitterhofer was not the inventor of *a* typewriter, but the inventor of *the* typewriter'—thereby staking Austria's claim.

To which the Frenchman, Sénéchal, responds indignantly: 'C'est à un Français, M. Progin, que doit revenir l'honneur . . .'[37]

The Germans have a harder time, but they try nevertheless. Some have championed the cause of a Baron Drais, rather unconvincingly, but when all else has failed, Ernst Martin[28] offers the following astonishing and unsubstantiated speculation: Germany 'probably' invented the writing machine but no one will ever know because patents were not granted then and no records were kept!

And so it goes on. The author was hardly surprised, then, when he took a shot in the dark and phoned the Cultural Attaché of the Russian Embassy, to be informed that the typewriter was, in fact, the invention of a comrade called Alissoff.

The myths proliferate, fed by patriotism, partisan interests, and partial ignorance. The most arbitrary measures are applied to the facts in order to make them fit the predetermined pattern, and the reader is expected to follow the prestidigitatious intellectual fantasy of the so-called historian he is studying. In truth, about the only thing these men (and others) share, apart from prejudice, is the common misconception not only that the typewriter is the most important single development of all time, but also that it is of the most staggering mechanical complexity. Both of these assumptions are sheer nonsense. It is as if early writers on the subject felt compelled to inflate their material beyond recognition in order to justify their efforts in recording its history.

The effective origins of the typewriter, like those of a multitude of other machines and devices, date back to that exciting fertile period in history loosely referred to as the Industrial Revolution. The word 'effective' is critical in this context: in fact, the typewriter was invented long before it eventually became a household word. But in the field of mechanical invention it is as useless to be before one's time as to be after it, and society must be ready for an innovation before it will accept and incorporate it. Herein lies the sole reason for the late arrival of the writing machine, and it is time to explode the myth that man lacked the mechanical knowledge and ability to develop it earlier.

In fact, from a mechanical point of view, there is no reason why a writing machine could not have been built successfully in the fourteenth century, or even earlier. A cursory examination of the history of horology should be sufficient to substantiate this, to say nothing of the development of the many engines of war and peace which date back beyond the Renaissance.

But a quick glance at clocks and watches will suffice to establish a little perspective on our subject. No one knows exactly when clocks were first invented, but their origins certainly go back to the thirteenth century or even earlier. In those days, clocks and watches were made to order. You told your local clockmaker what you expected your time-piece to tell you and he made it to your specifications. He would have designed and built you a writing machine, too, if you had ordered it, and if you had been prepared to pay his price, for the craftsman's products were expensive, the more so since he might work for years on a single

3. 'Employers very often married their typewriters.' A contemporary Bar-Lock advertisement. (AC)

4. The craftsman who made this late sixteenth-century watch (ex-Webster Collection) would have had no trouble building a writing machine, if a client or patron had ordered it and been prepared to pay the price. (AC)

complicated item. The widespread misconception that artisans worked for nothing in those days is totally unfounded.

The degree of skill required—and indeed the *number* of skills required—by the early clockmakers defies the comprehension of modern specialized man. He was first of all a master toolmaker, for each man made the tools with which he worked. He then needed to learn how to use them, for wheels were cut by hand and teeth had to mesh perfectly. He was a metallurgist expert in brass and steel, making his own alloys and bringing parts to the necessary degree of hardness and temper, blueing and burnishing as required. Astronomy held no secrets from him, and his mathematics and geometry were on a par with the best of his day. The result was that horology developed simultaneously in all directions: by the six-teenth century, miniaturization had been perfected and watches as small as a fingernail are recorded, while larger clocks offering all manner of astrological data and automatons were already common two centuries earlier. In fact, there were already hundreds of makers throughout the world accomplishing these feats by the sixteenth century, to say nothing of such masterpieces as the Prague clock which latest research dates at 1410: here, apart from astronomical and sideral information, at each hour a skeleton nods his head, other automatons shake theirs, bells are struck, the twelve apostles file past, each one briefly facing the crowd as it moves, a cock crows, and so on.

One example will suffice to give a clear insight into the degree to which mechanical skills had developed in the so-called Dark Ages: in the middle of the fourteenth century Giovanni de Dondi built an astronomical clock which not only showed

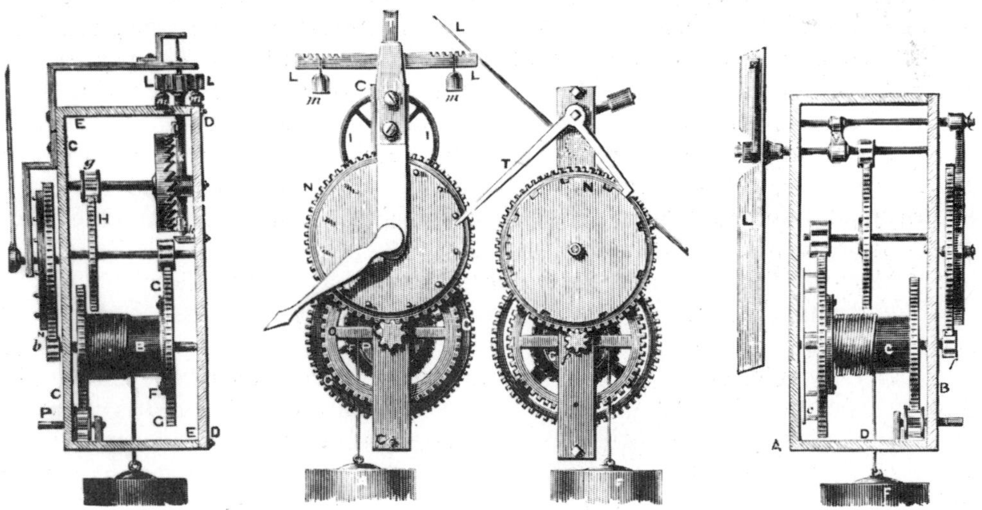

5. '...a common clock of which many examples existed'. De Vick's fourteenth-century time-piece, showing views of the going and striking trains (Britten, F. J.: *Old Clocks and Watches*, 1st Edition). (AC)

the time of day but also had separate dials and indications for each of the following: the fixed feast-days of the church, the movable feast-days, the nodes (i.e. the points of intersection of the orbits of sun and moon), the movements of sun, of moon, and of the five (known) planets, etc. The mere maths required to calculate the teeth on intermeshing wheels and pinions is staggering. In the contemporary manuscript describing the piece, de Dondi casually skipped over all but the astronomical part, stating that the rest was nothing but a common clock of which many examples existed.

Add to all this the fact that these machines suffered the severe limitation not merely that they had to work, but that they had to work at a very specific speed. And lest one be tempted to believe that these skills were the exclusive prerogative of a mere handful of men, Baillie's classic but incomplete record of clock- and watch-makers up to the year 1820 lists over 35,000 names.

By and large, all this happened because the demand for time-pieces was so great. Not only has man always been concerned with the passage of time, but also watches, then as now, were largely status symbols. They were worn exposed, hung around the neck or suspended at the waist, from a chatelaine pierced and engraved and highly decorative and destined to become ever more so as the art of the enameller and jeweller became fashionable. Later, when accuracy improved, greater attention was placed on the movement which was finished with exquisite attention to detail.

The watchmaker's client paid dearly for these objects—the prices some of them fetched are staggering. Records survive of single payments of hundreds of pounds (sterling) for the mere repair of clocks and watches in the early 1600s; at the end of the century, for instance, Tompion made a spring-driven clock which required winding only once a year and for which he was paid £1,500. Thomas Mudge charged 480 guineas for a minute repeating clockwatch set in a walking stick around the middle of the eighteenth century and John Arnold received 500 guineas from George III for a half-quarter repeating watch set in a ring (the movement was one third of an inch across and consisted of 120 parts). The Empress of Russia offered him 1,000 guineas for an identical piece but he declined the offer. At about the same time, John Harrison won a £20,000 award for a marine chronometer which proved accurate to within less than half a second a day.

Any watchmaker could have built his client a writing machine, too, if he had ordered it and had been prepared to pay the price, but what use did he have for it? Clearly, it could serve no decorative purpose (although at least one nineteenth-century typewriter was designed to be worn around the neck!) and it could do nothing for his status in a society in which beautiful script was the fashion. Even assuming that the client were educated and literate enough to use it, why should he bother when he had scribes and secretaries to take care of these chores?

Later, in the eighteenth century, society's approach to machinery began to undergo a change. Man became fascinated by the machine in much the same way that he is fascinated today by the computer: as a human substitute. He began to take machinery more seriously and was forever amazed at its potential. Fantastic what it could do! We tend to laugh at the naïveté of the average man of those days, much as future teachers will entertain their pupils with stories of how people in the twentieth century were so fascinated with the computer that they allowed it to regulate their lives, predict their fortunes, choose their spouses, declare their wars, provide their chess partners, and so on. A similar fascination in the eighteenth century led to the construction of a number of writing automatons which were displayed amid universal amazement, and a little fear. These robots had human figures and were programmed to write a particular text, which they did exceedingly well, even to the point of inking their pens automatically, changing lines, spacing, and punctuating. Their spelling was beyond reproach! But even though these robots were immeasurably more complex than a typewriter ever need be (and, incidentally, they were almost all built by watchmakers), the typewriter was still rejected by a hostile society.

Close to a hundred inventors were doomed to build writing machines before one of them 'caught on', late last century. Almost all these machines worked badly, in varying degrees of badness. They form a fascinating testimonial to the lonely efforts of close to a hundred doctors, lawyers, teachers, priests, artistocrats, journalists, even some mechanics and engineers, with limited means, incomplete knowledge, insufficient experience, an adverse or at best indifferent society, but a common vision of the future. And yet, some of them produced superb instruments, even though they typed badly or slowly by modern standards. Often the principles were sound, but the workmanship was deficient. After all, what means did these men have at their disposal? Only one or two of them, unfortunately, were watchmakers.

And then, in the pompous words of the Herkimer Historical Society,[21] '*The hour for the typewriter had struck*' (sic). Their book then proceeds to hail the efforts of their hero. A more sober report might have taken the wider view of embracing the hundreds of makes and models which were to flood the market in the few decades following that fateful 'hour'. It was a mad scramble of band-wagon-jumping. Inventors performed amazing mechanical somersaults in order to come up with a patentable design, and promoters anxious for a share of the growing market sank their money into projects whose chances of success were at best questionable. It was a period fraught with commercial battles and litigation, as interested parties defended their patents (and, thereby, their right to exist) as ferociously as others attacked them.

This was a feature of the times and, on a smaller scale, interesting parallels can be found in virtually every other field. During the early days of the phonograph, for instance, the Gramophone Company fought a losing ten-year legal battle merely for the exclusive right to use the word 'gramophone' to describe a machine that played music recorded on circular discs. And even as minor a development as the introduction of the tapered tone-arm to replace the cylindrical one was the subject of such acute litigation that a court order was eventually obtained to sequester rival machines from an international exhibition.

It is interesting to note how often the histories of the phonograph and type-writer run into each other. Thomas Edison, for instance, invented a couple of writing machines, one of which was to form the basis for the Stock Exchange ticker tape and another (Frontispiece) was primarily designed for cutting stencils. Later, around the turn of the century, both the phonograph and the typewriter were fully in vogue but there was still some doubt as to whether one or the other might not prove to be merely a passing craze. To play it safe, the Gramophone Company changed its name to the Gramophone and Typewriter Company and produced both talking machines and writing machines. Regrettably for our history, their Lambert typewriter failed as miserably as their His Master's Voice label succeeded. Who, at the time, would ever have believed that a little terrier . . .?

Diversification, then, is not a modern industrial phenomenon. In early days you went to a bicycle shop to buy your phonograph and cylinders, and many indeed are the cycle manufacturers who produced typewriters and sewing machines, as did the gunsmiths—not only Remington, of course, but also the equally famous French firm Saint-Etienne.

For all the similarities between the typewriter and other machines of the time, it is still rather special. It is in a class of its own, not so much because it had singular features as because it combined a number of features in a single instrument. Not only was it simple and cheap; above all, it was almost infinitely versatile.

Its simplicity has already been discussed. Typewriters were built with as few as half a dozen moving parts. Not only did these work, they often worked as well (or as badly) as those with hundreds of components. They sold as many, or more. In fact, the writing machine was so simple that virtually anyone could stumble on a potentially patentable brainwave. And on more than one occasion, just about anyone did!

And it was cheap. In fact, compared with other machines it could be remarkably cheap, not only to purchase but also, of course, to manufacture. For a dollar you could not only buy a typewriter but you could even take your choice of several competing makes. You could also spend a hundred dollars or more on a machine, of course, but even that was not excessively expensive. What this meant was that the typewriter enjoyed a vast popular market and that it was economical enough to allow a large number of people to give serious consideration to setting up a plant and manufacturing.

But above all, its versatility! There are perhaps a limited number of ways in which an engine can combust internally, or in which a motor can transmit power to a wheel, or in which a noise can be recorded on a surface, or transmitted on a wire, or without a wire, but the number of ways whereby an impression can be made on a sheet of paper appears to be virtually infinite. Writing machines were devised in every known geometric shape, and the principles they employed were based on pushing, pressing, dropping, lifting, spinning, sliding, striking, hammering. Virtually any system would work.

Let us establish a few points of reference to assist us in forming a historical perspective of this era. We are dealing, roughly, with the decades immediately before and after the turn of the century. After the Writing Ball, the Remington

6. The LAMBERT and the top-wind 'dog model': two simultaneous products of the Gramophone and Typewriter Co. at around the turn of the century. (AC)

7. The WRITING BALL, invented by the greatest Dane of them all. It was manufactured from 1870 onwards and was the first commercially successful typewriter. (DTM)

had been on the market longest, first as the Sholes and Glidden Type Writer and eventually under its own name. Neither the Type Writer nor the Model 1, introduced in 1874 and 1878 respectively, were good machines or commercially successful. The Model 2, also introduced in 1878, eventually proved to be both, although it wrote 'non-visibly', meaning that the printing was not visible to the operator without the carriage being raised. The printing-point was underneath the platen and the keys struck upwards from inside the machine. Remington defended this design until well into the twentieth century in a losing battle against competitors producing better 'visible' models. The first machine of modern 'conventional' design, which was to become the prototype of the instrument as we know it today, was the Underwood No. 1 introduced after 1895. This, of course, is called 'being wise after the event', for at the time it was far from certain what the eventual profile of the typewriter was destined to be, and many manufacturers continued to produce machines of the wildest designs until as late as the 1930s. Everyone appears to have believed in his own product, or at least to have hoped for its eventual acceptance. Those who did not believe in it made it sound as if they did, so as not to pay royalties for improvements which others had patented.

Remington's continued promotion of its inferior 'blind' (i.e. non-visible) design was thus motivated.

Consider briefly the bewildering alternatives which had to be considered by the potential client. He would first have to decide whether he required a machine that typed both upper and lower case or whether upper case sufficed, in which event a Longini might be the machine for him. If he wanted both, he would have to decide whether he wanted a keyboard machine or one with a letter index, and if he chose the latter he might be faced with a contraption in which the letters were stamped on an arc (Edelmann), a semi-circle (Globe) or a complete circle (Velograph). The circle itself might be horizontally (Liliput), obliquely (Edland)

8. The simple LILIPUT (1907) used a circular index which was turned by the knob to select the characters and then depressed for printing. A little slow, perhaps, but it worked well. (TMT)

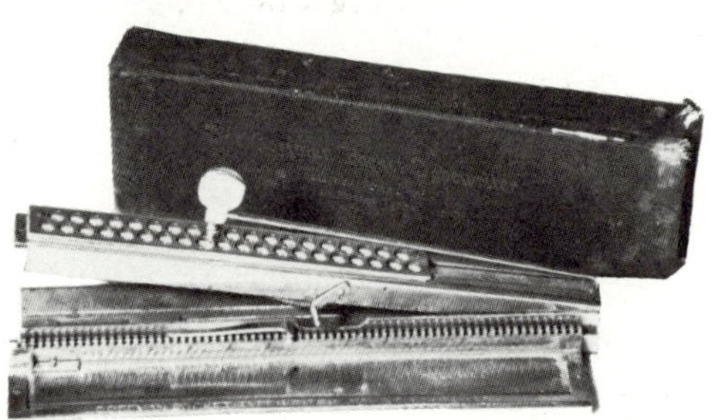

9. Good things come in small packages! The COFFMAN (1902) was designed to fit comfortably into any pocket, so long as it was ten inches deep. (GC)

or vertically (Niagara) disposed. But if he rejected the idea of a circular index, he could choose one that was square (Hall), rectangular (Mignon), eye-shaped (Eureka) or linear (Odell).

The type itself might be on a small wheel (Blickensderfer), or a large wheel (Columbia), a sleeve (Crandall), a small sector (Hammond), a large sector (World) or a straight line (Coffman). Or it could be at the end of a lever, whereupon our client might have a difficult time in assessing the conflicting claims of competing manufacturers who suspended the levers upside-down in a basket under the platen (Remington), or positioned them vertically in front (Salter), behind (Brooks) or to the sides of the platen (Oliver). He might prefer them obliquely at the rear (Fitch) or obliquely at the front (Ideal). If he wanted them horizontal, he could have them in front of the platen (Daugherty), or both front and rear (Williams). If he wanted a machine where a little hammer hit the paper against the type, rather than the type striking the paper, he might choose a Chicago. Or how about a machine that moved, keyboard and all, while the paper stayed still (Elliot Hatch)?

34

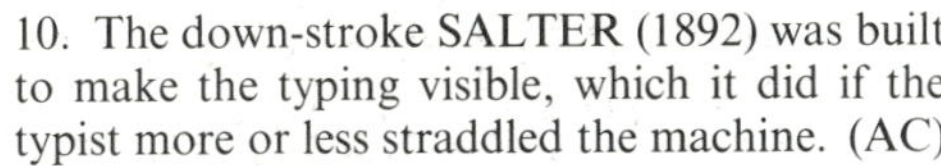

10. The down-stroke SALTER (1892) was built to make the typing visible, which it did if the typist more or less straddled the machine. (AC)

11. A little hammer flicking up from behind the CHICAGO (1898) struck the paper against the horizontal type-sleeve, with the ribbon between the two. (AC)

In fact, just how to dispose of the paper might present something of a problem. He might like the idea of rolling it up into a basket, whereupon he would have to decide whether he preferred to roll only the blank sheet (Hammond) or both that and the typed portion (North's). If he did not object to uncurling the paper after it had curled itself around the platen, he might consider the Lambert. He might prefer the platen to be vertical instead of horizontal (Hanson-Lee), or polygonal instead of cylindrical (Caligraph). Or perhaps he simply preferred the paper to lie flat (Fisher).

Of course he had already decided how large his paper, and hence his carriage, had to be. Most machines offered interchangeable lengths, but if he wanted to type on news-sheet or its equivalent, he had to buy an Electric Blick, which offered carriages up to a yard long. 'The convenience of this (36-inch carriage) is too great to need a single word of explanation' wrote a contemporary.[27] No family should be without one!

What about keys and keyboards? If the client wanted no keys at all, the American Visible was the machine for him, and he could not miss: printing was by depression of the entire letter index. If one key was acceptable, he might try a Boston; if two keys a Mignon; if three a Frolio, and he could play bingo with the number of keys until he reached a Duplex with 100 or even a Brackelsberg with 132. He could choose between any number of rows of keys from one (Saturn) or two (Helios) to eight (Yost) or more. Usually the rows were straight, but at times they were arcs (English) or complete circles (Daw and Tait).

Choosing the best available letter order might well prove a nightmare. Apart from a strictly alphabetical order, there was the so-called 'Universal' keyboard (QWERTYUIOP) and its chief rival the 'Ideal' arrangement (DHIATENSOR).

13. The entire 'keyboard' of the AMERICAN VISIBLE (1893) had to be pressed down. The type was on a rubber strip running parallel to the platen.   (AC)

12. NORTH'S (1892). One of the type-bars is shown striking down from the rear. On such designs, the paper had nowhere to go: it had to be coiled into the basket in front of the platen and it uncoiled itself into the one below as typing progressed.   (AC)

Manufacturers of index machines often preferred to have common combinations of letters within easy reach of each other; hence T H and E are next to each other on the American index; A N and D on the World; O F, A T, T H E, I N G on the Edison Mimeograph, and so on. But this is as close as one gets to seeing intelligence applied to keyboard arrangements, for, apart from the alphabetical, Universal and Ideal orders one finds ZQJBPFD (Hammond), ZPRCHMI (Crandall), KBFGNIA (Hall), XVGWSLZ (Morris), QCBLUIP (World), ZKPWMCR (Columbia), XZKGBVQ (Blickensderfer), CJPFUBL (American) and so on and so forth. Kind of made it tough to change makes after one had become accustomed to a particular machine!

Inking could present the next problem. 'Most modern machines use ribbons' announced the *Encyclopedia Britannica* of the day, although a Blickensderfer advertisement listed 'no ribbons' as one of the 'essential features of a first-class typewriter'. However, if our client nevertheless felt partial to them, he might be obliged to decide between ribbons all the way from a quarter of an inch wide to those over two inches wide. Or he could dispense with them altogether and take rollers (Junior) or pads (Yost). If he was still not satisfied he was in trouble, for printing through carbon paper went out a few years earlier.

Did he want a machine that strapped to his fore-arm? It had to be a Virotyp—there was no other. And he would have to choose the same make if he wanted to type while riding a horse, as he might well do if he happened to be a First World War correspondent attached to a cavalry unit—it was specifically designed for him. If he wanted one to carry in his pocket he had more choice (Trebla or Pocket)

14. The circular keyboard of the DAW and TAIT (1884) controlled a corresponding circle of type-bars hovering obliquely over the platen. (CSM)

15. The inventor of the AMERICAN (1893) obviously gave some thought to the problem of adapting the letter order to the characteristics of the design. (AC)

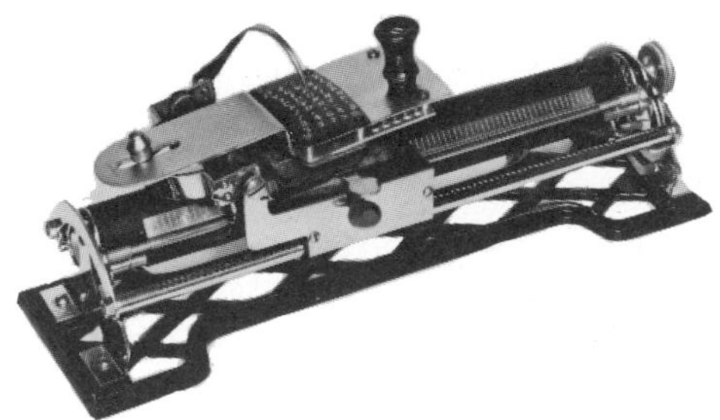

16. A very early MORRIS (serial no. 111), manufactured while the patent was still 'applied for', which dates it 1886. It used a rectangular rubber index, brought down on to the paper by the knob above. (AC)

17. The VIROTYP (1914) could be strapped to the fore-arm or else used suspended from the little belt on the left and the grips on the right. The paper was inserted below. (AC)

but if it had to look like a pocket watch as well, then there was only one (Taurus). On the other hand, if size was no object, he could get a book typewriter that covered half a desk, or perhaps one of those fancy Triple Typewriters which were essentially three Smith Premiers joined together one behind the other, all operated simultaneously from a single keyboard, or maybe a Megagraph which was 6 feet high, 5 feet 10 inches long, 3 feet 4 inches wide, weighing 400 lb.

If all this confusion overwhelmed him, our client might settle for a device that

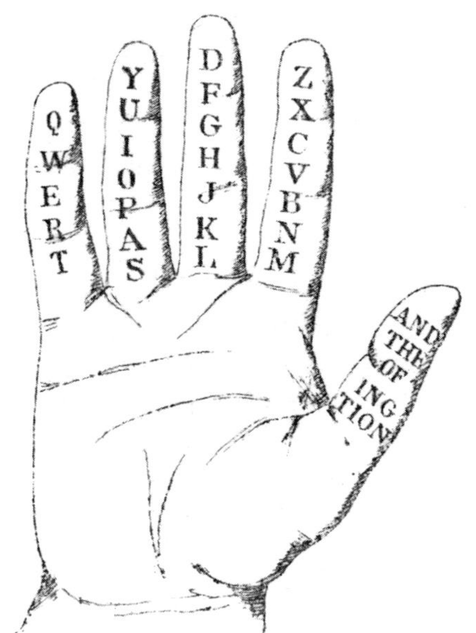

18. They did not come much smaller than the TREBLA, a hand-held machine made in 1910. The rack which held the paper moved a space to the left after each impression.   (TMV)

19. For those who were afraid of machines: the Cary WRITING GLOVE. It was probably better suited for boxing than for typing.   (35)

was not mechanical at all. In terms of sheer nonsense, the Cary Writing Glove must surely be one of the most mischievous inventions of all times. The secretary dons the rubber gloves, with the letters in relief as on the illustration, rubs her hands over an inked roller and is ready to begin typing her employer's correspondence.

What about these early contraptions? Did they really work? Did people actually use them? Actually take them seriously? Buy them? Affirmative. By and large, all the machines worked (ignoring the Writing Glove)—some better, some worse, but in the final reckoning they all fulfilled the purpose for which they were created, namely to print letters on paper one by one, by mechanical means. By today's standards, they worked badly, by which we mean, more than anything else, that they worked slowly. We are struck by their strangeness, by their often beautiful sculptural shapes, by their unorthodox mechanisms, their quaintness. They are perhaps among the last survivors of the now obsolete concept that an object must be not only functional but also aesthetically and artistically pleasing. But we tend not to take them seriously and find it impossible to believe that our ancestors did. We say they worked badly since they were slow and clumsy by modern standards. But that is all. They were less sophisticated; the touch was heavier; the movement more sluggish; consequently they were slower.

But rather than compare the old machines with modern ones, it might be more productive to compare the old machines among themselves. No one in his right mind would ridicule a Remington 2, for instance, because it looks something like a modern typewriter. But how did its performance compare with that of the early index machines which we today find so amusing?

The World, for example, was advertised as having a performance equal to Remington's. This may well be. It had no type-bars and therefore nothing to clash or jam. Or take the Mignon. This apparently ineffective plaything was a best-seller in its day, and was popular right up to the 1930s. The mere number manufactured and sold is staggering and the author himself has been able to trace serial numbers into hundreds of thousands. It continued to enjoy enormous popularity even after the conventional typewriter had long dominated the market. At the time of writing (1971), there is a Mignon in the Institute of Biblical Studies in Rome which the author has been unable to acquire because it is still in use. The Mignon instruction booklet claimed speeds from 250 to 350 characters a minute and the lower figure is quite feasible. But speed is only part of it. The mechanic at the small typewriter shop of Pilar Lorente in Barcelona, a man in his seventies and still at work, recalls that as a boy he was apprenticed to the Mignon representative, who sold thousands of machines in that city by teaching himself to type at the speed of dictation. If the potential client were still unimpressed by this demonstration, or in any way sceptical, he then performed his *coup de grâce*: he removed the letter index from the machine's baseplate, leaving only a gaping hole, and took another dictation directly on to the typewriter with nothing but experience to guide the pointer to the correct position. Touch typing indeed!

The Mignon still in use today is not the only recorded case of the longevity of an unconventional typewriter. A Hansen Writing Ball, manufactured from 1870 onwards, is said to have been in use in the Dutch Court until 'recently'[29] although it has not been possible to corroborate this information. An American newspaper article from 1954 reported Writing Balls in use during the Second World War and Fr Ferrerius v.d.Berg of the Tilburg Typewriter Museum himself

20. One of the most popular machines ever made was the type-sleeve MIGNON, which was produced without major change from 1903 to the 1930s. (AC)

21. Illustrating one of the testimonials in a Blickensderfer catalogue. (AC)

saw a Writing Ball being used by a grocer in his town in 1945. A couple of years ago a collector saw another grocer, this time an octogenarian in an English provincial village, invoicing on a Simplex: attempts to buy the machine were unsuccessful.

Or consider the Lambert, by all accounts a commercially unsuccessful machine. Destiny was unkind to it, for it was a fine instrument whose inventor spent something like seventeen years perfecting it. The Lambert illustrated in figure 6 had a typewritten sheet around the platen when the author purchased it. At the top was the heading *Garden Produce Records*, followed by the name of a British company and the address. The rest of the page was devoted to columns of vegetable harvests for the years 1940 to 1944 inclusive. The company turned out to be a manufacturer of electronic instruments which utilized spare ground around its factory for gardening in the wartime 'Dig for Victory' campaign, and the trusted Lambert was used, more than forty years after it had been built, by the gardener to record his production. As for speed, the Lambert could apparently hold its own with anything on the market. 'That it is fast may be inferred from the fact that several typists are said to have averaged 110 words a minute on it.'[27] It also has the dubious distinction of being one of the machines which a modern viewer considers to be least practicable.

Contemporary advertising abounds with glowing testimonials from satisfied owners. The Blickensderfer company was one that favoured this form of publicity. A drawing accompanying a testimonial from the Reuter's correspondent in the Omdurman Campaign shows the man in pith helmet atop a camel, peering through field glasses held in his left hand while with his right he hammers away on his Blick strapped to the camel's hump. Then there was General Baden-Powell, whom one presumes to have been well prepared, and who cried 'I feel quite stranded without my Blick' after a thief made off with it. Or try this: 'Capt. Gillespie, R.E., who was with the Thibet Expedition, writes from UMBALLA, Punjab, 1st Jan., 1905: ". . . it accompanied me on the march, travelling on a coolie's back. I used it under all sorts of conditions, in pouring rain, at over 13,000 feet altitude with the rain coming in all over the tent. At least once it was dropped and went rolling down the Khud. I hardly ever cleaned it and worked it very often unfairly . . . At the end it was writing just as well as at the beginning . . ."' Impressive indeed, even though one may wonder why the altitude should have affected the performance of any but the operator. On the other hand, what modern machine could have withstood such treatment, especially being dropped down the Khud—whatever that is.

But Blicks held no monopoly of testimonials. In a review which the Hall Type-Writer Co. reproduced in its advertisements, the respected *Times* said of that make (1884) 'The manner in which the machine is finished seems to leave nothing to be desired.' No mean boast, that. And the following year a certain Arthur Friske wrote a glowing testimonial advising the company that he had cured his pen paralysis with one of their machines. On the other hand, the Franklin Institute pronounced the Hammond the best, and the United States Government ordered seventy-five for one delivery, according to a Hammond advertisement of 1891. The 'wholly Visible Emerson Typewriter' gave you a free piano if you sold ten of their machines in three months. The scheme was outlined in a ninety-five-page booklet of which seventy pages were dedicated to literally hundreds of testimonials. And the American Writing Machine Co. published no less than 400 such statements for its Caligraph.

These advertisements sound effective when considered in the context of their times. Others tend to be less plausible: the Yost as an 'accessory to the motor',

23. On the 1886 WORLD TYPE WRITER, speeds of seventy-five words per minute were claimed. Impression was made by lowering the 'key' on the upper left. (AC)

22. A contemporary advertisement claimed that 'with a YOST TYPEWRITER it is perfectly easy' . . . except that the prudent entrepreneur had made sure the 'motor' was stationary. (WC)

for instance ('to write with a pen whilst a motor engine vibrates is next to impossible. With the YOST TYPEWRITER it is perfectly easy'), or Merritt insisting (1890) that their machine 'improves spelling and punctuation'. All manner of machines (the Bar Lock, to name but one) were promoted as being 'automatic' without any justification whatever, including one that was even called the Automatic.

The operating speed of unusual early machines has been mentioned on several occasions. The Mignon had 50 to 70 words per minute, Lambert 110, the World 75. Clearly, their worth is not dependent solely upon the velocity at which they operated, but this is an important consideration to a generation of users accustomed solely to conventional type-bar models. At the speeds cited, however, they were clearly useful by modern standards and these performances were creditable indeed. Conclusive evaluation, of course, is hampered by lack of authoritative and controlled testing—a little discretion is always required, together with the proverbial grain of salt.

Remingtons undoubtedly dominated those early years, giving rise to the fallacy that they were responsible for the invention of the machine itself, a misconception

41

they have persistently encouraged. For most of the 1870s, they were poor performers, packing up at maximum speeds in the vicinity of 30 to 40 words per minute. The type-bars clashed and jammed, apart from other problems, and although the company claimed 30 to 60 w.p.m., the lower figure was the realistic and even optimistic one. Performances were invariably related to the speed of longhand and so it is clear that this machine typed more slowly than writing, for which a reasonable figure of 40 w.p.m. is generally quoted.[35 etc.] Others consider it higher: Malling Hansen, who was qualified, reported in his 1870 British patent application the results of his studies, placing handwriting speed at 4 characters per second. That equals roughly 60 w.p.m., since spaces are obviously not to be counted in handwriting.

Index and other unconventional designs offered comparable performances, then. An independent writer[62] who was partial to Yosts nevertheless conceded that these 'one hand' machines were capable of 'nice-looking' work at speeds equal to or, with practice, a little faster than the pen.

After the introduction of the Remington 2, the speed of type-bar machines increased considerably and manufacturers began claiming terminal velocities for their products. In 1887, 99 words of familiar text per minute was supposed to have been achieved on a Remington in a publicity race against 'one of the most rapid longhand writers in the world' who managed the same text at the 'unprecedented' rate of 65 w.p.m.[54] Two years later, Remington claimed 151 memorized words a minute! Meanwhile, Caligraphs had beaten Remingtons in 1888 in a speed test which filled them with delight, but is likely to leave us sceptical as to

24. This is a facsimile of part of the typing produced by the winner of the 1888 speed championships in which the Caligraph beat Remington. The loser pointed out, however, that the winner had only succeeded in typing two correct sentences in the entire contest. (31)

the value of such contests (see Fig. 24). Both Remington and Caligraph claimed world records at the time, as did Hammond with 180 w.p.m. of selected material in 1891.[62]

By contrast, official World Championships were held under formal and controlled conditions from 1907 on. Twenty years had passed; the machines were better; visible designs had made 'blind' typewriters obsolete, yet speeds dropped dramatically. 87 w.p.m. was the first official world record, set on an Underwood. It became 95 w.p.m. two years later.

The discrepancy between the official and the earlier unofficial records is explained partly by cheating, and partly by the use of unfamiliar and random text as against selected and memorized copy. The latter, of course, goes much faster. Mark Twain, one of the Type Writer's first clients, was sold on the machine after watching a girl demonstrate it time after time at a fantastic speed. Only later, when he compared all the sheets she typed, did he realize that it was the same text repeated over and over again.

Statistics on sales are as inaccurate as early speed claims. A Californian typewriter authority[28] gave the following annual sale figures for the Remington 2: 1880: 704 units; 1882: 3,278; 1887: 27,000; 1890: 65,000. For the years 1882 and 1887 a different contemporary source[54] quoted half these figures: 1,400 and 14,000 respectively. Total sales for the Model Two were given as 100,000 units by a third writer,[33] while a fourth tells us there were altogether 150,000 typewriters in use in the United States in 1896,[35] and a fifth[62] reports 60,000 Remingtons in use by 1891 as against 30,000 Caligraphs, their closest competitor. It is all very confusing and proves that if the design of the typewriter has changed a lot since those days, the use of statistics has not. What emerges, however, is that unorthodox designs represented a large proportion of annual sales. 17,000 Worlds were sold in 1886, its first year,[27] while an Odell advertisement dated 1895 claimed 50,000 of that make in use. Both these figures may be high. On the other hand, total sales of 15,000 Halls[35] is probably close to correct, since the inventor is known to have made a good living off the machine. And as previously stated, the author has been able to trace Mignon serial numbers from four through to six figures. These quantities compare favourably with those of type-bar instruments.

Both from performance and popularity, then, the validity of the unorthodox designs can be established. But even the so-called conventional machine with its type-bars and four-row keyboard suffered the severe handicap of non-visible typing, and mention has already been made of the controversy between advocates of 'visible' and 'blind' machines. So serious was the dispute that even the contemporary *Encyclopedia Britannica* took a strictly partial stand: '. . . doubtless the novice who is learning the keyboard finds a natural satisfaction in being able to see at a glance that he has struck the key he was aiming at, but to the practical operator it is not a matter of great moment whether the writing is always in view or whether it is only to be seen by moving the carriage, for he should as little need to test the accuracy of his performance by constant inspection as the piano-player needs to look at the notes to discover whether he has struck the right one . . .' This absurd rationalization apparently ignored the fact that striking A and B on the piano produces two audibly different signals, whereas this hardly applies to a typewriter, except perhaps to an exquisitely trained ear. Furthermore it avoids the other side entirely, namely in what way 'writing in view' could possibly be construed as a *disadvantage*, especially with operations such as underlining.

Despite the fiercest efforts of Remington and other manufacturers of 'blind' machines, the eventual success of the 'visible' design was a foregone conclusion. Even the Union Typewriter Company, a rather ignominious trust formed in 1893 for the purpose of fixing the price of standard typewriters at $100, was unable to withstand the impact of the 'visible' attraction. The trust, promoted largely by Remington and originally including Caligraph, Yost, Smith Premier and Densmore —*all*, incidentally, 'blind' designs at the time—was difficult to compete against, but so firmly did visible typing establish itself that at least two additional makes (Brooks and Monarch) were added to cater to this growing market. The Union Typewriter Company was something of an *éminence gris* in those days: no one knew much about it, except that it did its utmost to establish a monopoly of the typewriter market. The various brands pooled production resources and facilities as well as sales and marketing outlets,[31] eventually simplifying their operation by dropping the Densmore, Caligraph and Brooks and concentrating on Remington, Smith Premier and (for a while) Yost, all of which were made to look like independent organizations. 'As one can see' a German writes bitterly 'the smart Yankees understood even then how to do business at all cost'.[28]

The Union Typewriter Company went completely 'visible' in 1908—Remington

25. The ODELL was the most popular of many similar machines using a sliding linear index at 90° to the platen. The patent was filed in 1887 and granted two years later.   (AC)

26. One of typewriter history's anachronisms was the GENIATUS which appeared in the 1920s. The type was cast on a rubber belt . . .   (AC)

and Smith Premier with their Models Ten and Yost with its Model Fifteen.

The era of standardization in typewriters had dawned but unconventional designs such as Geniatus continued to appear as late as the 1920s, and a host of old machines were resurrected under thin disguises, and at times under none at all. The Edelmann, for instance, was simplified, cheapened and brought out again in 1924 as the Gundka and Frolio, while the record belongs to Tip-Tip, a barely disguised Mignon, which was built as late as the 1940s. But it never worked out, this fairly frequent practice of remarketing old designs. The successful Remington 2 became the unsuccessful Manhattan after most of the important Remington 2 patents had expired; the basic Sun-Odell design went through over a dozen resurrections at different times; while one of the most successful machines of them all, the Blickensderfer—using the type-wheel which IBM claims to have invented—proved a dismal failure as the Rem Blick a mere decade after its original manufacture had ceased in 1917.

27. . . . and although the 1924 G & K (Gundka) used a type-wheel, the design was hardly better, given its late appearance on the scene.   (AC)

28. The BABY REM was nothing but a Blickensderfer Model Five, warmed up and served again in 1928 after the original patents had expired. Like a spaghetti bolognese, however, it was more successful when it was freshly made.   (AC)

Finally, we have an example of the typewriter as 'the lightning caricaturist, the pencil of nature, and the portrait painter *par excellence*'.[27] For as soon as our machine became popular, artists the world over (predictably enough) began using the new contraption to create their masterpieces. The following example from a contemporary source bears the title:

'QUEEN VICTORIA.
Executed on an Oliver Typewriter'.
. . . A reference, presumably, to the portrait, and not the monarch.

29. Painlessly, one hopes!   (27)

Typewriter Prehistory and Mechanical Monsters, 1714–1806

Some sources,[29] [etc.] attempting to give the typewriter a longer pedigree than it deserves, have traced its origins to the invention and use of engraved blocks in ancient times. This material, however, is related to the history of printing and not to the subject under consideration. A writing machine must possess two vital features in order to qualify here: it must first of all be a *mechanical* contrivance, and then it must print characters or groups of characters *in succession*, according to the will of the operator. There are controversial borderline cases which will be discussed in their due context; obvious errors made by otherwise careful and conscientious researchers,[15, 37] and others who merely plagiarized their work, must be rectified. These errors include compositors, printing presses, and an assortment of non-mechanical appliances.

Blocks reportedly[29] invented in 1575 by the Italian Rampazzetto, in this case as an aid to the blind, are disqualified from the genealogical chart of the typewriter. There are, however, tantalizing but unsubstantiated rumours of writing machines in the seventeenth century. An engine permitting several copies of a writing to be made simultaneously is one of those mentioned. More important, perhaps, is the report that a type-bar machine was constructed by a watchmaker called Le Roy and presented to Louis XIV. All attempts at research on this machine have proved fruitless and no confirmation of the existence of such an instrument has been located. This is a great pity, for the Le Roys were famous watchmakers; which of them could have dedicated himself to this thankless (at the time) task? Hardly the eminent Julien, who was a mere thirteen years old by the end of the seventeenth century, unless the date mentioned in the reference is inaccurate.

Discarding speculation for concrete evidence, we find the first record of a writing machine in the early eighteenth century. A British engineer, Henry Mill, was granted British patent no. 395 in 1714 for a device capable of impressing letters on paper or parchment one after another 'as in writing', the product being so neat and exact as to be indistinguishable from printing. No small claim this, if it is true; unfortunately neither mechanical specifications nor drawings of this machine have survived, for the presentation of these details was not a prerequisite for the granting of a patent until much later, when a model had to be produced to prove that the invention worked.

The operative paragraph of Mill's patent is as follows:
'WHEREAS our trusty and wellbeloved subiect, HENRY MILL, hath, by

his humble petičon, represented vnto vs, that he has, by his great study, paines, and expence, lately invented and brought to perfection "AN ARTIFICIAL MACHINE OR METHOD FOR THE IMPRESSING OR TRANSCRIBING OF LETTERS SINGLY OR PROGRESSIVELY ONE AFTER ANOTHER, AS IN WRITING, WHEREBY ALL WRITINGS WHATSOEVER MAY BE ENGROSSED IN PAPER OR PARCHMENT SO NEAT AND EXACT AS NOT TO BE DISTINGUISHED FROM PRINT; THAT THE SAID MACHINE OR METHOD MAY BE OF GREAT VSE IN SETTLEMENTS AND PUBLICK RECORS, THE IMPRESSION BEING DEEPER AND MORE LASTING THAN ANY OTHER WRITING, AND NOT TO BE ERASED OR COUNTERFEITED WITHOUT MANIFEST DISCOVERY;" and having, therefore, humbly prayed vs to grant him our Royall Letters Patents for the sole vse of his said Invention for the term of fourteen yeares,'.

From the wording 'the impression being deeper . . .', it has been deduced[35] that the machine was designed to serve the blind. This does not necessarily follow; given the thick spongy paper commonly used in those times, any instrument at all was likely to make a deeper mark than 'other writing' which was normally performed with a quill pen. Furthermore, some primitive early machines which used the blow of a mechanical hammer against the type left a three-dimensional impression whether this was the inventor's intention or not. Some of these hammers banged so hard, as we shall see later, that the machine began to disintegrate under the blow! And it has been pointed out quite correctly[15] that a machine destined for operation by the blind is hardly likely to be of 'great vse in Settlements and Publick recors'.

But if it was not Mill, then it was another Englishman, a blind mathematician called N. Sanderson, who is claimed[28] to have built the first writing machine for the blind in 1730. The man in question was Cambridge don Nicolas Saunderson, inventor of a computing board, not a writing machine. The blind, however, continued to inspire the development of the typewriter for over a hundred years.

So did the musicians, although it took almost twice as long for a successful musical writing machine to evolve. The correlation between the finger movements of a pianist producing musical notes and those of a typist printing characters inspired a host of inventions and provided a starting point for the many typewriters which relied upon a piano keyboard. The first recorded attempt at the mechanical writing of music dates from 1745: Johann Friedrich Unger conceived a device which he later presented to the Berliner Akademie, and in the same year a man called Cred was supposed[28] to have invented a Melograph, of which we know only that it was an apparatus to be attached to a harpischord or clavichord whereby every note played was committed to paper.

It is believed to have been similar to Unger's 'Machina ad Sonos et Concentus quoscunque ope Clavichordii productos in ipso cantationis actu chartae tradendos', according to his impressive Latin dissertation to the Berliner Akademie in 1752. It consisted of a supplementary keyboard attached to that of an existing instrument

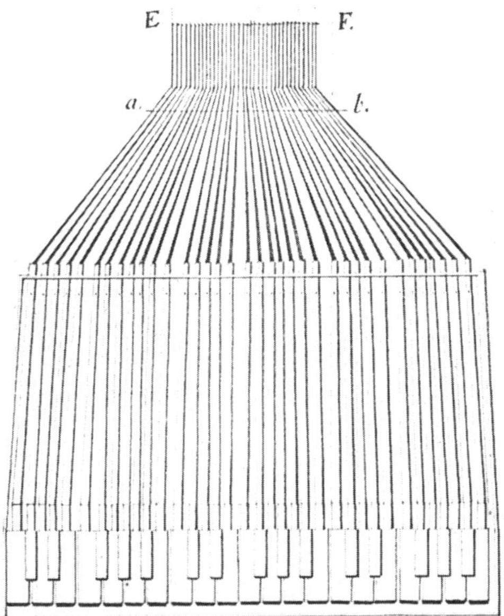

30. The keyboard of the 'Machina ad Sonos et Concentus' invented by Johann Friedrich Unger in 1745. The mechanical complexities of the design were left unresolved. (BA)

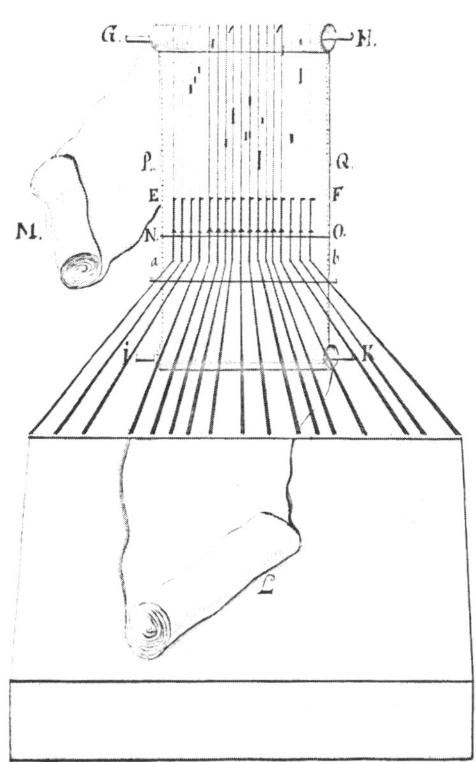

31. Optimistically, the Machina printed a musical code of dots and dashes; in devising the means of making the impression on the paper, Unger accidentally invented the felt pen. (BA)

in such a way that depression of a key actuated a series of rods and levers and resulted in a note being drawn on an endless roll of paper. The printed note took the form of a straight line, the length of which was determined by the duration of key depression. The inventor, or theoretician as he might more accurately be called—for it is highly improbable that he ever actually built his machine— realized that the success of his creation depended principally on moving the paper at a constant rate past the printing point; how this was to be accomplished, how- ever, was left largely to the reader's imagination. In fact, Unger devoted more attention to the description and justification of the resulting musical code than to the machine itself, the problems of which could 'easily' be solved. Quoth he!

An interesting by-product of this device may yet immortalize its inventor, even if the 'Machina' itself does not. The impression was to be made by ink, using as many pens as there were notes on the keyboard; the complexity of such an inking system, however, was more than Unger could handle. Regular pauses for inking would not do it: pens operated by notes which were rarely used, he said, would be dry by the time they reached the paper. This is indeed hard to understand, and why these pens should dry faster than those used frequently is likely to remain a mystery, for Unger made no attempt at revealing the secret. What is more to the point is that he discarded one inking system after another, settling eventually for a series of special pens with spongy absorbent points fitted to tubes and receptacles to retain the ink, thus giving him no small claim to the invention of the felt pen, if not of the writing machine.

Back, momentarily, to pure speculation: a Frenchman called Pierre Carmien was reported[28] to have built a writing machine with a piano keyboard in 1749. It was claimed that the instrument embodied all the advantages of the modern typewriter and that it had been patented, although no records of the patent have been located. American engineers were reputed to have seen the machine and copied it, an accusation repeated with greater or lesser justification in later years, Carmien's invention being the first recorded instance of this phenomenon.

The year 1753 saw the birth of the first of a series of veritable mechanical monsters. The German Friedrich von Knauss exhibited his first model which is sometimes[6] considered the earliest true writing machine. It was, in fact, a writing automaton, the movement enclosed in a metal globe from which protruded two hands, one holding a quill pen and the other an inkwell. The contraption had to be programmed to reproduce a predetermined message, thereby making it an ancestor of the computer. Its hand was automated not only to print the five-word text but also to dip the pen into the ink. Von Knauss created a sensation with his monster, and he immediately proceeded to build another which he completed five years later. This improved model composed its nine-word message in a mere five minutes, while a third machine reduced the time even further. The fourth and last model, completed in 1760, was the inventor's *magnum opus*; it was topped by a seated figure which wrote texts of up to 107 words in 3 mm letters. The beast had reached maturity: it had grown to a height of $7\frac{1}{2}$ feet, standing on a three-foot high pedestal, while the globe containing the mechanism was a yard in diameter.

While they cannot be considered typewriters, von Knauss's creations were decidedly mechanical and they spawned a mass of sophisticated automatons which were miracles of misplaced mechanical genius. They were built by watch- and clock-makers who, as we have seen, had already enjoyed hundreds of years of experience in constructing instruments of staggering complexity; seen in this context, they offered no significant contribution to mechanical history. Their frivolity of purpose, for they were built solely to amuse, does not, however, detract from the excellence of their mechanical conception and construction.

Von Knauss's contraptions were the first writing automatons, but they were not androids and these, performing different tricks, were becoming commonplace. There were those that danced, played musical instruments, drew, spoke, and so on. Of the writing variety, the first robot after von Knauss's final model was one by Payen presented to the Académie des Sciences in 1771 and favourably reported

33. The most famous writing robot of them all was L'Ecrivain of 1772, made by the Swiss watchmaker Pierre Jaquet-Droz.   (MHN)

32. Friedrich von Knauss's masterpiece, 1760. The attractive lady seated upper right does the writing; she is dwarfed by the globe which contains the machinery.   (TMV)

on in the archives of that institution. The doll performed its movements by means of concealed internal levers connected to an external mechanism. With prophetic insight and a clear bid for late twentieth-century favour, it wrote the word LOVE! in upper case and in the imperative!

Then there was L'Ecrivain, by Pierre Jaquet-Droz. Some dispute appears to surround the exact date of its manufacture; the most reliable authority on the subject[9] says that it was completed in 1772. It was the first self-contained writing automaton; where von Knauss had encased his mechanism in a large globe, and Payen in a small box, Jaquet-Droz miniaturized it and embodied it in a wooden doll some thirty inches tall. The clockwork movement inside operates the robot by a complex system of cams and levers. The doll holds the pen in its right hand with its left resting on a small writing-desk, and when it is set in motion it dips the pen in the ink, shakes it twice, then places it at the top of the page; upon depression of a lever, it begins to write its message in neat lines, with head and eyes following the written word, finishing with a period. Now, two hundred years after it was built, the doll still performs regularly for visitors to the museum.

This automaton, together with others built by Jaquet-Droz, quickly established him as the leading maker of his time. With his son Henri-Louis and a number of well-known associates including Jean Frédéric Leschot, Henri Maillardet and others, he began taking orders for similar machines, including two writer-designers, between 1782 and 1787. One of them, resembling a Chinaman, was fitted to a

51

34. The 'guts' of the automaton is a highly sophisticated mechanical 'computer' where the text is programmed. Miniaturizing this complex piece of equipment into a thirty-inch doll was no mean feat. (MHN)

36. Henri Maillardet's bilingual (French and English) android writes four different verses and draws three landscapes. Chapuis believes it may have been made by Jaquet-Droz, but the doll is not self-contained—the movement is in the base. The ball-point pen is an unfortunate 'gaucherie'. (FIP)

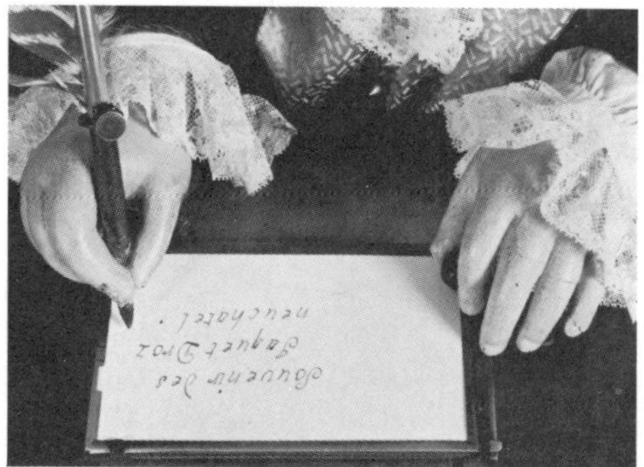

35. Two hundred years after it was built, the automaton still writes its message every week for visitors to the Museum. (MHN)

37. Writing frames and guides such as this one by an Englishman called Hawkins were plentiful, but they were not machines. (33)

clock and traced[8] to the Peking Museum; another was found in poor condition in Philadelphia but, when it was meticulously restored, it revealed its identity by writing a couplet in French, after which it signed off with the words: 'Written by the Automaton of Maillardet'. This message may well be false, for Maillardet is known to have toured with a robot show in 1814 and 1815 and may have re-programmed a Jaquet-Droz doll; on the other hand, it may be a copy built by him from one of the master's originals. As for the one in China, the Emperor was apparently so pleased with it that in 1790 he ordered his technician and watchmaker in Peking, Charles Paris, to build him another. Not to be outdone, Paris's android was polylingual and wrote in Chinese, Tartar, Mongol and Tibetian.

Meanwhile, in 1783, a writing and drawing automaton by Joseph Neussner of Vienna was described in full technical detail in a sort of do-it-yourself writing automaton instruction manual, obviously written to satisfy the widespread curiosity aroused by the machines. His book appeared none too soon: fickle public enthusiasm had already begun to tire of robots by the turn of the century, although they continued to be made for several decades.

The discussion of automatons has been pursued to completion although it has upset the strict chronological sequence of other writing-machine developments. Not that anything of importance was offered in the meantime—in fact, most of the devices listed in this period are either speculative or not writing machines at all, and secondary sources that have not bothered to check their subject have thrown it into utter confusion. Thus Count Leopold Josef von Neipperg built a 'copiste secret' in 1762, a kind of pantograph consisting of a guide frame and a series of pens with two or three nibs allowing a corresponding number of copies to be written simultaneously. Many such devices were designed. The one by Hawkins[33] in fig. 37 is typical; all are disqualified on the grounds that they are non-mechanical.

Nine years later a German called Hohlfeld invented a musical writing machine along the lines of Unger's invention with which he is said to have become familiar.[28] The same source reports that a portable embossing machine was mentioned in correspondence as having been invented by Abbé Rochon in 1779, but no substantiation of this information could be located. Similarly speculative is the writing device for a blind noblewoman invented in the same year by one Wolfgang von Kempelen, of which no details are available. Further research has revealed that this man achieved a certain degree of fame as the inventor and builder of talking machines, and of a chess-playing robot which was a fraud: it was seated at a hollow desk concealing the real flesh and blood chess player!

These were hard times for the earlier typewriter historians who seem all too prone to throw in names in order to fill chronological gaps. Thus a Frenchman by the name of Pingeron is invariably listed as having developed a writing frame for the blind in 1780, despite the fact that such a device is no machine; an improvement on the same instrument by L'Hermina four years later offered no mechanical advantages and is likewise disqualified. An Englishman by the name of Jenkins came up with the same thing in the same year, we are told, and can therefore be dismissed in the same way. The record is surely held, however, by yet another Frenchman in 1784: not only have all details of the man's machine been lost, but even his name has failed to survive! How a piece of information can even be located under such adverse circumstances is something that the source[28] fails to reveal and it has stimulated the present author to reconstruct this whole affair as a typical example of how shabbily and carelessly typewriter history has been handled. It turns out that a man called Jenkins, in an 1894 lecture,[59] mentioned in vague terms the invention of a machine in France in 1784. Two years later Rochefort-Luçay[35] repeated this information, adding that the inventor's name and machine were lost. Sénéchal,[37] in 1903, tracked down the details, discovering that the machine was Pingeron's 1780 frame perfected in 1784 by L'Hermina. But Herkimer[21] ignored Sénéchal, perhaps because no one in the County Historical Society understood French, and copied Jenkins. *Typewriter Topics*[42] perpetuated the mistakes, adding a few of its own, such as mis-spellings. Bliven,[5] Tilghman Richards[40] and others repeat them. Martin,[28] however, takes the prize: he lists everything—Pingeron's machine, L'Hermina's, the supposedly anonymous one, plus another by someone called Jenkins, which clearly arose out of misreading of Dupont and Sénéchal.[15] This Jenkins was in fact the same man who delivered the lecture two centuries after the invention with which he was erroneously credited. And the whole affair concerns a device that was not even a machine!

All is not lost, however. Joseph Bramah, a famous English engineer and the prolific inventor of an assortment of unrelated devices, had the idea of putting concentric wheels to a number of uses. The combination lock is one of his inventions. Conceived in 1784, its principles are still used to this day. It is also interesting in that it may well have inspired Thomas Jefferson to make his wheel-cipher device which achieved fame a century or so after his death. Jefferson was in France from 1784–9; here he may well have become familiar with Bramah's lock, inventing his wheel cipher in the decade from 1790 to 1800[24]. Essentially, Bramah's

38. Joseph Bramah invented just about everything from flushing toilets to this 1784 combination padlock, which has remained substantially unchanged to this day. (AC)

39. Adapating the padlock's concentric wheels to printing, Bramah first devised this banknote numbering machine, and later a similar writing machine which he patented in 1806. (33)

and Jefferson's devices are identical and both served the same basic purpose, namely secrecy. Bramah's wheels had to be turned to produce a message and thereby open a lock. Jefferson dispensed with the lock and used all other rows of letters as alternative encipherments of the plaintext.

Bramah also invented a banknote-numbering device and the writing machine he patented in 1806 follows similar lines (the 1794 date sometimes quoted[36] is incorrect). It consisted of a series of twenty-six rings or discs, with the characters around their perimeters, stacked one after the other. A line of text was first composed by spinning each disc in turn and selecting the desired letter, whereupon the whole line was brought into contact with the paper. According to the wording of the patent, more discs could be added *ad infinitum* to the printing roller, thereby covering just about all contingencies and then, with a view to posterity, improvement upon improvement would be suggested 'to the end of time'! Inking was by an open-bottomed trough fitted on to the printing roller in such a water- (or ink-) tight way that the sector of the roller provided the bottom of the trough, a leather 'wiper' hopefully taking care of excess. By building divisions into the trough, simultaneous printing in different colours could be achieved.

This was the first of several similar appliances which appeared in the nineteenth century (see page 79). The device is certainly mechanical but rests somewhere on the border between the writing machine and the printing press, the major objection being that an entire line of text had to be composed in advance. It cannot therefore be considered a thoroughbred typewriter.

Chapter 3

Evolutionary Progress, 1808–50

The new century brought a change in approach, as we have already seen. The potentials of the machine began to be recognized, and its role changed from that of a contraption designed to inspire awe and perhaps a little fear to that of a functional device with universal applications. This process did not occur overnight, of course, but considering merely the acceleration in the numbers of inventions as the nineteenth century progressed, to say nothing of their nature and purpose, one realizes that although society was not yet ready for mechanization, it was at least being prepared.

More specifically, the writing machine began to evolve as a potential reality, and one inventor after another promoted his creation by enumerating its advantages over handwriting. The uniformity of the message was listed, as were ease of legibility, facility of making copies, and above all speed. Not unnaturally, instruments were evaluated by arbitrary comparisons between their speed of typing and the speed of writing by longhand. This issue was mentioned briefly in Chapter One and will be considered in further detail later.

But early in the nineteenth century the world was not clamouring for the typewriter. Far from it! Resistance to the machine was ubiquitous, and the task of the inventor was a thankless one, as the diaries and letters of many of them indicate. Society was in fact not fully ready for the typewriter until the last quarter of the century, when suddenly the demand was there, and a frenzy of commercial activity set about filling it.

Meanwhile, the special requirements of the blind continued to occupy inventors' minds for the better part of the century. With machines for this purpose, speed of operation was a secondary consideration and usually was not even mentioned. Certainly it was unimportant to thc Italian nobleman Pellegrino Turri who, in 1808, built a writing machine for the blind Countess Carolina Fantoni, whose plight he hoped to alleviate. This is the first device on record from which the typewriter as we know it may be said to descend, and it has been called[42] the first to which the designation 'machine' may be applied. The meaning is clear, for while the robots and automatons of the eighteenth century were decidedly mechanical, they were more closely related to computers than to typewriters.

Turri's machine was put to hard use over a period of many years, during the course of which the inventor made several modifications, including change of type. Impression was first by means of 'black paper', then by ink itself. The former was the more satisfactory but was apparently abandoned for logistical reasons: the inventor, who lived in another town, was the sole source of this carbon paper and when his girl-friend ran out of it and could no longer use the machine, she was obliged to resort to the services of a secretary. This meant, basically, that she could not say what she wanted, since by then both she and he were married —and not to each other. All of which has come to light from some of their correspondence which has fortunately survived, and which is dealt with more fully in Chapter Seven.

A brace of unimportant machines followed: still in 1808, a Pastor Schönfeld is reported[61] to have devised yet another musical writing machine to be attached to a piano keyboard, and fourteen years later, a short article[50] records that a Claude Lechet, in collaboration with his blind master Franz Huber, developed a mechanical device for the use of the sightless. It was in the following year, however, that the next decisive step was taken: the Italian Pietro Conti's Tachigrafo or Tachitipo of 1823 printed on paper or any other material at a speed equivalent to that of the spoken word, according to the inventor who, failing to specify how fast he spoke, continued by extolling the virtues of his invention for those with poor eye-sight or with badly trembling hands. Conti's Tachigrafo was portable and was housed in a wooden box. It consisted of a frame with a marble or iron plate under which the sheets of paper were attached. This plate moved backwards and forwards for line spacing; underneath it, a circular type-basket containing tempered steel type-bars moved from left to right for letter spacing. In front of it all was a keyboard with a separate key for each character so arranged as to permit the operator to manipulate all keys without moving his hands, this being the main indication of the compactness of the instrument. Upon depression of a key, the type-bar swung upwards to the printing-point, inking itself on the way (just how is not explained). It was a highly sophisticated instrument by any standards and was the first known up-stroke machine, meaning that it printed by striking upwards at the paper as against the conventional typewriter with which we are familiar, which is of front-stroke design. The up-stroke principle found much favour throughout the nineteenth century and was the movement used on early Remingtons, among others, until as late as 1908. Furthermore, the Tachigrafo was the first of a number of machines in which the type moved while the paper remained stationary.

Conti's invention enjoyed considerable recognition, if little practical success. In 1827, the Arcadia di Scienze, Lettere ed Arti published a detailed description in its journal (from which the above account was taken) and notified its intention to purchase one of the machines. In the same year, the Tachigrafo was demonstrated to the Société d'Encouragement pour l'Industrie Nationale, having already been approved by the Académie Française, which was reported to have purchased it, although no record of an actual transaction survives in the archives.

These details are mentioned because they are indicative of the way in which mechanical information was collected, stored and transmitted in those times.

Few people worked in a vacuum; inventors began studying past efforts and borrowing from them. The above-mentioned societies and academies were but a few of those which began to open their doors to inventors; society as a whole was not ready, but these isolated pockets of intellectuals and visionaries were. Conti's machine was in fact the first typewriter to have received extensive dissemination. A later inventor who learned of it was his compatriot, Giuseppe Ravizza, the man whom Italian historians consider to be the inventor of the typewriter. He began corresponding with Conti in 1832, and the first model of his later Cembalo Scrivano borrowed so heavily from the Tachigrafo that he has been directly accused of copying it.[28] Italian chroniclers[1] prefer to call it 'inspiration'.

While Conti's Tachigrafo was being demonstrated in Paris, a French librarian named Gonod quietly invented the shorthand typewriter. It was the year 1827, and Gonod's inspiration came from the ease with which pianists manipulated a keyboard. His machine had twenty keys in four groups arranged in such a way that two were for opening consonants, the third for vowels, and the last for closing consonants. Each hand controlled ten keys. The operator played 'chords', meaning that more than one key could be depressed at the same time; thus, Gonod's

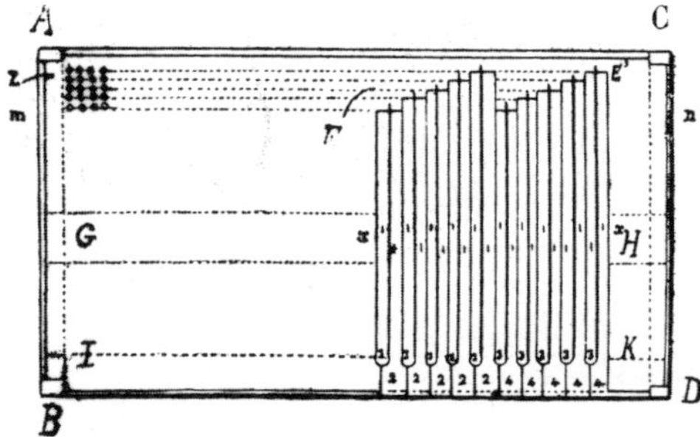

40. The first shorthand machine was one with a piano keyboard invented by a Frenchman called Gonod in 1827. His system of depressing keys in 'chords' has survived to this day.   (16)

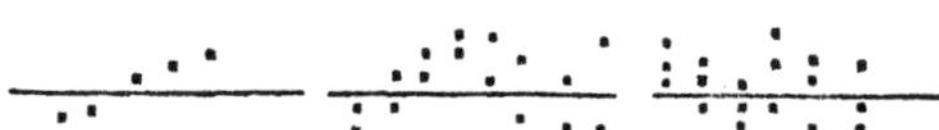

41. Gonod's shorthand code consisted of dots differently positioned on both sides of a dividing line.   (16)

inventive genius established a system which has survived to this day. Paper was wound around two cylinders and moved from right to left after each impression. The instrument printed in a shorthand code of dots on either side of a dividing line, in a system designed by the inventor, and speed is reported to have been equal to that of dictation.

But Gonod's ingenuity deserted him when it came to devising the means by which the impression was made on the paper. It is hard to understand why he did not use carbon paper, for it had been invented and patented more than twenty years before. Or he could have chosen a pencil or a fountain pen, both of which he specified as alternative means for drawing the dividing line on the paper. Instead, the best method he could think of was to place the sheet of paper against another on which a greasy substance had been smeared; when the character hit the clean sheet, it pressed it against the greasy one, leaving a greasy imprint over which coloured powder was subsequently sprinkled. Details both of the stenographic system as well as of the machine itself were sent by the inventor to the Société Académique de Clermont-Ferrand. He claimed that he had built a test model of his device, but the final instrument appears not to have been presented.

Two years later, across the Atlantic, came the first American entry into the writing-machine field. William Burt was granted a patent in 1829 for his Typographer, as he reluctantly christened his invention after canvassing far and wide for a more pleasing alternative. (Burt's Family Letter Press was second on a photo

43. A reconstruction of Burt's improved machine was undertaken by his great-grandson in 1893. The original had been destroyed by fire in 1836. (CSM)

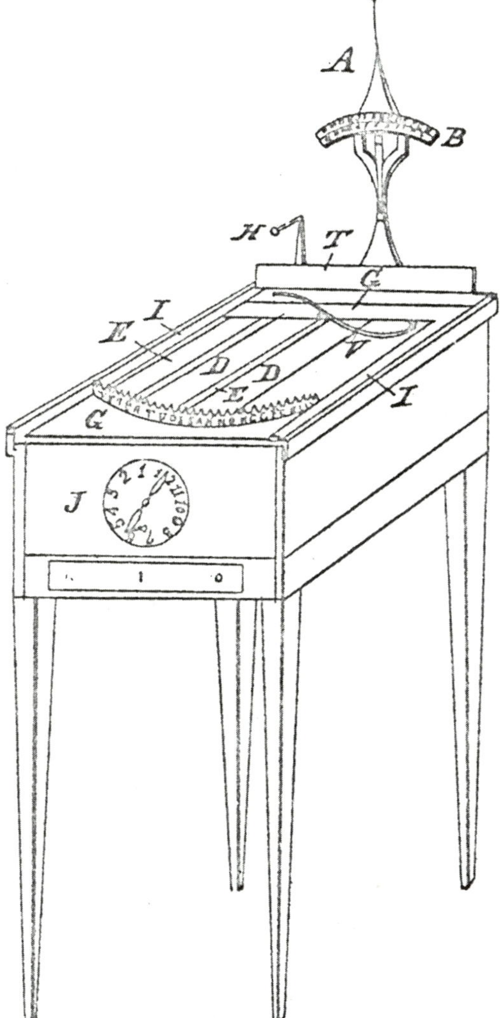

42. The first American typewriter was the Typographer, patented by William Burt in 1829. The type was on the underside of a swinging sector, raised in the illustration as was necessary to permit insertion of paper. (USP)

finish.) Like his contemporary Conti, he succeeded in stirring up considerable interest in the machine, none of which, however, translated itself into commercial or financial advantages.

Initially, the Typographer was constructed entirely of wood and, considering the rough home-made job it was, it did surprisingly good work. It consisted of a box at one end of which was a swinging lever. The type was mounted on a short sector attached to the underside of the lever, the other end of which carried an indicator for selecting the letter along a curved scale. Downward pressure on the lever brought the type into contact with the paper, making the corresponding impression. Pads on both sides of the printing point provided the ink and the machine printed on an endless roll of paper held in a flat horizontal frame; when a page was filled, it was torn off at the end, like paper towelling. A clock-gauge on the front provided an accurate measure of the length of the page and a primitive escapement moved the paper a space after each impression. Printing was in upper as well as lower case: this was the first machine ever to have been so equipped and it was precocious indeed when one considers that fifty years later, when type-writers were already being manufactured commercially, only upper-case models were available. But the machine was slow and clumsy and Burt was unhappy that it failed to fulfil the prime function for which he had initially designed it, namely to assist him with his voluminous correspondence. He was ready to pack it all up, but an enthusiastic associate of his, one John Sheldon, induced him to build an improved model which he eventually completed six months later. (This phenomenon of the disillusioned inventor and the exhortative associate is a recurrent theme in writing-machine history and, in fact, in the history of invention in general.) The new and improved model was little better than its predecessor, however, and the so-called improvements were limited mainly to the appearance of the Typographer, which now cut a better figure. Even so, it failed to attract the interest of the investors for whom it was principally made and the project was dropped. Burt went on to fame and fortune as the successful inventor of surveying and navigational equipment, among other things, and his Typographer was destroyed by fire in 1836. A model was built from original manuscripts by the inventor's great-grandson and exhibited in the World Columbian Exposition in 1893, and several reproductions of the Typographer are displayed in museums in various countries, together with copies of a letter typed on the original machine by the inventor. Many rash claims have been and are still being made for it, among them the assertion that it was the first machine known to have been capable of practical work, and that the accompanying letter was the first ever to have been written on a typewriter. Both these claims are false and unfounded.

Burt's failure to attract commercial interest appears to have discouraged his compatriots, for no further attempts were made in the United States for more than ten years. In Europe, however, things were buzzing. The 1830 decade began with a stenographic machine by an Italian living in London called Celestino Galli. He has been variously described as a language teacher[28] and as an artist[16]—in view of his wild imagination, the latter seems more probable. No one knows for sure whether his Potenografo ever really existed; the inventor, of course, said it did, and contemporary reviews and descriptions of the machine are not lacking. The Société d'Encouragement carried one of them in 1831[48] with a footnote, however, that the author knew of the machine by description only. To confuse the issue further, no one actually came out and said that he had seen or tested the machine itself, and the Italian inventor Ravizza later claimed to have a letter from Galli confessing that the Potenografo was never built. Ravizza at one time was pressed to produce this letter and failed to do so, thereby throwing even more ambiguity into the matter, the more so since he might be considered to have had more than a passing interest in discrediting a rival invention.

In any case, whether it had been built or not, details of the design survive. It was a radial-plunger machine with a keyboard of two concentric circles of keys so arranged that one hand operated the circle of consonant keys and the other that of the vowels. Depression of a key brought the type at the other end of the plunger into contact with an endless strip of paper wound around a roller, the paper uncoiling as printing proceeded. This made it possible, among countless other things, to write sixty (yes, sixty) times faster than by hand, and ten times faster than by any other shorthand method, according to a circular distributed by the inventor. Other more sober sources[48] quote a speed equivalent to that of dictation, and even that with some scepticism.

Hyperbole was not a monopoly of the Italians, however, for the next inventor was German and his claims make Galli sound like the personification of modesty. Baron Karl Friedrich Christian Ludwig Drais von Sauerbronn, for that is the man's improbable name, developed a shorthand machine which he exhibited in Frankfurt in 1831 and which was reported by the local newspaper[53] the same year. Describing it as the first German machine, the proud Martin[28] devotes seven pages to it while, in the same book, Burt warrants only two and Conti less than one; on the other hand, the English writer Mares[27] does not even mention the Baron. Such is the strength of national passions. What is certain is that the Baron's contribution to typewriter development is considerably less important than another of his inventions which bears his name, the Draisine, so that history is destined to remember him better as one who contributed to the bicycle rather than to the writing machine.

Drais's claims far surpassed Galli's. Among countless other things, he stated that not only was his instrument capable of keeping pace with, and even overtaking, the fastest dictation, but that also speeds of over a thousand characters a minute (or *a million a day*) were 'often' achieved. Not to be sneezed at—and all this on a relatively small machine in a wooden box, a little more than a foot cube, with a square keyboard consisting of sixteen keys in four-by-four arrangement. As in Gonod's case, the 'chord' principle was used, the impression being made on a strip of paper which was advanced by clockwork. The inventor evolved a shorthand code to go with the machine, and one that was arbitrary and flexible enough to be adaptable to any speed of dictation: if your boss spoke quickly and you had trouble keeping up with him, you simply abbreviated words more brutally!

Contemporaries of the Baron's referred to the device as 'eine mechanische Narrheit und Alberne Erfindung'—a mechanical madness and an absurd invention. The more sober *Mechanic's Magazine* editorialized[63]: 'We have seen Baron Drais write by his machine, which is undoubtedly very ingenious; but we think that a degree of pains has been taken to make it difficult of use, which might well have been spared.' It is then criticized for producing only shorthand characters, and at a speed no faster than by writing, whereupon a suggestion is made that it might well be adapted for printing the letters of the alphabet 'especially if more than one copy could be produced at a time, which (if *one* can be produced) might, we think, be easily managed' (original italics). This must certainly be one of the earliest references to the possibility of using a typewriter for making several copies of the same writing.

While the wild Baron was touring with his bag full of inventions, a Frenchman was granted the world's first patent for a down-stroke machine in 1833. Xavier Progin, who has suffered the indignity of having his name mis-spelt in more ways than any other writing-machine inventor, called his instrument 'Plume Ktypographique'. The 'k' spelling is hard to understand, the more so since there are

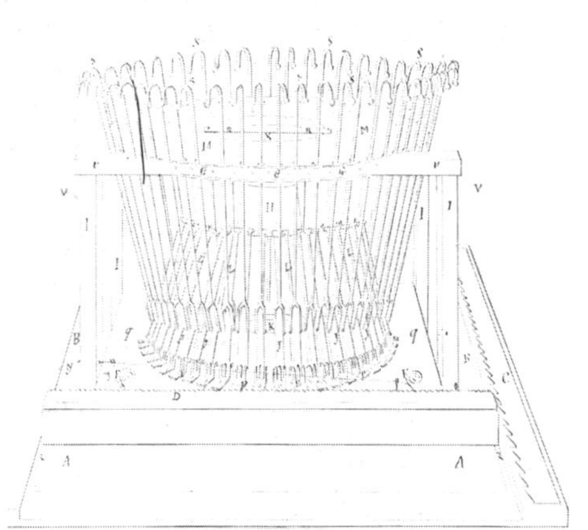

44. Progin's 'Plume Ktypographique' (1833) was the first type-bar machine and the first to offer such revolutionary features as visible typing and differential spacing. (FP)

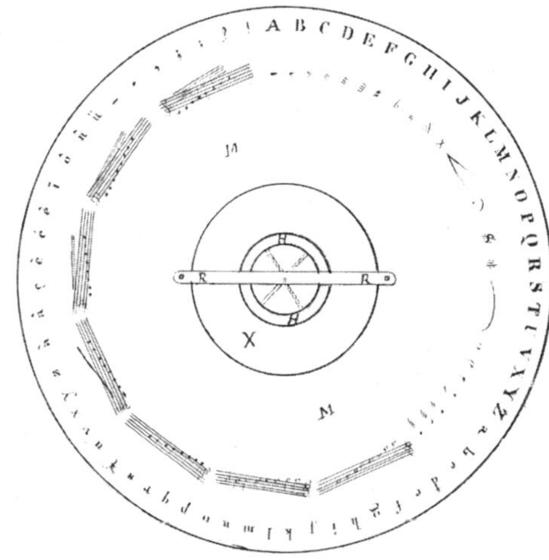

45. The circular letter index of the Progin machine. Interchangeable type was offered to permit the printing of music. (FP)

several more conventional references in the patent to the machine as a 'plume typographique'. Type-levers were arranged in a circular basket inside a square frame, the tops of the levers protruding above it being bent to facilitate gripping with the fingers. A circular index of figures, characters and upper- and lower-case letters in alphabetical order was positioned inside the circle of levers, the lower ends of which were pinned to type-bars hinged to the baseplate and designed to strike at a central printing-point. Inking was effected automatically from a removable pad whose capacity was sufficient for typing two or three pages. Paper was held flat on a plate under the type-basket and stayed still while the body of the machine moved, letter and line spacing being performed manually along lateral and frontal racks. The machine is usually listed as having sixty-six type levers in all, but the patent-drawings of the circular letter index show sixty-seven characters, and the specifications merely call for a separate lever for each character.

Sophisticated indeed! But Progin's importance lies as much in the lesser details of the machine as in its overall design. For instance, he recognized the importance of being able to read what was being typed, a revolutionary concept in the light of later controversies referred to in Chapter One. Even more to Progin's credit is that his design was not one which merely happened to offer visible typing, *malgré lui*: he made a special point of including this feature by cutting a large circular hole in the centre of the letter index, thereby permitting the operator to look straight down through the machine to the printing-point.

But there were more surprises yet: two spaces were allotted to capital letters, thereby making the Plume the first machine to offer a form of differential spacing (see p. 210). Some[28, 40, 42] say that Progin was a printer by trade and that such a practice would have seemed natural to him; this is apparently incorrect and originated from a faulty translation of the patent which refers to the inventor as an *homme de lettres*. Progin also proposed removability of the type and its replacement by musical or other characters, and the adaptability of the machine for stereotyping. The design was a direct lineal ancestor of the book typewriters which made their appearance towards the end of the century. Its big failing, however, was that it proved slow to operate: the modest Progin admitted that it required practice to write merely as quickly as with a pen.

It was a long time before anything else as good as the Plume Ktypographique was developed; meanwhile, around the world, work was proceeding on an invention which was truly to change the course of history on a scale that the typewriter could never hope to equal. Man had always suffered from his inability to communicate with others who were out of sight or hearing and society was more than ready for a practical device to serve this purpose. A history of the telegraph is beyond the scope of this book, however, except where the two fields overlap; it was in the United States that the first decisive step was being taken in the happy marriage between the writing machine and telegraphy. The type-printing telegraph, born to this marriage, posed 'one of the prettiest problems in telegraphy'[66] and is discussed more fully in Chapter Eleven. In 1835, however, Samuel Morse first showed friends his primitive telegraphic embossing device; contrary to popular belief, Morse did not invent the telegraph, nor was his the first practical system to be used. He does, however, appear to have been first with a permanent record device, although even this has on occasion been challenged. In depositions before the Supreme Court, Morse claimed that he first thought up the idea in 1832, and witnesses testified to having seen drawings of his device in that year. He had not actually built a functioning apparatus until 1835, when friends confirmed having seen it, and it was publicly shown the following year in New York. Experiments and improvements followed: the first instrument embossed, and Morse tried printing instead, using various inks and inking systems but with little success. In 1837, however, the first practical public demonstration of the instrument was given over a ten-mile circuit; by then, Vail in England and Steinheil in Germany were ready with their own machines. Morse is given priority, however, on the strength of evidence in favour of his 1835 date. He filed his caveat in the US Patent Office in 1837, his specifications the following year, and the patent was granted in 1840.

One might add, enviously, that the telegraph was an almost instant world-wide success. By the mid-1840s, overland lines had been strung or laid in virtually every major country and submarine cables were also being tested. How different was the attitude of the world to the telegraph from its reception of the typewriter!

SIW   The Morse embosser consisted of a train of wheels, driven by clockwork, which advanced the paper past an electromagnetically operated stylus. This, essentially, was the receiver; the transmitter worked as follows: the code which Morse elaborated for each letter of the alphabet was cut out on an edge of a small strip of wood in the form of teeth and flats, the dots being the teeth and the dashes the

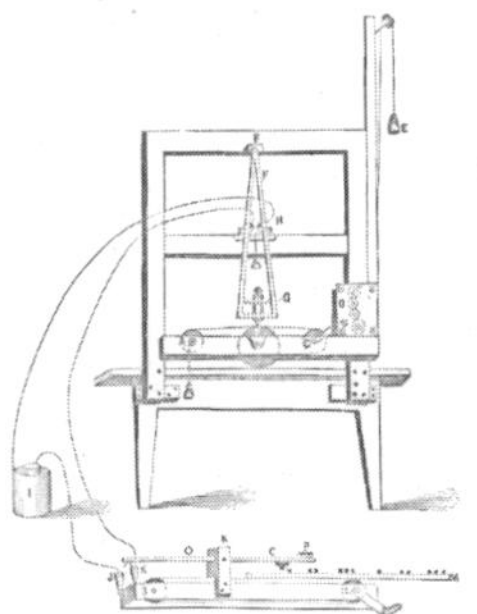

46. Morse's electric telegraph receiver used an electromagnet to emboss paper tape advanced by clockwork. Details of the transmitter (lower unit in the illustration) are shown in Fig. 47.

(38)

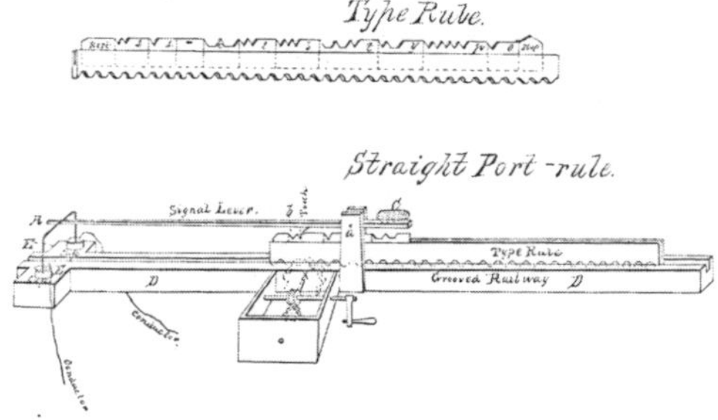

47. An enlarged view of details of the first Morse transmitter. The 'type-rule' with the message cut into its upper edge was cranked along a groove in the 'port-rule' beneath a lever which closed a circuit as it rode over flats of varying lengths to send impulses to the receiver, embossing corresponding dots and dashes on paper tape.

(USP)

flats. These strips (called type-rules) were then placed in a carrier (port-rule) according to the text to be transmitted, and were passed beneath a lever which closed the circuit as it rode over teeth and flats, actuating the stylus and embossing dots and dashes on the paper, if all went well, which was not always the case. Morse later substituted for this transmitter a plate with long and short metal bars for the dots and dashes; a metal pointer passing over them performed the task of closing the circuit. Hordes of other inventors offered further improvements. However, the decisive step in printing telegraphy had been taken, only to be retracted by the inventor soon afterwards when he realized that an audible signal could be decoded faster and just as easily. The Morze buzzer was the result.

Although Morse was able to prove that his was the first device completed, his telegraph was not put into commercial use until 1844. Cooke and Wheatstone's telegraph, 'The Great Western Railway Telegraph', had by then been working between London and Drayton for five years.

Meanwhile, in 1836, another attempt was made at perfecting a musical typewriter to be attached to a piano keyboard, this one patented by an Englishman called Miles Berry. Once again, both notes and their values were recorded as shorthand

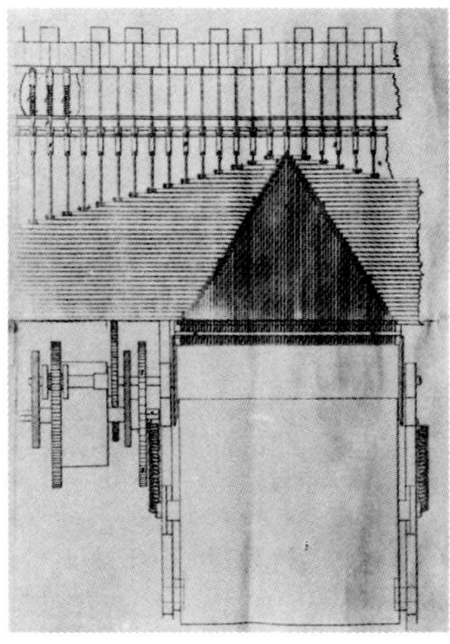

48. Berry's 1836 musical writing machine was only marginally better than Unger's. It printed a similar musical code, but at least the inventor illustrated the linkage between key and stylus and gave details of the clockwork paper advance.
(BP)

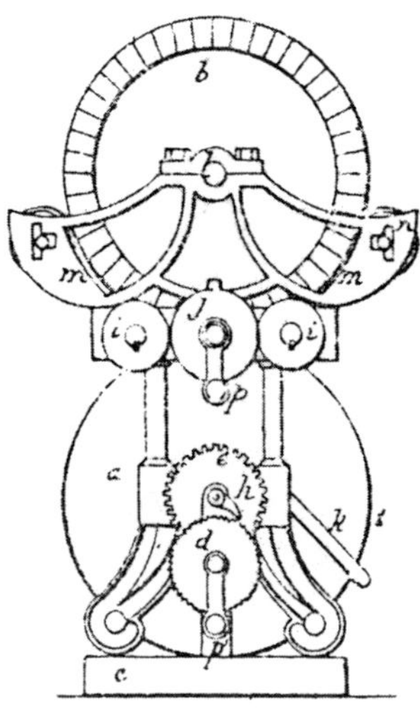

49. Bramah may have invented flushing toilets, but a Frenchman called Bidet invented the typewriter platen. It is represented by the lower of the two cylinders in the illustration; the upper circle is the type-wheel. Manual differential spacing was incorporated.
(15)

symbols on a strip of paper advanced by clockwork; this was virtually the only improvement offered by Berry over Unger's design, apart from his use of a stylus point to make the impression through carbon paper. According to the inventor, however, it worked perfectly. The only other offering from the same year is an anonymous circular index machine described and illustrated[27] and faithfully reproduced time after time,[31 etc.] erroneously, because the instrument is Cooper's of 1856 (Fig. 87).

1837: although Drais, as we have seen, gave his name to the Draisine, a Frenchman called Gustave Jean Emile Bidet is known only to have designed not bathroom

C

fixtures but a Mechanical Typographic Compositor. A slow and cumbersome apparatus, it consisted of a type-wheel mounted on a horizontal shaft above a cylinder around which the paper was fastened. After selection of the desired character, a lever was lowered to bring the cylinder up against the type-wheel, after which it fell back in place by gravity. Inking was by roller. The cylinder was rotated manually for letter spacing and the type-wheel moved along its axis for line spacing.

Bidet deserves credit as the inventor of the platen and is sometimes[15] also credited with the invention of differential spacing, for the platen was turned one or more clicks on a ratchet according to the width of the letter being printed. However, since this operation was not performed mechanically, it is doubtful whether he is entitled to this honour.

The first printing telegraph to use the alphabet rather than a code was invented by Alfred Vail in 1837, the same year that Morse was demonstrating his own device. Vail's should have been the better machine, for the ultimate goal was always to print directly in Roman characters, but the design was plagued with troubles and it never quite made it. There were apparently[44] three models, the last of which was the well-known one using twenty-six type-plungers disposed radially around the perimeter of a vertical wheel. Paper, as before, was advanced by clockwork.

A Johann Knie is reported[28] to have developed a workable machine for the blind in 1838, but no details have survived, and in the same year, an Englishman called Edward Davy invented the electro-chemical telegraph. This system developed from the discovery that electricity could be used to decompose chemical compounds; by treating a piece of paper (or cloth, in this case) with such substances and subsequently subjecting it to electric impulses, decomposition of the compounds took place and a permanent discolouration at the point of electrical contact was the result. Davy's invention was not successful, although it was later perfected by Bain (see page 74) and quite extensively used.

The German Professor C. A. Steinheil made a much greater contribution to telegraphy in the same year. The machine he invented is less noteworthy, however, than his discovery of the earth return: the 'grandest discovery ever made in practical telegraphy'.[38] This has become so much a part of our lives that we tend to overlook the advantages of being able to complete a circuit merely by leading it into the ground; the alternative, namely the use of a separate return line, was both cumbersome and costly. A few years later Bain discovered that a body of water could be similarly used and in his 1843 patent advocated a further refinement that has been repeated ever since: 'Attach the wire to a good water-pipe.'

Steinheil's printing telegraph consisted of two arms, fitted with cups of ink and capillary tubes at one end and attached to electro-magnets at the other. Actuating the circuit brought the tips of one or other of the tubes in contact with paper advanced by weight-driven clockwork, producing a two-row code of dots thus: A = .·., B = .¨., etc. This code was well conceived and was for a time a rival of

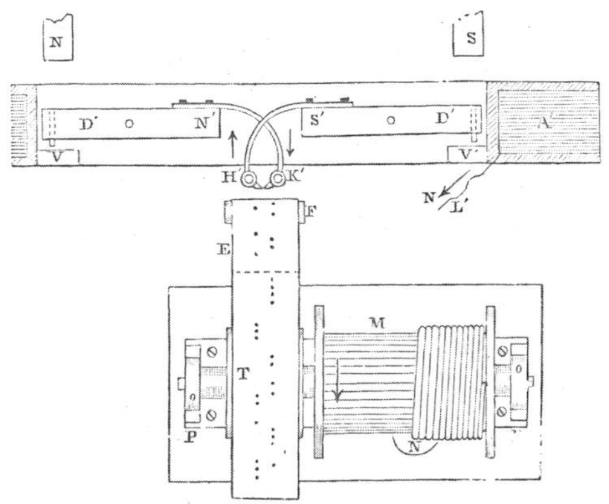

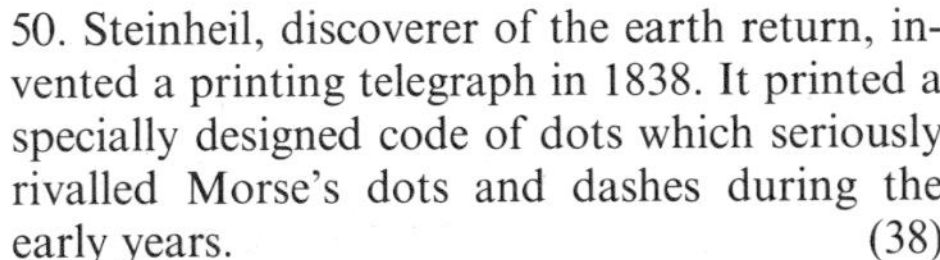

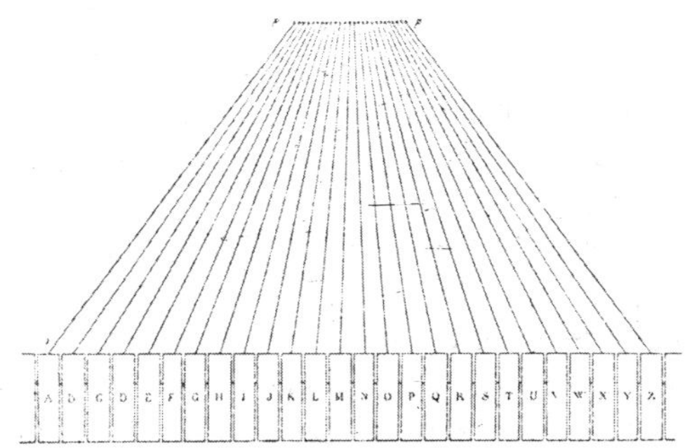

50. Steinheil, discoverer of the earth return, invented a printing telegraph in 1838. It printed a specially designed code of dots which seriously rivalled Morse's dots and dashes during the early years. (38)

51. Each key of Dujardin's Tachygraphe printed its own dot on a paper tape advanced by clockwork. A scale of the alphabet was subsequently passed over the tape, the location of the dots indicating the characters. (16)

Morse's, but gave way beneath the advantages of the latter's adaptability to audible signals.

Still in 1838, a Frenchman called Antoine Dujardin presented a rather remarkable piano-type instrument which printed dots over a width of 10 cm on a roll of paper. The Tachygraphe, as he named his invention, had twenty-six keys in alphabetical order, connected to levers with the same number of pens on the ends; the strip of paper was advanced by means of a weight, and after it was all over, a 10 cm-long scale with the same alphabetical order of letters on it was superimposed over the dots in order to decipher what had been written. Such a scale had long been known in cryptography and Dujardin may thereby be given some credit as the inventor of the printing cipher machine, although he is not mentioned[24] in this context. Some later stenographic machines also used a similar system.

Dujardin then thought of obviating the use of the scale by charging the pens with coloured ink, but as he was unable to mix enough distinguishable colours, he proceeded to divide the alphabet (minus k and w) into six colour-groups as follows:

BLUE	:	a	g	n	t
RED	:	b	h	o	u
YELLOW	:	c	i	p	v
VIOLET	:	d	j	q	x
BLACK	:	e	l	r	y
GREEN	:	f	m	s	z

He then ruled his paper into four columns, and the rest is self-evident. A green dot in column one can only be F, and so on. The Tachygraphe, however, was a flop, and a larger model for printing music proved no more successful than its predecessor. Dujardin is last heard of in 1847, patenting a printing telegraph which made dots on paper around a cylinder by means of an electromagnetically controlled pen in an ink reservoir.

These were good years for France, which continued to harvest healthy crops of writing machines. In 1839, Louis Jérome Perrot patented a 'machine tachygraphique': he was either a very undecided man, or else of many minds about his invention, for over the years he added some fifteen supplements to the original specifications. They did little good, for the device was never warmly received. Each hand operated half the keyboard with its own corresponding type-wheel; depression of a key first selected the desired character and then brought it into contact with the paper by means of a hammer striking the type-wheel spindle. Both type-wheels could be made to print simultaneously. Carbon paper made the impression, but this was wound around the platen face *upwards* under the sheet of writing paper with results that inexperienced typists know only too well— except that Perrot's characters were not mirror images of the impressions they were to leave, so that the printing turned out correctly after all. In the absence of carbon paper, the inventor suggested the messy alternative of blackening the platen itself!

In the same year, another Frenchman, Pierre Foucauld or Foucault or Foucaux invented the first of three important machines for the use of the blind. His name is indiscriminately spelt in all three ways; in fact, in a July 1843 source[48] it terminates in a 'd' throughout, while in the February 1850 issue of the same publication the 't' form is used, and his death certificate uses the terminal 'x'. Himself blind, he was at first concerned with developing a machine which would permit the blind to correspond with those who could see. The machine had to be capable of small print, he felt, because paper was heavy, letters were weighed, and postage was expensive. His first effort was a radially-striking plunger device called a Raphigraphe which used only ten keys to compose all the characters used in printing. Foucauld had broken down the letters of the alphabet into their components, one of which was at the extremity of each key. By simultaneous depression of several keys, the letters were built up of their vertical, horizontal and oblique portions. The ten keys struck downwards at a common printing-point—the first three handled accents and tall strokes as in 'h' and 'b'; the following four keys were used for all letters contained within the lines such as 'a' and 'm', and the last three for those with strokes below the line, like 'g' 'p' and so on. At first, the plungers terminated in points perforating the paper, later in pencil lead, allowing the writer to use both sides of the page, with a consequent saving in weight. Printing was slow; although several keys could be depressed simultaneously, the paper still had to be back-spaced for additional impressions. Nevertheless, a competent blind typist was reported[48] to be capable of printing fifty alexandrine verses (i.e. fifty lines of twelve syllables each) per hour.

68

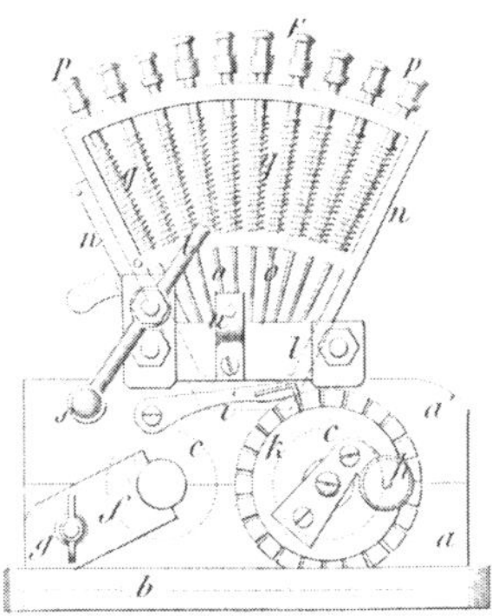

52. On the Raphigraphe, which dates back to 1839, Foucauld placed ten plungers in a row and fanned them out radially from the printing point. The simultaneous depression of keys in different combinations built up the alphabet by a series of dots. (46)

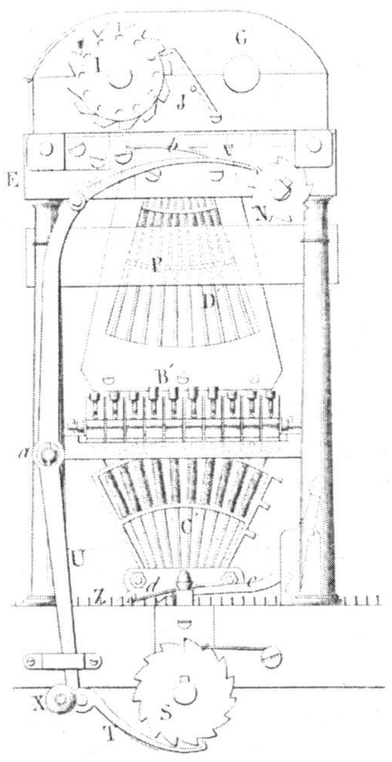

54. ... so he devised a machine consisting of one Raphigraphe mounted upside-down over another, with a common set of keys operating both machines simultaneously. The lower one printed while the upper one embossed, permitting the blind operator to 'read' what he had written. (46)

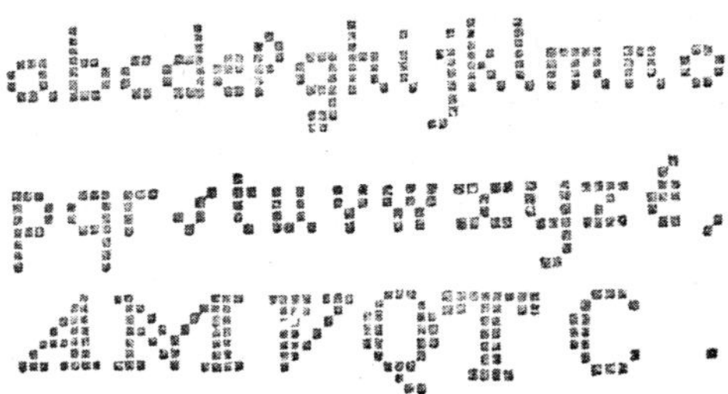

53. A sample of the Raphigraphe's printing. But the inventor, who was blind, was dissatisfied because he could not read what he had written ... (15)

The machine was well received despite its handicaps and earned its inventor quite a reputation. Braille himself seems to have entered the picture at this point although no two sources agree on the extent of his participation in future efforts. According to some,[50] 'Braille himself constructed a machine with his assistant Foucault', this being the most extreme, and probably erroneous, reference.

Others[34] have the two blind men 'collaborating' on the Raphigraphe, but even this would appear to give Braille more credit than he deserved. On the strength of original sources, it seems reasonable to assume that Foucauld did the inventing on his own and later on brought the more famous Braille in for prestige reasons. Thus, Braille is not even mentioned in early descriptions[48], while by 1862[58], it has become 'the embossing and printing frames of Monsieur Braille and Monsieur Foucaut' (*sic*). By then, Braille had already been dead for ten years and his reputation was enhanced by the passage of time.

Foucauld himself dismissed his first machine as suffering from the serious deficiency that it did not allow the blind to read what they had written. The consequent limitations were threefold: the communication had to be short, otherwise the writer forgot what he had already written; it had to be completed in one sitting, otherwise he forgot where he left off; and above all, it had to be written with extreme care since it did not allow him to make corrections. The inventor therefore developed an improved model to overcome these handicaps. It is best described as one Raphigraphe turned upside down and mounted over another, with ten keys connecting the two sets of radial plungers. Depression of a key caused one plunger to make a small impression on a sheet of paper below at the same time as another plunger perforated the character four times larger on paper held by a roller above. The writer could therefore feel what he had written on the upper page, while the lower one served for postage. Pencil lead was abandoned because of the difficulty of alignment after wear, and was replaced by blunt characters printing through carbon paper.

Foucauld's final machine was a departure from this intermediate design and once again printed on one surface only. The Clavier Imprimeur of 1849, reported in 1850,[48] employed the same radial striking principle: sixty plungers fanning laterally from the common printing-point were disposed in two parallel rows, the rear being vertical and the front obliquely inclined towards it. A guide was used to achieve alignment and each plunger was now equipped with a separate character so that only one key had to be struck to print each letter. Paper was held in a flat frame; there was carbon-paper impression; a bell indicated the end of the line; there was a space bar between the two rows of keys, and an increase in speed to seventy alexandrines per hour. It is immediately evident, of course, that the machine was no longer designed solely for the blind, although the inventor suggested that, with the substitution of different type it could be equally useful for embossing or perforating. Its versatility and adaptability to other uses was also convincingly advocated, for the machine earned its inventor a gold medal at the London Exhibition of 1851, and was soon manufactured on a small scale (see Chapter Four). It was generally well received everywhere.

The invention of Jaan Jaackson of Riga was not. He is reported[5] to have designed a contraption that looked like a permanent-wave machine and consisted of a semi-sphere of curved ribs, each of which bore a series of characters. This semi-sphere was rolled in a concave inking-pad by means of a handle, and then turned over until the desired character appeared underneath at the printing-point. 'It was an appliance absolutely without value, an impractical invention for which we are unable to locate the date as to when it was devised.'[42] 1840 has been mentioned[5] and for lack of substantiation will have to do, although both sources are often notoriously inaccurate.

Inaccurate reporting is also responsible for the historical injustices suffered by Alexander Bain, an inventive genius who is rarely given due credit for the work he did. In works on writing machines, he is usually[28, 40 etc.] dismissed in a sentence, sometimes in a few sentences,[15] and is invariably overshadowed by his more famous contemporary and bitter rival, Professor (later Sir) Charles Wheatstone. And yet the 'bible' of early telegraphy,[38] printed in 1859, spoke of 'the many ingenious contrivances invented by Mr Alexander Bain. He was not a commercial man, but his inventive powers were most wonderful. He has given the world some invaluable inventions.'

Bain was a watchmaker's apprentice in his late teens when he first became interested in electricity. Son of a provincial farmer, and a self-made man, he moved to London and, despite lack of money, managed to experiment with numerous applications of the 'electric fluid',[18] in the process of which he invented, among other things, an electric clock and an electric printing telegraph. In need of a patron, he went to Wheatstone in August of 1840 with models of both these; the professor patted him on the head, gave him £5 and promises of more, and promptly plagiarized his work. He advised Bain to keep his inventions secret, presumably with the intention of beating him to the Patent Office. However, Bain found another willing patron in a prestigious London watchmaker called Barwise and together they filed a patent for the electric clock in October 1840; the following month Wheatstone exhibited 'his' electric clock, which he was later obliged to withdraw after Barwise successfully served an injunction against him. Bain and Wheatstone became bitter enemies and when Bain, after a legal battle, was paid £7,500 by the Electric Telegraph Company in 1846 for his patents, Wheatstone, who was associated with that company, resigned from his post.

The controversy over the priority of invention of these two men will probably never be solved. It is of course true that Wheatstone's contributions to telegraphy precede Bain's and it is similarly true that, by a matter of a few months, Wheatstone patented a printing telegraph prior to his rival. For a number of reasons, however, one is inclined to reverse this priority and concede to Bain the benefit of the doubt. He is known to have possessed the printing telegraph machine in 1840 and to have shown it to Wheatstone, but one needed influence to get patents granted in a hurry and influence was one thing which Wheatstone had and Bain lacked. One also needed a certain amount of money—either money or a patron. Barwise, the watchmaker, filled this role with the clock patent but is hardly likely to have been interested in printing telegraphs, in the patenting of which Bain associated with a Lieutenant Wright. All the more credit to Bain, then, if he was able to win his legal battles despite Wheatstone's financial and professional influence.

Wheatstone was undoubtedly a great man, but his ethics and procedures at times appear to have been questionable and Bain was not the only inventor with whom he fought. He also had a serious dispute with his associate William Fothergill Cooke, with whom he shared many inventions and patents. He appears to have suffered from a failing common to many men of outstanding genius: an inability to concede credit to contemporaries for even minor contributions. Hooke from the seventeenth century comes quickly to mind, as does Edison (see page 146).

The details of the dispute between Wheatstone and Cooke fall outside the scope of this book, since they relate more closely to telegraphy in general than to the printing telegraph. Briefly, it appears certain[38] that Cooke built his first telegraph device within weeks of seeing an electrical experiment in Germany in 1836. He used a mechanical escapement tripped electrically, rightly deeming it preferable because it required only one circuit. Wheatstone, meanwhile, was working independently on electrical pursuits, but as a theoretician with little concern for the practical application of his experiments. The unknown Cooke was referred to the well-known Wheatstone and the two agreed to collaborate—Wheatstone's prestige was a desirable commodity. The result was the 1837 patent filed by 'Cooke and Wheatstone' and four years later, when independent judges were called upon to adjudicate the dispute between the two men, their award stated that 'Mr. Cooke is entitled to stand alone, as the gentleman to whom this country (England) is indebted, for having practically introduced and carried out, the electric telegraph . . . and Prof. Wheatstone is acknowledged as the scientific man whose profound and successful researches have already prepared the public to receive it.'[11] Wheatstone accepted this judgement conceding him the minor role, and both he and Cooke agreed to the wording and signed it. Shaffner[38] considered Cooke the inventor and Wheatstone merely the discoverer, uninterested in practical matters; but this, written in 1859, is hardly commensurate with Wheatstone's career as a whole. Rather, he appears to have taken advantage of his professional standing to plagiarize lesser men's inventions, many of which, it is true, he ultimately improved upon.

Considerable space has been devoted to these three men—Bain, Cooke and Wheatstone—not only because they were important but also because they have been so poorly studied in the past in connection with the printing telegraph and writing machines in general. As for Bain, an examination of his relevant patents discloses a number of important facts, not the least of which is the revelation that in 1841 he invented the typewriter ribbon, an honour till now invariably[1, 28, 40, etc] bestowed on Ravizza with his Cembalo Scrivano of 1855. There is no question about Bain's rights in this matter, for his patent (No. 9204 of 1841) is quite unambiguous: '. . . a roll of half-inch-wide ribbon, which is rubbed over the surface with a composition of two parts of oil, four parts of lamp black, and one part of spirits of turpentine, which will enable the pressure of any form thereon, as the type, to communicate the same to the paper'. In case that may appear too primitive, the matter is further resolved two years later (No. 9745 of 1843) and dismissed almost contemptuously: 'ZZ are rollers carrying an endless silk ribbon which has been previously saturated with printers' ink and dried.' And for those who object to the 'endless' element, here is the typewriter ribbon as it was when it reached finality in Bain's Pat. 10838 of 1845: there are two rollers, F & G, of which F is operated by clockwork. And there is a silk ribbon. 'This ribbon is first wound on the roller G, and in working the telegraph it is gradually wound off on to the roller F by the action of the clockwork.'

As for the instruments themselves, Bain's 1841 patent specifies two alternatives. On one, there are eight keys numbered 0 to 6 plus a blank for spaces. Depression

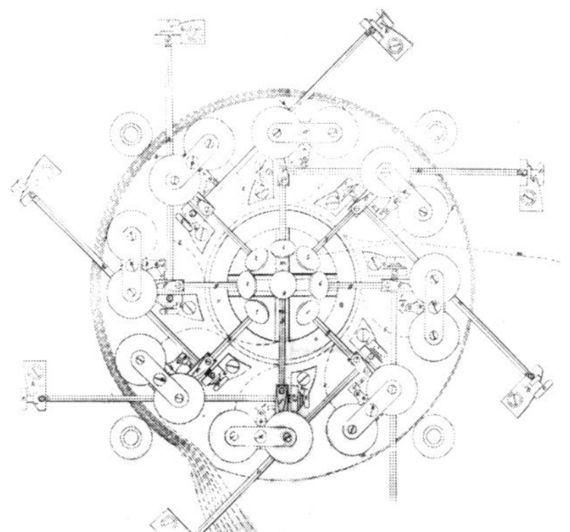

55. The first of the two alternative printing telegraphs patented in 1841 by Alexander Bain, the inventor of the typewriter ribbon, was this clumsy device using a separate line and electromagnet for each character. (BP)

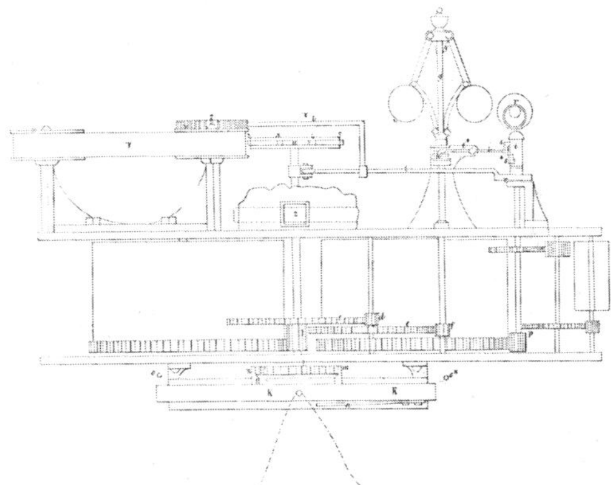

56. Bain's second design was excellent: it operated according to the 'step by step' principle to release a train of clockwork a tooth at a time. A type-wheel did the printing. (BP)

of a key closed a circuit, causing the corresponding type-bar to be drawn electromagnetically to a common printing point on an endless strip of paper. This was advanced by means of a ratchet controlled by its own electromagnet. The system, therefore, required the use of as many lines as there were characters to be printed, with an extra line for spacing. Not particularly good, and rather anachronistic. The other device was far better. It printed mechanically, using one clockwork train for character selection and another for the printing action. Electricity was used only for the escapement, and required one line and earth return, though actually Bain preferred a body of water. The instrument used a type-wheel mounted co-axially with a ratchet, and advanced by a means comparable to an anchor escapement on clocks, the ratchet being released electrically a tooth at a time by the 'step by step' principle, as a pointer on a dial passed successively over conductive and insulated sections, thus breaking and closing the circuit.

Improvements followed in later years. The design became more compact and the synchronization of different sets was simplified. A sheet of paper was sub-

C*

stituted for the paper strip by being placed around a cylinder which wound itself gently up a worm screw as printing progressed, thereby leaving a single sheet of copy with the typing in slightly inclined rows. Wheatstone used the same system.

In the 1843 patent, Bain presented details of a chemical recorder of such ingenuity that if he had done nothing else in his life, he would have ensured his place in history with this alone. He is usually credited with the invention of the electro-chemical telegraph itself, although this honour was secured by Davy some years earlier (see page 66). What Bain did was to take a metal frame and fill it with insulated wires with the live ends protruding; this he suspended in a wooden framework so that it formed the weight for a train of clockwork. A pendulum with a metal feeler swung alongside, so that the path of the feeler just cleared the field of live wires. When type or any other metallic object was placed over the wires and the pendulum was swung, the feeler passed over the type and broke and closed the circuit, while the frame dropped steadily as the pendulum swung and the clockwork unwound. These electrical impulses were transmitted to another identical machine which recorded them as discolourations on a sheet of paper previously soaked in prussiate of potash (potassium ferro-cyanide) and nitrate of soda.

Later patents perfected and simplified the system and it was used in England

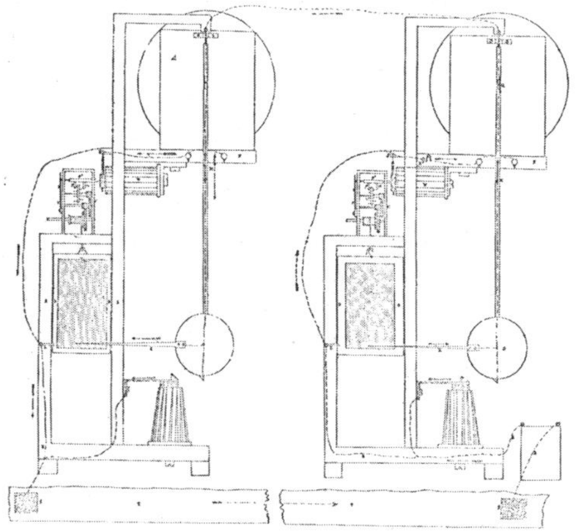

57. An ingenious electro-chemical telegraph (1843) was perhaps Bain's greatest contribution. As the pendulum swung on the transmitter, its pointer passed over whatever metallic object had been attached to the 'live' frame, which served as the weight for the clockwork. With each swing of the pendulum the 'weight' therefore dropped a little, and, in so doing, presented a fresh line to the pointer. The impulses produced corresponding discolourations on treated paper on the identical receiver. (BP)

58. Wheatstone's first printing telegraph (1841) used 'step by step' clockwork like Bain's but instead of a type-wheel, it had a flexible steel disk cut into segmental teeth, like slices of a cake, with a character on each tooth. An electro-magnetic hammer did the printing. (SML)

until the 1850s and even later. At first it printed the Steinheil code, later the Morse, and was the 'most prominent'[38] of all such telegraphs. It was introduced into the United States, but Morse interests secured an injunction claiming infringement of a Morse patent which had been shelved as inferior to the embossing and printing recorders. The advantage of the electro-chemical system, however, is that it operates without electromagnets, the opening and closing of which slowed down transmission speeds. Bain's alternative was 'so rapid and sensitive in action that it may again come into favour'.[13]

SML   Charles Wheatstone ultimately made greater contributions to the printing telegraph than did Bain, and three of his machines have been preserved. His earliest efforts date back to an 1841 patent for an instrument which 'borrowed' heavily from Bain's of the same year. This patent (No. 9022 of 1841) is the only one in which Wheatstone filed specifications of his printing instruments although, as will be seen below, he continued to experiment with them for another fifteen or twenty years. In the 1841 patent, carbon paper was specified for the impression; in later years ribbon was substituted, and printing was by means of a circular steel disk with slots cut radially into it, the type being arranged on the resulting segments, one character per slice. An electromagnetically controlled hammer struck the type against the paper, the steel strip springing back after the blow. The desired character was selected by the step-by-step method, and paper was attached to a cylinder which travelled linearly by revolving along a spirally cut screw. This same system appeared in Bain's 1843 patent. Alternatively, a sheet of lead or other metal could be used to receive the imprint, or a flat paper table moved by rack and pinion could be substituted for the cylinder.

A more sophisticated writing machine, departing from this principle and using a square metal comb with the type at the end of the teeth, is known to have existed prior to 1850, but the inventor appears not to have been satisfied, for it was not made public, not even the following year at the famous London Exhibition. It

59. The flexible teeth were retained on Wheatstone's next machine, dated 1851, but a linear comb now replaced the disk and a piano keyboard was added. (CSM)

75

has, however, been preserved and dated 1851[40] and several of its essentials were retained, with some refinements, in Wheatstone's later models. The metal comb was one of these: each character was mounted on the end of a flexible metal tooth and the comb, of sectoral profile, was swung until the desired letter was opposite the printing-point, whereupon a hammer pressed it against the paper to make the impression. The comb returned to its neutral position after the key was released. Numerous patents by later inventors borrowed elements from this design. Printing was on a narrow paper tape in upper case only, and inking was by pad.

Wheatstone's next machine, dated 1856 and produced in collaboration with a Mr Pickler of Buda Pesth, was a radical departure from the first. It has not

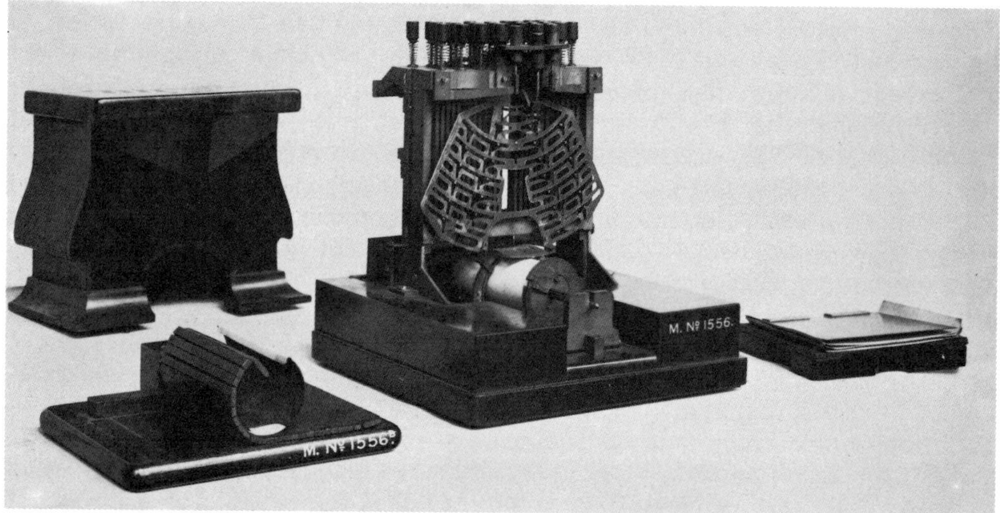

60. The profile had changed on this Wheatstone-Pickler instrument of 1856. The keys were now at the top of the unit and operated a complex guide plate; the principle of the printing comb was retained but it was modified into a swinging sector. (CSM)

been possible to determine the relative contributions of Pickler and Wheatstone to this machine; if past performance is any guide, however, one might well speculate that such a dramatic change in the machine meant once again that the design was originally Pickler's. The appearance was completely different, for it had acquired a vertical rather than a horizontal profile. The piano keyboard had given way to cylindrical buttons located on the upper surface of the instrument, now printing both upper and lower case. A cylinder was used to transport paper of larger dimensions; the cylinder rotated mechanically for letter spacing as keys were depressed, sliding along its axis for line spacing—in other words the arrangement was the opposite of that on modern machines. The principle of the type comb was retained, as was pad inking, which was perhaps the extent of Wheatstone's contribution.

A final machine to have survived is dated 1855–60. It appears to be the last of some six models which Wheatstone made and it combined the essential features of earlier machines: the piano keyboard from the 1851 model, change of case and the platen from the other one. It required too much effort; inking was still deficient, and in general it apparently worked badly, which may explain why the inventor abandoned the project after this last attempt.

It seems strange that Wheatstone should choose pad inking (and later inventors often preferred carbon paper) even though the ribbon had already been invented and Wheatstone himself used it on early instruments. Clearly, the ribbon is so efficient and its use so universal today that anyone attempting to adopt an alternative system might be accused of defective judgement. But it was not always so, and in the days before mass production, making a ribbon at home could be something of a nightmare. More of this on page 202.

Back to chronology and 1841, when Baillet de Sondalo and Coré, two Frenchmen, patented a Universal Compositor which was particularly recommended for people who could read but not write; for the blind, who could learn its operation in two hours; for cryptographers, and for a host of other worthy causes, such as making all typographical material, the pen, and the study of shorthand, instantly obsolete. An indicator performed these miracles: it was manipulated by the right hand and selected the desired character, while a pedal brought the paper into contact with a type-wheel and the left hand took care of manual line and letter spacing. Essentially a compositor, it was, the inventors insisted, a versatile device. Add a further pedal arrangement and it printed a letter other than the one selected, thereby turning out a simple substitution cipher which presumably only the initiated could decode. Or, alternatively, by merely inserting a large letter index and type-wheel containing all 500 syllables in the French language, it permitted

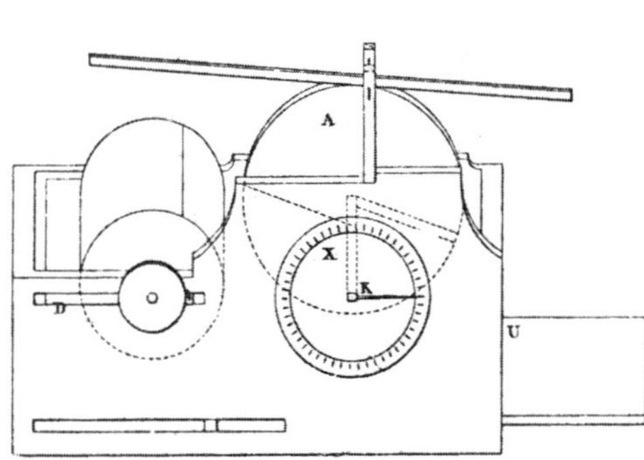

61. Baillet de Sondalo and Coré patented this primitive machine in 1841. It was offered optionally as a printing cipher device. (FP)

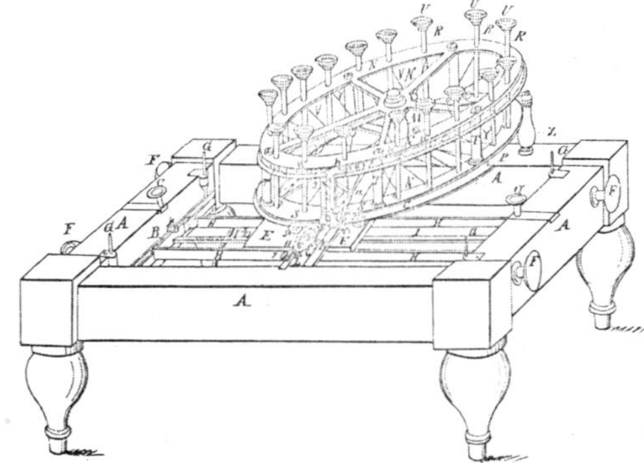

62. Charles Thurber built his Patent Printer in 1843 for a 'gentleman . . . to whom writing was a very irksome employment'. It specified a flat paper-table . . . (USP)

up to six letters to be printed simultaneously. According to the inventors, this gave the instrument a speed equal to that of the pen and even allowed dictation to be taken directly—'highly doubtful',[15] given the clumsiness of this enormous design.

Two years later, a man named Kohl is reported by an unreliable writer[29] to have invented a writing ball, the famous machine later produced by Malling Hansen (see Chapter Six). This information is almost certainly wrong. A cipher machine by Alexis Kohl of writing-ball design, dating from the 1880s, has been preserved (see Fig. 210) and this undoubtedly is the instrument referred to. Furthermore we might be excused for taking this writer's information with a grain of salt since he persists, among other things, in calling our next inventor Carlo Thurbero.

Charles Thurber of Mass. took out a patent in 1843 for a machine he unimaginatively named 'Patent Printer', the specifications of which provided the basis for a later more sophisticated instrument which has fortunately survived. In the idiom of the times, 'the work was taken up at the request of a Southern gentleman of means, to whom writing was a very irksome employment'.[67] Thurber made it less irksome by mounting a wheel on a vertical shaft and disposing forty-five plungers (fewer are illustrated in the patent specifications) in three groups around its perimeter. Each plunger terminated in a cupped key with the character in relief on the upper end, and the type at the lower. 'To use this machine,' suggested

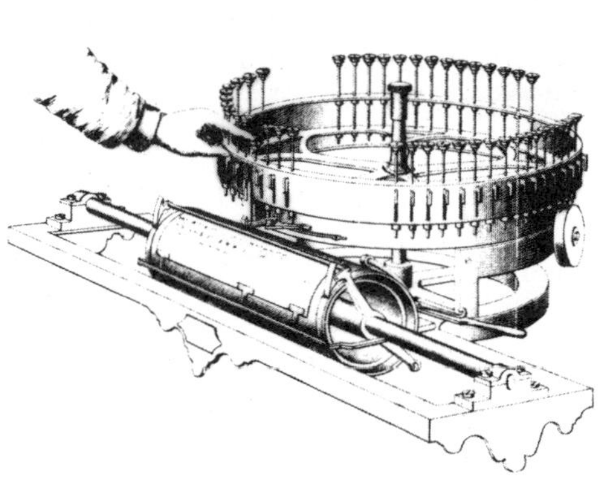

63. . . . but on this later model, it sported a very precocious carriage and platen. Note the peripheral type-plungers, passing over the inking roller on the right. (35)

64. He should not have done it! Thurber abandoned his typewriter and built this handwriting machine which he patented in 1845. The only thing that can be said in its favour is that it employed differential spacing. (SIW)

the inventor in his patent, 'the first thing to be done is to whirl around the horizontal wheel . . . till the ink is properly distributed.' Inking rollers were provided for this purpose. The wheel was then rotated (more slowly presumably) until the desired plunger was above the printing-point, whereupon it was pressed down to make the impression on the paper. A guide provided correct alignment. Thurber's patent specified a flat paper-carrier; in the later model, this had matured into a prototype of the present-day carriage, complete with cylindrical platen, paper-feed system and scale, and the necessary escapement for revolving the platen and providing it with lateral movement. Letter spacing was automatic, a key controlled spaces between words, and the rotation of the platen was accomplished manually. Bidet had provided the platen, and Thurber had modified and perfected it.

It is disturbing, however, to think that the inventor of the modern typewriter carriage should have felt obliged to return to a flat paper-carrier in his later writing monstrosity for the blind. This was an enormous machine which reproduced all the motions of the hand so that the final production was a letter mechanically handwritten in pencil. The device was called a Mechanical Chirographer and was patented in 1845. It is tersely reported to have proved a failure.

Thurber's contributions to writing-machine history were many, however, and his importance should not be minimized. Not only did he provide us with a carriage and platen on his early machines, but also he proceeded to embody full differential spacing in his handwriting device. Differential spacing has been mentioned several times and is discussed fully on page 210; briefly, it concerns mechanical recognition of the printing principle that all letters do not occupy the same space. The first incomplete application of this principle was in the Progin machine: capital letters were assigned twice the space allowed for lower-case letters. Bidet made manual provision for full differential spacing while Thurber made it mechanical, applying the entire principle for the first time well over a century ago; Jones ingeniously incorporated it into his Mechanical Typographer seven years later, as did Harger and others, and it has been used on numerous occasions from then until the present day. With apologies to IBM who, in their *Informal History of the Typewriter*, are informal enough to have claimed credit for this invention.

Our next inventor went mad towards the end of his life and committed suicide shortly after his release from an asylum, although there is no evidence that his writing machine was responsible. Labrunie de Nerval, a French writer, took out a patent in 1844 for a quaint device called a Stéréographe, the purpose of which was to print an entire line in one swipe. A series of type-wheels—as many as there were spaces on the line—was mounted side-by-side along an axle. Each wheel contained two separate sets of characters on opposite sides, the upper set serving for composition, and the lower for printing. The line was first composed, then inked by means of a hand roller, and finally brought down on to the flat paper by depressing a lever. It was, of course, essentially Bramah's machine of 1806.

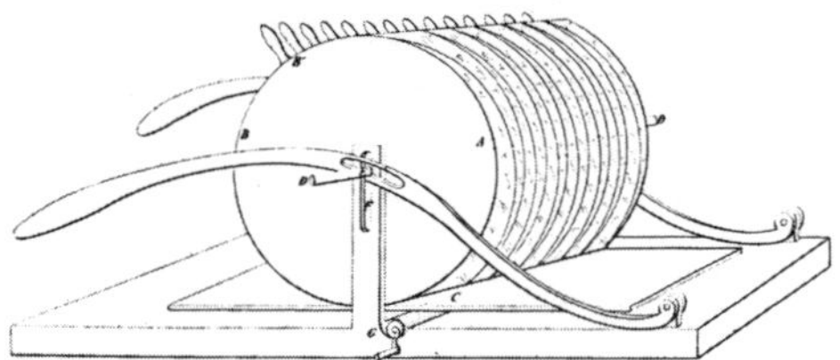

65. Labrunie de Nerval virtually re-invented Bramah's concentric wheels in this 1844 patent . . . (FP)

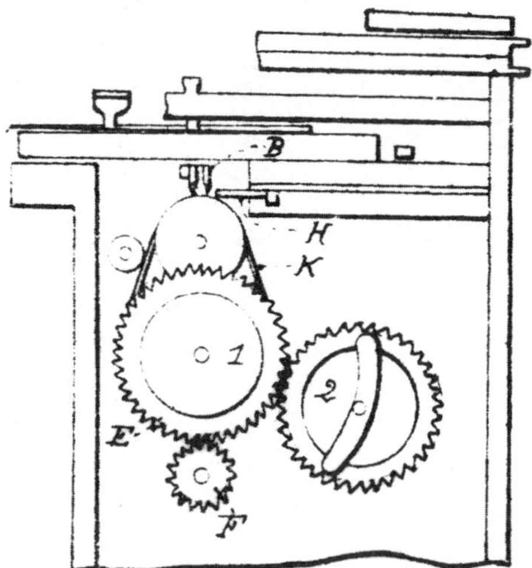

66. . . . and Pape, in the same year, more or less re-invented Dujardin's Tachygraphe. (37)

A French piano manufacturer called Henri Pape also appears to have borrowed heavily from a predecessor. His 1844 invention was Dujardin's Tachygraphe all over again. Twenty-four keys made marks across a strip of paper or cloth which wound off one cylinder on to another. After writing was completed, the keyboard was hinged back to reveal a fixed scale divided into the twenty-four letters represented on the keyboard; the paper or cloth, having first been rewound on to the original cylinder, was then moved past this scale for decipherment, the dots indicating the letters that had been printed. Old hash.

In the same year, the famous Herbert Spencer is reported[28] to have conceived the idea of a writing machine; his autobiography, quoted as the source, reveals, however, that it was not a typewriter at all: 'To make type by compression instead of by casting, was the idea.'[39]

A device for the blind invented by a Mr Littledale was described[47] by the Revd W. Taylor. It consisted of a single row of wooden type, sliding in a box a yard long and three or four inches square. At one end of the box was a hammer which clobbered the characters that appeared beneath it 'by the application of an ingenious contrivance at the opposite end of the case'. Embossed letters for the blind were printed, these being made visible by prior insertion of carbon paper. But the 'ingenious contrivance' appears to have been the chief weakness of the device, for it caused the type to disintegrate under the blows of the hammer. Wooden type

80

67. The 'ingenious contrivance at the opposite end of the case' appears to have been the principal weakness of Littledale's writing machine.

was used to withstand the impact and a cloth damper was added as a shock absorber, but to little avail. It all sounds vaguely terrifying and in the absence of more specific information one can only speculate as to the size and weight of the hammer!

1845 was another bad year. A printing telegraph utilizing a type-wheel was the only thing that saved it. Patented by Jacob Brett in England in 1845, and by Royal E. House in the US the following year, it was developed from then until the 1850s, but it was plagued with troubles and eventually abandoned. The circumstances of this invention are somewhat mysterious: House and the Brett brothers are reported[25, 41 etc.] to have developed the instrument together; the former patented it in the United States while the latter is said to have patented it 'on communication' in England. Some sources[38] even offer a short biographical sketch of the inventors but without giving any details of just how, when and where these men co-operated, and it sounds, once again, very much like another case of each writer merely repeating information from a secondary source. Specifically, what ought to be explained, is why Brett's patent fails to specify that it was filed 'on communication from abroad', as is standard procedure in such cases.

The drawings from the original patent specification bear a striking resemblance to the skeleton clocks of that period, which appear to have offered the initial inspiration; a family of watchmakers called Brett are recorded in London until the early nineteenth century. The device consisted of separate transmitters and receivers, each powered by clockwork, with electricity used only to regulate the motion by means of an escapement. The transmitter had a piano keyboard with twenty-eight to forty keys, parallel to which was a revolving cylinder with pins spirally disposed around it—something like the barrel on a music box. On the receiver a type-wheel revolving at sixty r.p.m. was synchronized with this cylinder; depression of a key stopped the cylinder at the pin corresponding to that character, the circuit was broken, and the type-wheel stopped at the same letter. Several future inventors borrowed this design, including Devincenzi, D. E. Hughes and Edison.

Further patents were granted on both sides of the Atlantic for improvements

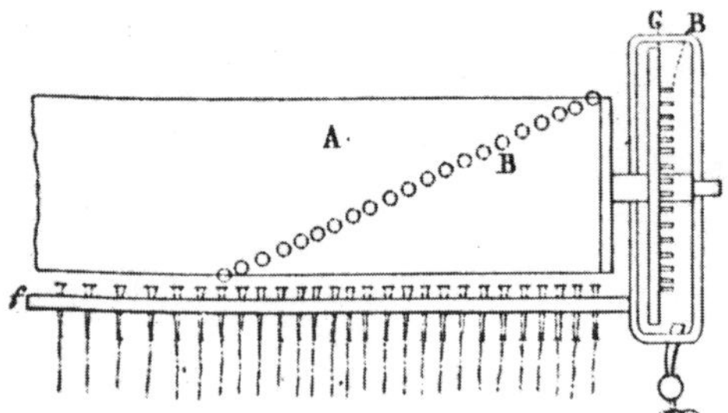

68. A comb and pin barrel not unlike that used in musical-box movements was at the heart of Brett's 1845 printing telegraph. Depression of a key stopped the barrel revolving by bringing the corresponding tooth of the comb into line with one of the pins. This idea was used time and again by future inventors.   (BP)

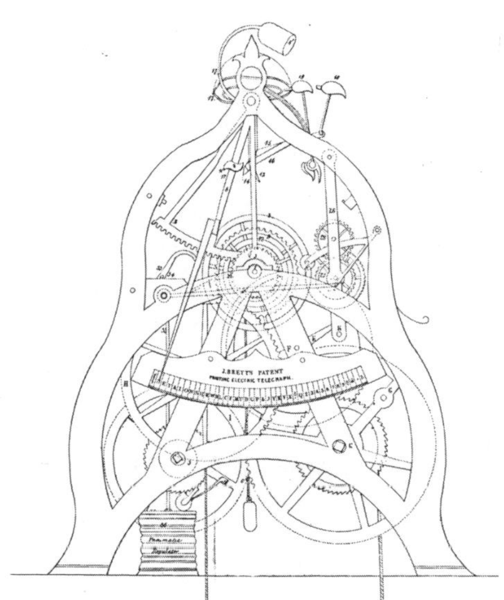

69. The design of Brett's printing telegraph was clearly inspired by the skeleton clocks popular at the time this 1848 patent was granted.   (BP)

to this basic design. The original alphabetical letter order on the type-wheel was later changed to one based on frequency, although it was still retained on the keyboard; in order to achieve correct selection, the pins on the cylinder had to be staggered. Brett had a limited measure of success with his invention, which was used for a time both in England and Europe, including the Dover-Calais cable in 1850, but it gave too much trouble and was dropped in favour of more successful instruments. House, on the other hand, appears to have persevered longer than his colleague and patented improvements continued to be filed as late as the 1860s. He also achieved a greater measure of success and his machine was used quite extensively throughout the United States.

70. One of the later models of Royal E. House's printing telegraph. It enjoyed greater success than Brett's. (38)

Apart from that, in 1845 a simple writing frame for the blind invented by the French Dr Saintard and his mechanic Saint-Gilles was later developed into a primitive writing machine through additions to the original patent specification of a sliding circular index bearing twenty-eight characters. A mere reference in a friend's letter is all that survives of a typewriter invented by Dr Leavitt[27] or Leavitte[14] of Kentucky, which must have worked, however, as a specimen of the typing was enclosed. No other details. Nor any of a machine erroneously reported[40, 42 etc.] to have been built in 1847 by an American called Prentice: this man was in fact editor of the *Louisville Journal* and author of the letter which referred to Leavitt's invention.

No less confusion surrounds the next inventor whose machine was not even a typewriter. Gustave Froment was the Frenchman's name; one source[28] insists that this is incorrect and that it should be Franet, and other writers have picked either spelling at random and added a few of their own. The correct name, once and for all, is Froment, for what it is worth, for it matters little, since his pantograph machine is not a typewriter at all. It was used for engraving tiny messages on glass, metal, etc. and came complete with its own microscope. And before we leave that fateful year another invention, by the Frenchmen Rohlfs and Schmidt (Frenchmen?) is commonly quoted[37 etc.] it is a device with text on a rotating rectangular prism and is thus a printing, not a writing machine and so fails to qualify. In the search for authentication of this information, a French patent

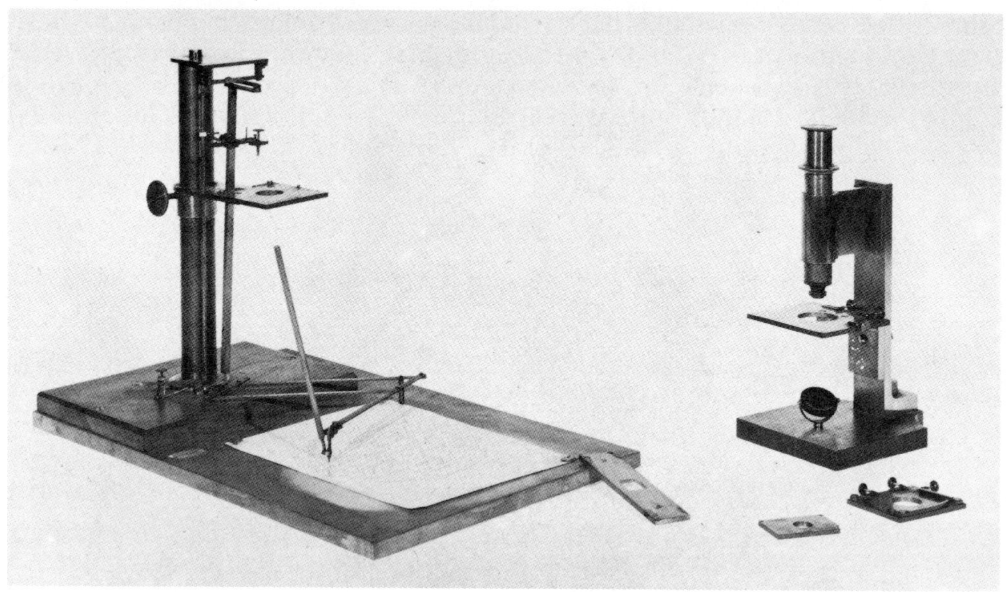

71. Froment's pantograph was designed for making miniature engravings and was not mechanical. It is thus disqualified . . .   (AMP)

72. . . . as is Rohlfs' and Schmidt's invention which, although mechanical, was a printing press and not a writing machine.   (FP)

issued to a German called Rolffs in 1849 was discovered, but it is not for a typewriter and therefore not of interest to us.

Scanty documentation, however, was not a problem from which Alfred Ely Beach suffered. As one of the owners and editors of *Scientific American*, he had reason to be confident that his writing machines would be sympathetically reviewed. His first typewriter appeared in 1847; 'the machine worked very well' according to the article,[67] 'but the quality of the printing did not satisfy the inventor's critical eye. So he laid it aside for improvement at a future time.' Which is a reporter's euphemistic way of saying that his boss's machine was a flop! It employed a basket of type-bars, printing at a common point on the sheet of paper supported on a platen. It had a keyboard, with letter- and line-spacing keys, a weight-driven carriage, warning bell, and carbon paper for the impression.

Beach's second machine, patented in 1856, was intended for the use of the blind, and embossed characters on a narrow strip of paper by clamping it between

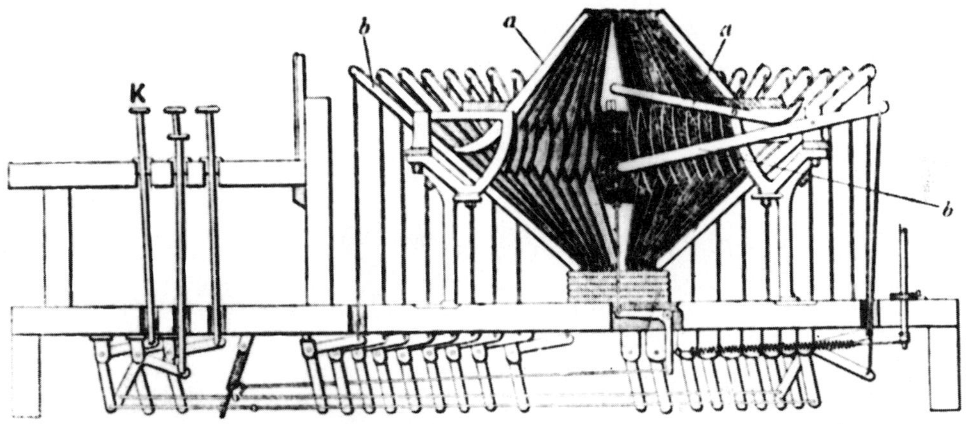

73. The only up-stroke down-stroke machine ever devised was this invention by Beach (1856). There were two type-bars for each character and they converged upon depression of a key, clamping the paper tape between them.   (59)

two type-bars, one male and the other female. These type-bars were arranged in two circular baskets, one above and the other below the printing point, the machine was thus an up-stroke down-stroke machine, the only one of its kind. Depression of a key caused the corresponding type-bars to converge simultaneously on the printing-point from above and below. This design might well have doomed the instrument to the ranks of historical curiosities and nothing more; in other respects, however, it was precocious indeed. For one thing, the linkage between keys and type-bars was sophisticated, and the escapement was operated by clockwork and tripped by a type of universal bar which later became standard equipment on typewriters. A three-row keyboard that Beach later added had a distinctly modern look to it, although, obviously right-handed, he had disposed the most frequently used letters to the right. This is not really much worse than the terrible arrangement we are at present burdened with, which will be discussed at greater length in Chapter Ten.

To return to 1847, or some time around that date and before 1849, when the book[4] was published, we find the sole reference to a 'Gravicembalo' (harpsichord)

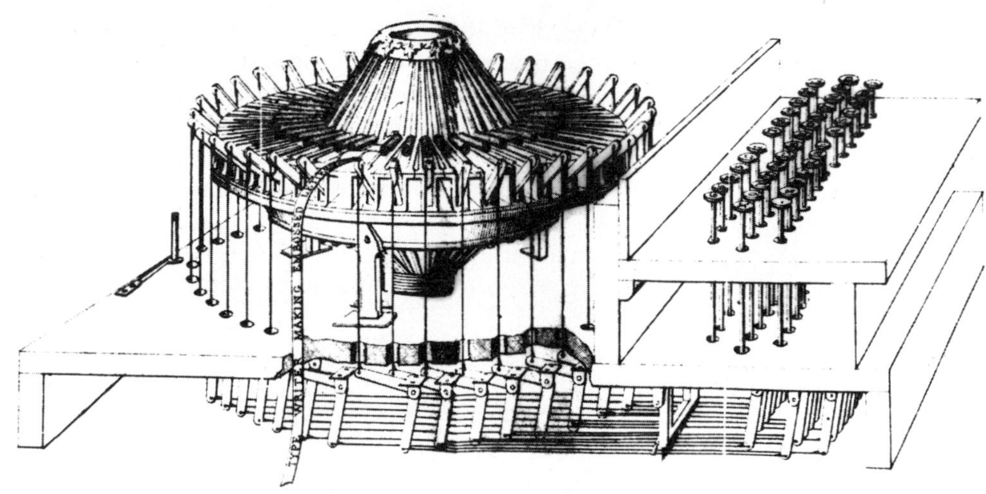

74. Despite its peculiar action, his excellent
design has earned Beach credit for the invention
of the universal bar.   (35)

invented 'some years ago' by an Italian called Bianchi. According to this report,
the machine as presented to the Paris Academy had a piano keyboard, type-bars,
and a cylindrical platen around which specially prepared paper was wrapped.
This was apparently made to rotate for line spacing, a feature to which we have
become so accustomed on modern machines that it seems impossible that anyone
should ever have built otherwise. In early days, however, it was by no means
certain that this design would ultimately evolve. Many machines, as we have
seen, used a flat paper-carrier; when the platen was incorporated, this was usually
made to rotate for letter spacing and move laterally for line spacing, clearly because
this system is easier to design. Harnessing the printing movement to rotate a
cylinder by engaging the teeth of a wheel is a simpler mechanical proposition than
devising an escapement that permits lateral motion. But apart from the platen
design, Bianchi's writing harpsichord was apparently a good machine, capable of
fast, clear work and performing 'with the speed and precision of the Thalberg
variations',[4] a reference to the virtuoso pianist who was the idol of his day.

This was precisely the kind of praise which Christian Sörensen of Denmark
was fated never to hear. He is on record in 1849 as requesting financial assistance
from the authorities for the construction of a piano-type machine which he claimed
was capable of 500 strokes a minute—on the drawing board, that is, for his
request was refused and this speed demon never materialized. Some years later,
however, having abandoned the typewriter for more acceptable inventions, he
constructed a printing press with the very financial aid which was refused to his
typewriter project. It appears to have worked well, and he won a decoration and
a medal for it, awarded in person by Napoleon III, no less! But ultimately this

DTM

did him little good, although he was later able to pawn it to ward off starvation.

Oliver T. Eddy, of Baltimore, was even less fortunate, for he was left with nothing to pawn at all. He built a monumental writing machine in 1850 and literally dedicated his life to promoting it. He received nothing but setbacks; a cry for help was turned down, and he died, as he had lived, in abject poverty. He has been nominated[28] as the first of many professional typewriter inventors whose lives ended in tragedy as a direct result of their inventions.

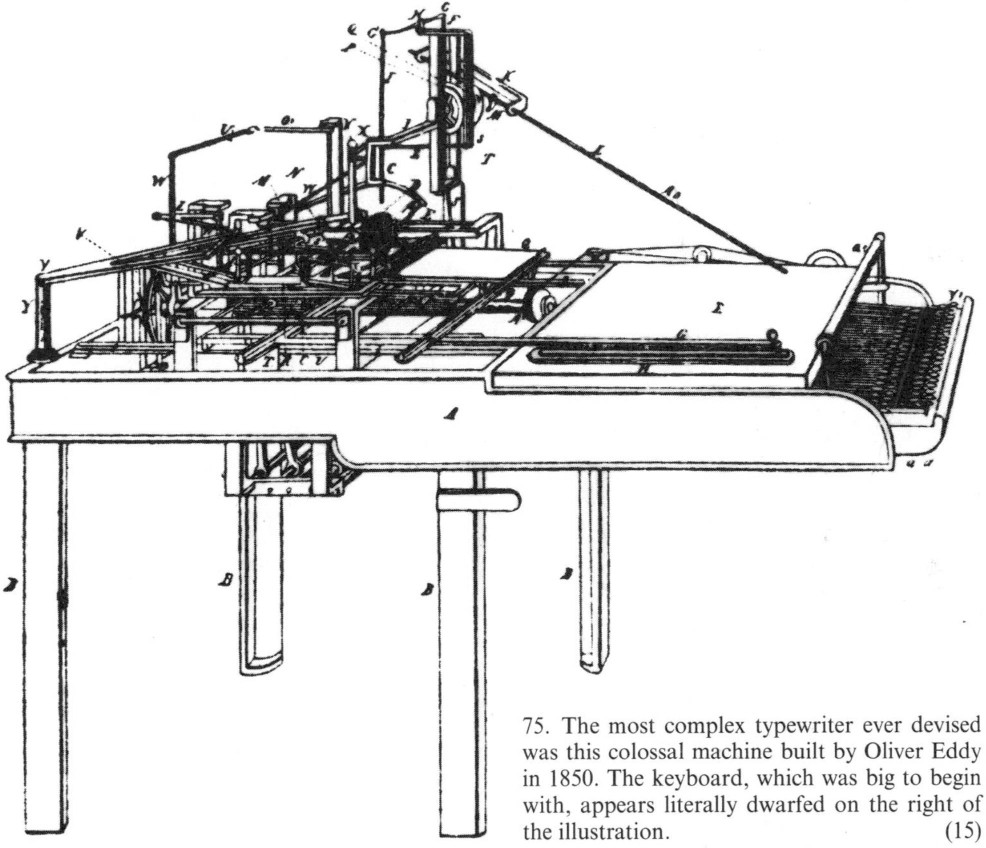

75. The most complex typewriter ever devised was this colossal machine built by Oliver Eddy in 1850. The keyboard, which was big to begin with, appears literally dwarfed on the right of the illustration. (15)

Eddy's apparatus was a huge and highly sophisticated device, as complicated as a description of its operative elements. 'The upper part of the standard is an oblong rectangular frame so constructed as to support the horizontal shaft, U, and having across the top of it the projecting arm, R, which sustains the standards, 0′ 0″, which are the boxes of the shaft, H, which shaft is the fulcrum of the lever, G, the long end of which is connected with the plunger, C, by the connecting rod, I, and the short end of which lever, G, is connected with the lifting rod by the connecting rod, F . . .'[27] and so on. The essence of the machine was a keyboard of seventy-eight keys in four rows dwarfed at one end (on the right in fig. 75) and actuating a similar number of plungers containing the characters, bunched together in the middle of the monster. The paper in its flat carrier moved a space after each impression; there was a warning bell, and line spacing was automatic. The type-plungers printed at a common point, inking themselves on the way to the paper by one of the most complex linkages ever designed for that purpose.

The year 1850 was important, as will be shown in the next chapter, and Eddy's was only one of many machines which put in a simultaneous appearance. In Germany, Werner Siemens transformed his pointer telegraph into a printing instrument by gearing into it a type-wheel of radial segments similar to Wheatstone's. An electromagnetic hammer brought paper and type-wheel together. There was nothing revolutionary in this well-tried design; however, Siemens later made more sophisticated contributions and he and his business associate Halske built up one of the largest European organizations manufacturing telegraphic and electrical equipment.

The same year, an Italian called B. G. Marchesi invented a machine for the blind: it was circular in shape and permitted visible or embossed printing in three kinds of type. The Blind Institute in Milan used it, and it was entered in the 1851 London Exhibition, where it won a medal: this was apparently about the most a writing machine inventor could hope to get for his troubles! An American, Joel Wight, also built a writing machine for the blind, consisting of a flat paper frame with a rack for line spacing and manual cell spacing. Nine keys embossed pin-pricks in a ⋮ format, the small keyboard riding over the frame on two rails.[57]

SIW   There was also a novel phonetic printer by a compatriot, J. B. Fairbank, whose name is invariably mis-spelt by the addition of a terminal 's'. His device was of radial plunger design and was essentially a rectangular box standing vertically ('somewhat resembling in its outward appearance an upright piano case', according

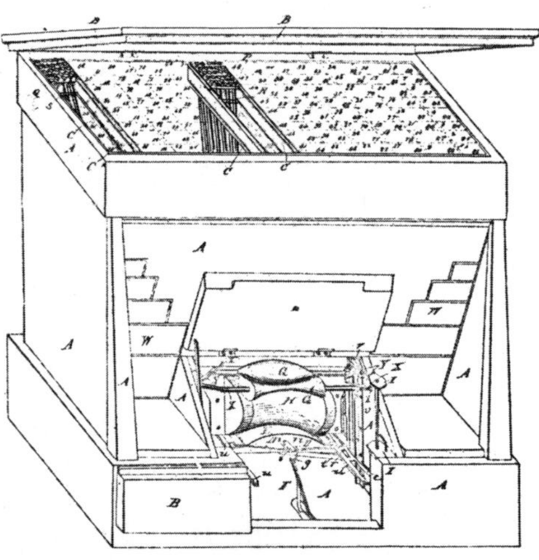

76. Fairbank's 1850 invention specified no less than forty-eight complete phonetic keyboards— row after row of them! Paper was stationary so that the first character to be printed was selected from the first keyboard, the second from the next, and so on. Spacing was by merely skipping a keyboard. In the illustration, the raised cover reveals two of the keyboards.　(USP)

to the patent specifications), with the keys at the top, paper carriage beneath, and a guide for the type plungers between. Inking was either by a roller applied to the type face at the end of every line, or else by carbon paper. The carriage was moved by hand or foot at the end of the line, but there was no lateral movement of either the type font or the paper-table for letter spacing: instead, Fairbank's unique solution to the problem was a keyboard consisting of no less than forty-eight complete phonetic alphabets from which the sounds to be printed were selected in succession. In other words, the first sound on a line was chosen from alphabet one, the second from two, and so on. Spaces between words were made, of course, by skipping an alphabet or two. At their lower end, the type plungers described an arc as they radiated upwards: both the inking roller and the paper-

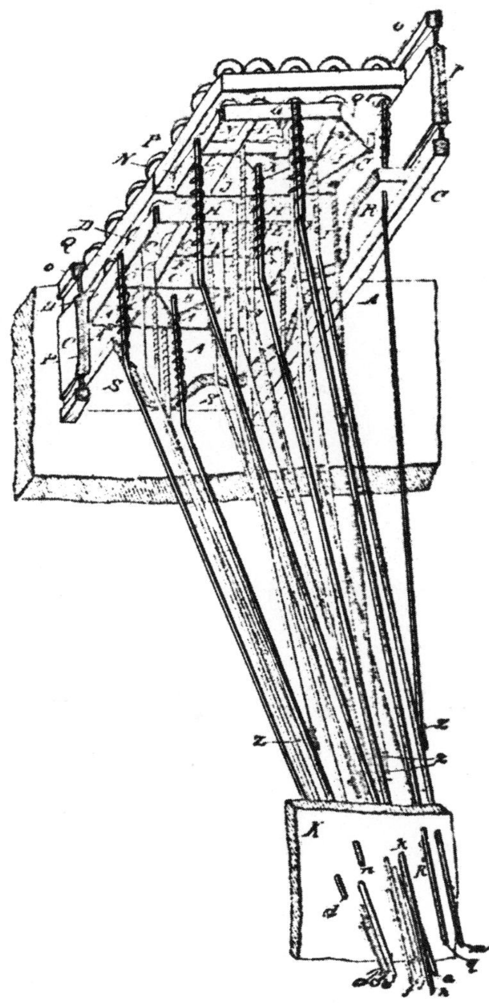

77. It was rather busy inside the Fairbank box, with a separate type-plunger for each key of each of the forty-eight complete alphabets, all of them radiating upwards from the curved platen. (USP)

table were similarly convex. Since it was stationary, in a lateral sense, the patent suggested the use of a continuous roll of paper, a feature destined to become quite an issue in later years (see page 147).

Virtually every secondary source[27, 28, 40, 43 etc.] blithely perpetuates a mistake once made as to the nature of Fairbank's invention. Referring to the machine as a Calico Printer, they describe it as one designed primarily for printing patterns on to cloth and thus they summarily dismiss it as a writing machine, quoting as their authority for this verdict Fairbank's patent no. 7652 dated 17 September 1850. It so happens, however, that this patent contains the description and illustrations included above. Clearly, earlier writers simply borrowed from each other without ever bothering to refer to the original source, namely the very patent they claimed to be describing.

Chapter 4

The Beginnings of Production, 1851–67

Up to this point, writing machines were virtually one-off models built either by the inventor himself or to his specifications by a competent machinist or carpenter. Whenever a patent was sought the inventor lost his original instrument right away, for it had to be submitted with the application. If he could afford it, he then had another one made. At times two or even three were built, but usually a single instrument was then toted from place to place for demonstration and display. However well it may have worked, the typewriter nevertheless failed as a commercial enterprise and few are the early inventors who were rewarded at all for their troubles; none of them was adequately recompensed.

The half-century, however, ushered in a new era: the manufacture of writing machines actually began, at first on a small scale but with increasing momentum as the decades passed. The choice of the year 1851 is not arbitrary, then, for the date may well be considered a turning point in mechanical history. The great London Exhibition was opened in the spectacular Crystal Palace, built specially for the occasion—the Exhibition of the Works of Industry of All Nations—and this and the many similar events all over the world that followed its successful lead provided the impetus that had previously been lacking. Almost 14,000 exhibitors took part on that first occasion and total attendance exceeded six million.

Several typewriters were manufactured in the 1850s. One of these was Foucauld's last machine, first reported[48] in 1850 and successfully entered in the Exhibition

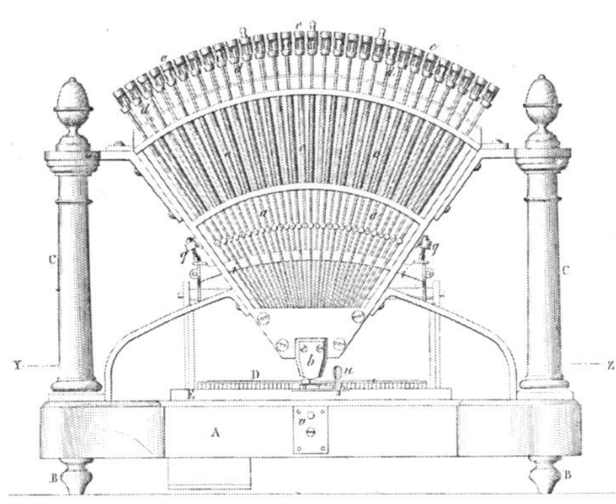

78. Foucauld's Clavier Imprimeur of 1849 perfected the radial plunger design of his earlier inventions. It was manufactured in small quantities during the 1850s.   (48)

soon afterwards (see page 70). Eleven years later, it was still 'generally used in France and on the Continent'.[58] However, exactly in what year production began and how many units were made is not known.

SML   Another instrument with a virtually identical history is the Hughes Typograph. Invented in 1850, it, too, was entered in the Exhibition where it was warmly praised by the Juries and was decorated with the highest awards; at the 1862 Exhibition it was again decorated and deemed the best machine of its kind, thereby providing another remarkable example of longevity. Once again, dates of manufacture, number of units built and selling price are not known.

Apparently, however, the Typograph was heavily copied on the Continent under the flimsiest disguises: the Frenchman Levitte merely provided it with lower case while his compatriot Larivière put the letters in two rows. The Swiss Hirzel and German Ehlwein also brought out virtually identical copies. There is only one source for this information[28] which the author could only partially substantiate. A patent granted to Georges Henri Larivière in 1850, for example, dealt only with colour printing of cloth and lingerie—whether this is the same Larivière mentioned above is unknown. On the other hand, the 1862 Exhibition Reports[58] indirectly confirm the information by saying of the Hughes instrument: 'So useful do the directors of the French schools for the blind consider this writing machine that they have adapted it to the *système braille* in use amongst their pupils, and have exhibited their improved machine in case No. 2807, in the French department.'

It is as well to dispel a confusion that originated in the Reports of the two exhibitions[52, 58] and that has been perpetuated in several sources[40 etc.] since that date. The 1851 Catalogue and Reports speak of a machine for the blind invented by G. A. Hughes, while the 1862 Reports refer to one by W. Hughes. They are somewhat differently described, as well, leading to the mistaken conclusion that they are in fact two separate inventions by different men. This is not the case: the two references are to slightly different models of the same machine—in twelve years, it had obviously been improved. A phrase in the 1862 Reports confirms

79. The fine Hughes Typograph for the use of the blind was also manufactured during the 1850s. Of peripheral plunger design, it was considered the best machine of its kind. (CSM)

this: 'The Typograph . . . an instrument admitted by all the instructors of the blind to be the most perfect as yet produced, which received the Prize Medal at the Exhibition of 1851 . . .'[58] According to London Science Museum records, the correct initials of the inventor are G. A.

Hughes originally devised his typewriter to print embossed characters; later, carbon paper was used for visible printing. Forty-three plungers were positioned around the perimeter of a rotating disc, with a corresponding character in relief and a guide hole for each plunger; the disc was spun until the desired letter was above the printing-point whereupon a lever was depressed and, after locking itself in the guide hole, forced the plunger into contact with the paper. The unit was attached to a vertical toothed wheel that travelled along a spiral groove in a horizontal rod; depression of the lever therefore advanced the unit a space at a time. The paper was held flat on the base of the machine, at one end of which was a screw for line spacing. It was a good simple machine, effective and beautifully made, a little slow perhaps, but that was never a factor of primary importance to inventors of instruments for the blind.

Several other machines were displayed at the 1851 Exhibition along with the Foucauld and Hughes. Marchesi's was described in the last chapter; another,

80. The third invention to have reached production stage during the 1850s was the Mechanical Typographer by John Jones of New York (1852), but the first consignment of 130 units was destroyed when the factory manufacturing them burned to the ground. The design incorporated differential spacing.   (MPM)

by Tollputt, was merely described[52] as an 'apparatus' for the blind. It clearly failed to make much of an impression, for it received no award nor was a description of the device offered.

MPM   In the following year, John Jones of New York invented a Mechanical Typographer, also presented as a Printing Instrument for the Blind. He contracted with a factory for manufacture and 130 of his Typographers were either built or in the process of completion when the factory was burnt to the ground and all machines were destroyed. He left us, however, with history's first production figure (for in the case of Hughes and Foucauld this number is not known). It seems pathetically low, a mere 130 machines! On the other hand, when it began production more than twenty years later and with a far more sophisticated device, Remington was manufacturing almost exactly that number per annum, and it was laconically reported that there were more machines than potential clients that could be found for them. In the light of this information, 130 Jones Typographers as an initial batch is a creditable number indeed.

The machine, as presented in 1852, was pretty, simple, and capable of excellent work, predating similar index machines (World, etc.) by over thirty years. Jones took pains to decorate it and make it attractive to potential buyers, believing that it could be manufactured so cheaply that 'everybody' could afford it. As well they might, for it consisted mainly of a wheel on a vertical axis with the type underneath around its periphery, each character corresponding to a similar one visible on a ring surrounding the wheel. This was revolved till its pointer corresponded to the desired character, whereupon it was depressed by means of a lever, grooves providing correct alignment. The paper was on a platen of large diameter held in a carriage, the platen rotating for line spacing and moving linearly in the direction of its axis for letter spacing, as it does on modern machines. Depression of the type-wheel forced the carriage outwards, by means of levers, against the pull of a spiral spring; upon completion of a line of typing, the carriage was automatically returned and rotated a notch for line spacing. Inking was by roller.

Most important of all however, was the incorporation of mechanical differential spacing. 'It will be borne in mind,' states the patent specification, 'that the types on the horizontal wheel will vary in thickness according to the letters on them, and if some provision is not made for their inequality printing will not be even, the large letters will crowd or overlap the smaller ones. The arrangement just shown obviates this . . .' The 'arrangement' was not brilliant, but was good enough. Since printing was in a linear direction, wide letters (such as 'm') stuck out further than narrow ones. As these were brought down to the printing-point they brushed past a roller at the end of a lever, causing the carriage to be correspondingly displaced.

Jones continued to work on his invention and was granted another patent, for an improved Domestic Writing Machine, in 1856. It was essentially the same instrument as the earlier one except that it now printed both upper and lower case, and the handle which brought the type-wheel down on to the paper also turned the wheel for letter selection. The cylindrical platen was abandoned in favour of an endless belt serving a similar purpose; inking was still by roller, and full differential spacing was retained.

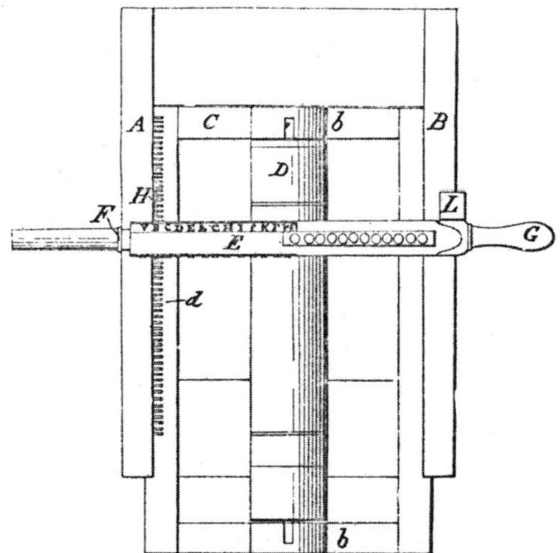

81. The primitive Thomas machine of 1854 can be built on a do-it-yourself basis by using 'two rolling pins and a knife tray'. But even a design as primitive as this made a contribution: locking holes in the type sleeve.   (USP)

94

A pity about that fire in 1852! If only it had spared the fine Jones machine and consumed Robert Thomas's Typograph instead! Even by 1854 standards, an associate of the inventor denounced this contraption as very primitive and little more than a toy. Poor Thomas has been ignominiously treated indeed: one source[28] hesitates even to call his invention a machine, while another[27] delivers the most unkindest cut of all: anyone can build a Typograph for himself by using a couple of rolling-pins and a knife tray! Merely arrange the characters in several rows around the centre of one of the rolling-pins and drill the necessary number of locking holes (to ensure alignment) at the end. Mount this rolling-pin across the wooden knife-tray so that it can be manually revolved and slid along its axis. It is now at right angles to the other rolling-pin which is inside the tray. In this way, the upper one is the type-sleeve while the lower one serves as the platen. Add a few sophistications such as a locking pin to engage the holes you have

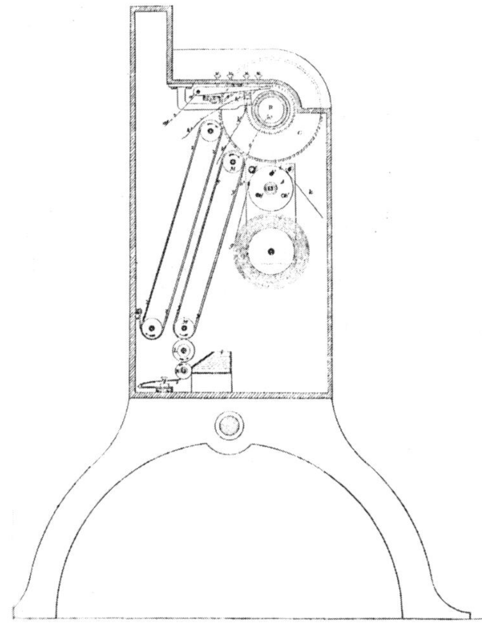

82. The first electric typewriter, Devincenzi's of 1854, borrowed the comb and pin barrel principle from the printing telegraph. The large wheel at the top of the drawing carried the type plungers, with four rows of keys above its shaft. Ink was transported from the trough at the bottom by means of a conveyor belt.   (BP)

drilled and an inking pad or two, and there you have the Thomas Typograph. Give it credit for one thing, however: the type-sleeve with locking holes is a direct ancestor of that used on the famous Crandall twenty-five years later.

The same year produced a most interesting design for an electric typewriter. Giuseppe Devincenzi (*sic*), described in his 1854 British patent merely as a 'gentleman', drew his inspiration from the comb and barrel of a music box, a system first used in 1845 by Brett with whose machine the inventor was certainly familiar. He disposed a row of pins spirally around a cylinder and by means of a treadle

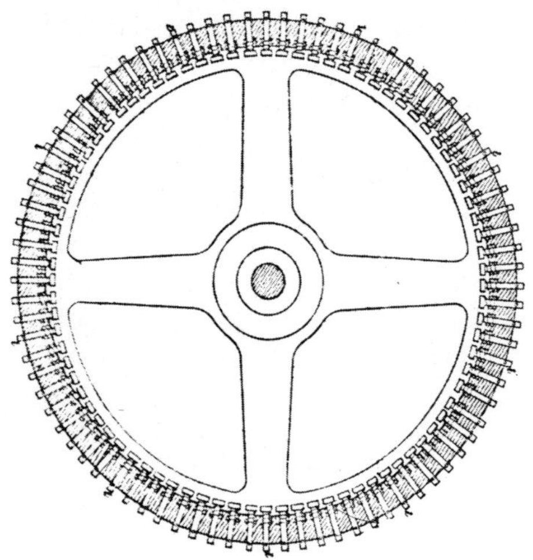

83. Details of Devincenzi's radial plunger design.
A hammer struck the type from inside the wheel.
(BP)

rotated this cylinder near the comb, above which was the keyboard. Depression of a key brought the corresponding tooth of the comb in line with a pin and stopped the rotation of the cylinder. A wheel, with type-plungers fitted radially around its perimeter, was mounted co-axially with the cylinder, and under the type-wheel were the platen and carriage. An electromagnetic hammer struck the plungers against the paper (from the inside of the wheel outwards) and letter spacing was also electrically operated, occurring automatically upon release of a key. The machine printed full upper and lower case, and inking was by means of a leather conveyor belt replenished from a trough. A wiper took care of excess. All in all, a beautiful design.

STM   Giuseppe Ravizza of Novara, Italy, who had been concerned with the problems of mechanical writing from the 1830s, was granted a patent in 1855 for a Cembalo Scrivano, a Writing Harpsichord, the first of sixteen (or perhaps more) models which he was fated to produce before his death some thirty years later. He is the inventor who corresponded with Conti about the latter's Tachigrafo from 1832 onwards and who reportedly possessed a confession from Galli that the Potenografo was never constructed (see page 61). His developmental work on the writing machine was amongst the most important to his day and the Cembalo Scrivano has been vociferously proclaimed[1] etc. the world's first typewriter. Ravizza's tribulations and the development of his machine are discussed in detail in Chapter Five. A few models have survived; most, however, have been destroyed.

Destruction, in fact, was exactly what obsessed Count d'Aunay, a French provincial politician of the Second Empire, who delivered a broadside not only against a machine invented by Abbé Clément but also against writing machines in general. It seems that in 1855 the poor Abbé built one such device, which worked, and in his initial excitement at the success of the instrument he typed out a list of the names of all his pupils. This in itself would hardly have kicked up a storm

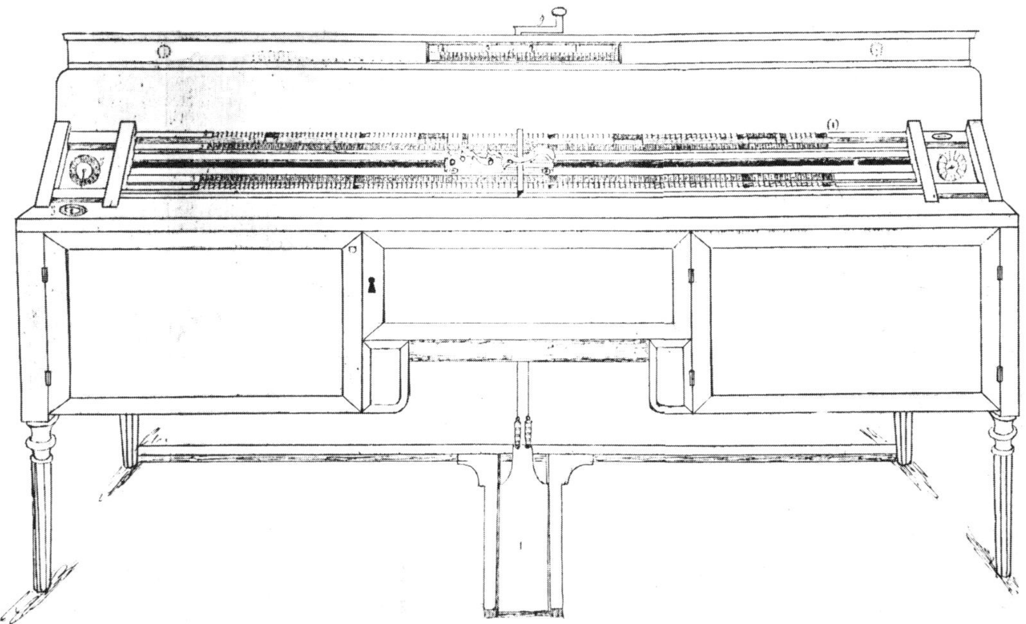

84. Abbé Clément's secret weapon against 'Morality and the Régime' was this linear plunger device built in 1855 and patented two years later. (FP)

except that, pride and human frailty being what they are, he could not resist the temptation of showing the list to the children, one of whom squealed on him. The Count d'Aunay was horrified: writing machines were highly dangerous inventions, he raged, because anybody without even the help of a printer could produce and distribute pamphlets and flyleafs not only against morality but even against the Régime! And so the poor priest paid for his sins by sinking into oblivion, but not before applying for a patent which was granted in 1857. His table-sized machine was of the plunger type, with a manual selector and a pedal for making the impression on the flat paper-carriage. The type-carrier ran on rails and its length was determined by the number of alphabets, placed end to end,

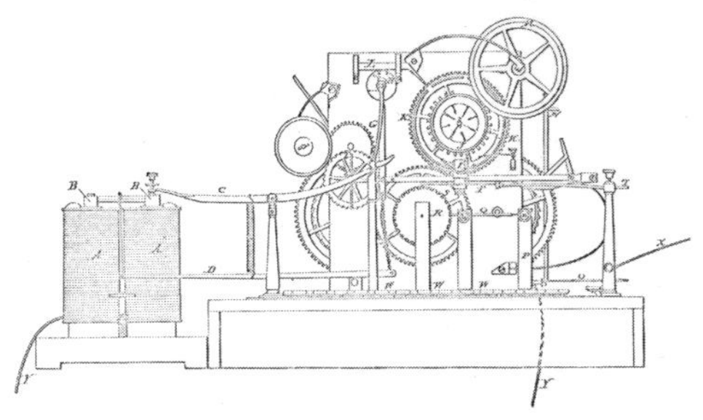

85. One of the most famous printing telegraphs ever designed was the one by D. E. Hughes of Kentucky, patented in 1856. It was of comb and pin barrel design. (USP)

D

that the operator desired. A dial on the left of the machine indicated the number of lines typed and another, on the right, the length of these lines. All in all, it was hardly a machine with which to assail the Régime let alone morality!

It was while the Abbé was sinking into oblivion that another inventor in the more liberal United States was rising out of it. D. E. Hughes of Kentucky (no relation to G. A.) invented his important printing telegraph in 1855 and patented

87. The carriage and platen rocked forward to strike the type on Cooper's 1856 invention.
(SIW)

86. The Hughes machine was extensively used throughout the world and different models evolved from country to country. The one illustrated was the British version.　　(CSM)

it the following year. As in Brett's and House's machines of a decade before, the Hughes used clockwork to drive a constantly revolving pin barrel which controlled type-wheels rotating 'in harmony' on both transmitter and receiver, if all went well—for such systems always suffered from difficulties in maintaining accurate synchronization. Nevertheless, the machine was used throughout the world and, over the years, many improved models were developed in different countries: by Froment in France, Siemens in Germany, and so on. There was nothing revolutionary about Hughes's design, of course, but the Jury of the 1867 Paris Exhibition considered his 'the best of all the type printing telegraphs'.

MPM　　John Cooper of Philadelphia made less of an impact. Like Hughes, he borrowed heavily from his contemporaries on the machine which he invented in 1856, but the product was less noteworthy. A wheel bearing the type was mounted on a vertical shaft above which a lever selected the character on a circular index. Alignment was by conical countersinks, and inking by roller. The paper was transported in a carriage automatically advanced by a ratchet escapement: this carriage

88. Dr Samuel Francis built an advanced up-stroke machine in 1857. Among its important features was a free-swinging type-bar which had no direct linkage with its key. (SIW)

was not rigid but rocked forward upon depression of the selector lever, thereby bringing the paper into contact with the type-wheel. Cooper's machine was most noteworthy for its sophisticated combination of cylindrical platen and feed roller, very like that in modern typewriters.

SIW   Dr Samuel Francis of New York provided another such element: a carriage escapement operated by a spring which was wound up automatically on carriage return. His Printing Machine, patented in 1857, was a sophisticated device of the up-stroke class, declared by the inventor to be capable of a speed double that of

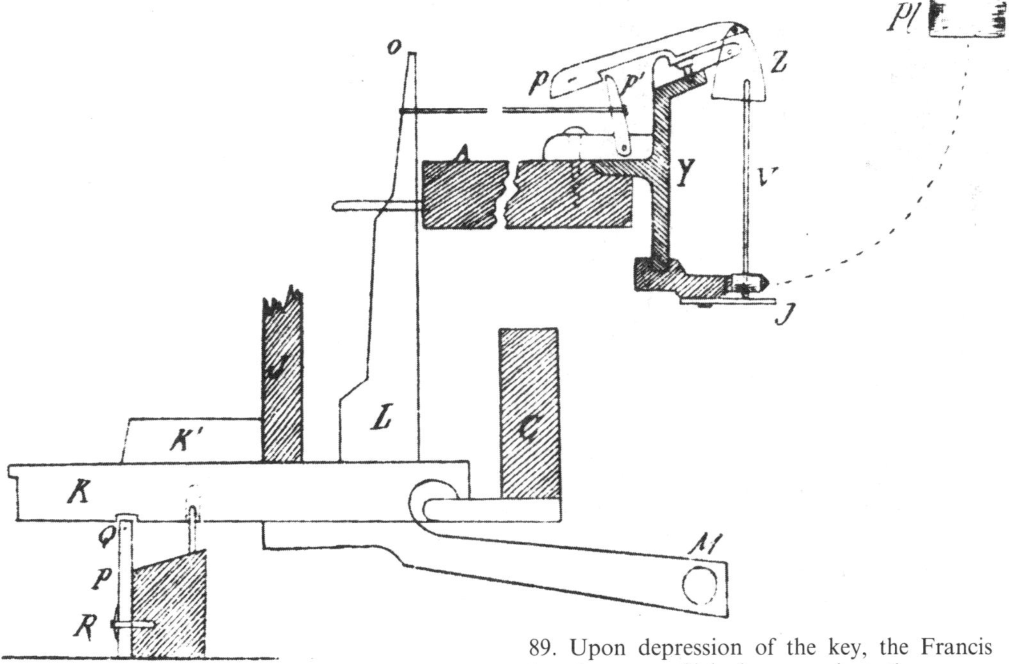

89. Upon depression of the key, the Francis type-bar was flicked up against the paper, being free to fall back into place by gravity.

(14)

handwriting, a claim which was only marginally unrealistic. A piano keyboard of thirty-six keys controlled the type-bars by a novel method which avoided direct linkage. The type-bars hung freely in a circular basket, depression of a key causing the corresponding bar to be hit and swung up to the common printing point, whereupon it was released and fell back into place by gravity. Printing was by means of a silk ribbon saturated with dye; a warning bell sounded four spaces before the end of a line, and a special device prevented jamming if more than one key was depressed at a time. Francis's patent offered two alternatives for making the impression: either the ribbon was to be placed between the type and the paper, whereupon the type would have to be reversed, or else the key hit the paper against the ribbon, in which case this would not be necessary. He favoured the latter as being more readily visible to the operator, and he suggested, furthermore, that if a sheet of transparent paper was then placed on the other side of the ribbon, a copy would simultaneously be made. This would be legible by reading through the transparent sheet. Add more ribbons, he said, and you got more copies.

The machine was considered[56] to be capable of printing cleanly and faster than handwriting but it was too bulky, too intricate, too delicate, and, in some of its elements, too costly for commercial exploitation. This appraisal, of course, was written in 1886; the criticisms are unconvincing and further development might well have remedied the defects. It was a fine machine which bore more than a passing resemblance to Ravizza's Cembalo Scrivano, and the inevitable charges of plagiarism are not lacking.

Of course, development of the writing machine had now reached a level at which most of the essential features had already appeared. It becomes increasingly rare to find a completely original design or even part of a design; generally, it

90. Hood (1857) geared a vertical type-wheel to a horizontal indicator, formulating a mechanical principle made famous a quarter of a century later on the Columbia Type Writer.   (SML)

100

may be said that the components had become the same jigsaw pieces which inventors merely fitted together to form different or better pictures. There are, of course, exceptions, although these are now hard to find and there is little in later machines which is not in some way strongly reminiscent of what had already been tried. The Francis machine looked like the Cembalo Scrivano and, later, an early Sholes model clearly borrowed heavily from them both. The ribbon had long been invented, as had the carriage, its escapement, platen and feed systems, the type-bar, type-wheel, universal bar, keyboard, space key, change of case, the printing telegraph and shorthand machines, electric assistance, differential spacing —all of this, and much more, had already been tried and was now merely being perfected.

ML   Peter Hood of Scotland, for instance, had an idea in 1857 which appeared a quarter of a century later on the successful Columbia Type Writer. His machine consisted of a vertical type-wheel geared at right angles to a horizontal index-wheel on which letters appeared in alphabetical order. A knob and indicator selected the desired character on the horizontal wheel which turned the vertical one till the corresponding letter was above the printing point, whereupon it was brought into contact with the paper by depression of the knob. A spring returned the type-wheel to its original position. Locking holes were provided in the index for correct alignment, and an escapement automatically advanced the carriage

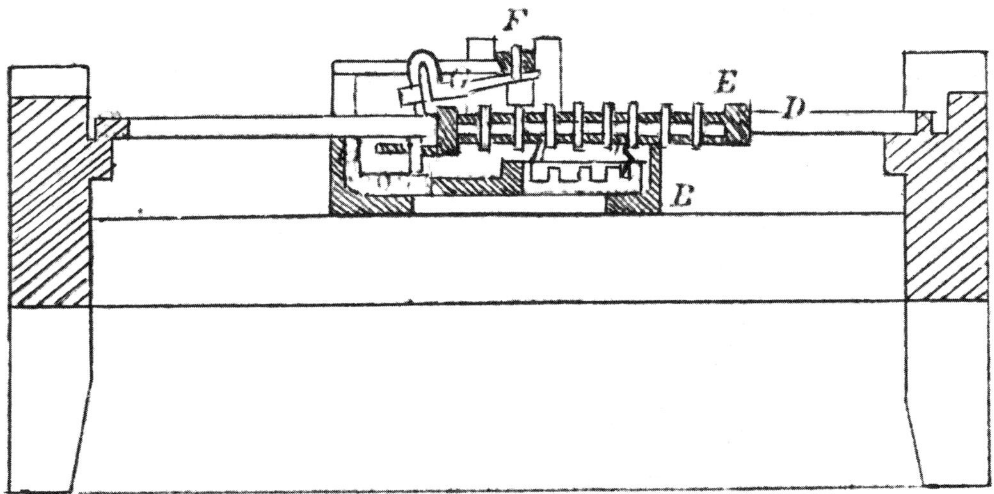

91. History has been cruel to Henry Harger, whose ingenious type-plunger device with dif- ferential spacing, patented in 1858, has been overlooked by every other secondary source.

(USP)

for letter spacing. Inking was by means of a pad. The inventor is also reported[28] to have built a machine for the blind, of which no details have survived.

The same year, across the Atlantic, Daniel Johnson is reported[28] to have taken out a patent (which could not be located) on yet another machine which printed embossed characters for the blind, made visible by means of carbon paper. On the other hand, the same source states that all details have been lost of a machine by Henry Harger who did, in fact, take out a US patent on his device in 1858. And an ingenious machine it was, too! Depression of a hand-operated lever advanced the carriage and brought the selected type-plunger into contact with the paper—in that order, i.e. the paper was advanced *before* the impression was made. The rest is brilliantly simple: by varying the lengths of the type-plungers, and hence the travel of the lever, full differential spacing was incorporated: the shorter the plunger (e.g. 'm'), the longer the distance the lever had to travel in order to bring that plunger into contact with the paper; the longer the lever travelled, the further it pushed the paper, leaving more space for that wide character. Fontaine used a more sophisticated application of the same principle a decade later (see page 121).

AMP Adolphe Charles Guillemot—and *not* Guillemont[40, 42 etc.]—employed considerably less ingenuity. Born into a family of inventors and himself holder of a host of mechanical patents, this Parisian instrument-maker began to occupy himself with the construction of a writing machine in 1855, completing and patenting it four years later. It was a type-wheel device with a piano keyboard and flat paper

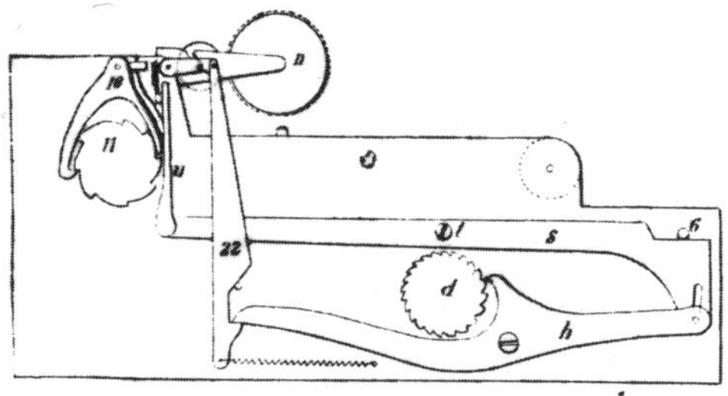

92. Guillemot's first model, 1859, featured a type-wheel and piano keyboard, but a later version used three type-wheels without, however, tripling the inventor's chances of success.

(37)

carriage; inking was by pad. A second and improved model was entered in the 1862 London Exhibition whence it was subsequently lost, much to the inventor's dismay, having failed to attract the interest of a potential manufacturer. Small wonder: Guillemot's piano keyboard now controlled no less than *three* type-wheels, one each for upper case, lower case and figures. According to the claims of the inventor, the machine was particularly recommended for the incapacitated, the infirm, the blind and the illiterate. The illiterate? Some years later, he is reported to have tried his hand at a musical typewriter with which he enjoyed similar lack of success.

SML In fact, these were lean years in typewriter history. John Cox built a device in 1860 (or so) which might have warranted historical attention had the inventor lived two centuries earlier. It consisted of a vertical type-wheel fitted with a knob by which it was turned, and a button by which it was pressed against paper resting on an ebony strip on the base board. The type-wheel moved linearly along an

102

93. The type-wheel was turned by means of a knob to select the character and the key above was pressed down for printing. Primitive though it was, the Cox of around 1860 served as a model for the later Herrington. (CSM)

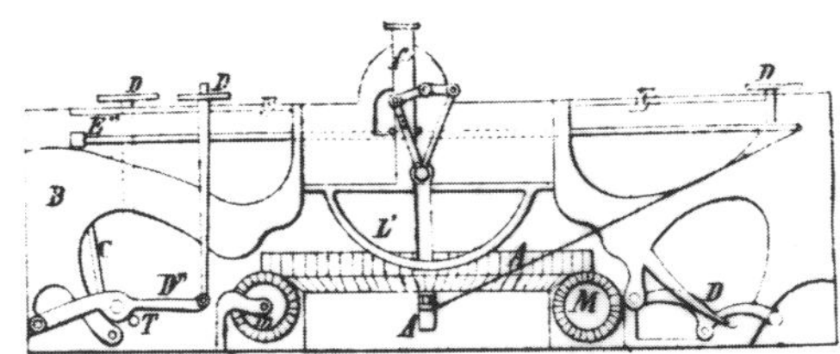

94. Codvelle designed this machine in 1861 to print capitals by decorating lower case letters.

(15)

arbor for letter spacing; inking was by roller. There was no provision for holding the paper in place nor for line spacing; in general the design was so primitive that the inventor might well have tried to fix the paper with saliva!

The following year, a Frenchman called Valentin Codvelle—not Codville[27, 40] etc.—produced L'Ecriveur, with forty-eight keys, the chief peculiarity of which was that it printed upper case by the simultaneous depression of the corresponding lower case key, plus a special ornamental key which decorated the small letter and made a capital out of it. Not bad. Still in 1861, a stenographic typewriter came from Brazil, of all places. Francisco João de Avezedo, a priest, displayed a wooden instrument at an Exhibition in Recife, receiving an enthusiastic press and creating quite a stir which soon, unfortunately, petered out without producing concrete results. It seems[46] that the priest first planned to build a straight typewriter, but the Legislature, to whom he applied for assistance, were more interested in a shorthand machine which they immediately patronized. According to the report of the Recife exposition, there were fourteen keys, eight for the right and eight for the left hand (*sic*)! A picture of the instrument accompanying the report is not clear enough to permit solution of this arithmetical riddle. The keys operated type-bars with characters in relief and, as in Beach's machine, produced an embossed impression by clamping the paper between male and female type. In de Avezedo's case, however, the negatives were on a fixed metal sheet across which a strip of paper moved: the component letters of a word were therefore embossed in the correct order but separated by longer or shorter gaps. Despite the sound of it all, it was apparently a good machine; after Recife, it won a gold

96. Martin's machine for the blind was reported to have left a feeble impression, not only on the paper but also on the public.　　(SML)

95. The first Brazilian typewriter, invented by de Avezedo in 1861. 8+8=14 . . .　　(46)

medal at the Exhibition in Rio de Janeiro the same year, plus the honour of selection as one of Brazil's exhibits in London in 1862. It appears never to have made the voyage, however, for it was not submitted.

SML　　John Martin's writing machine for the blind was, but it did him little good. It produced a feeble impression not only on a sheet of paper but also on the Exhibition Juries who[58] deemed the machine inferior to Hughes's. It was a small and compact device: type-plungers were located vertically around the perimeter of a horizontal wheel with a circular index on its upper surface for identification of letters. After the selection of the desired character, a handle pressed the plunger through to the flat paper table beneath. Rollers advanced the paper.

Still in 1862, a historically important development came from an Italian, Antonio Michela, whose fine stenographic machine was one of 'real merit . . . the first to allow practical results to be obtained',[16] an evaluation the more noteworthy since it was not an Italian but two Frenchmen who made it. An oft-expressed view is that Italian inventors were never taken seriously either at home or abroad— this has repeatedly been said of Ravizza both during his lifetime and after. In Michela's case, as far back as 1879 a local newspaper commented cynically that his shorthand machine would have been used in the entire civilized world, had the inventor been born in London, Paris or New York instead of in a small Piemontese village.[28] There is perhaps some truth in the assertion, for Italy in

104

those days was so preoccupied with its own internal political strife and struggles for unification that it had little time or energy to spare for industrial development. And, of course, there has always been that inherent Italian contempt for anything practical and not artistic.[7] To this day, addressing an Italian engineer as 'Ingegnere' is mildly offensive and contemptuous; you should call him 'Dottore'—unless he happens to be a Count or a Prince.

Michela was a drop-out and a failure in his youth, and was dismissed even by his distressed parents as a crazy dreamer. One of his schemes concerned a universal alphabet, to the development of which he dedicated many years of his life. In the process, he studied speech organs with minute care, reducing speech itself to all its component syllables. With these qualifications, he built his machine,

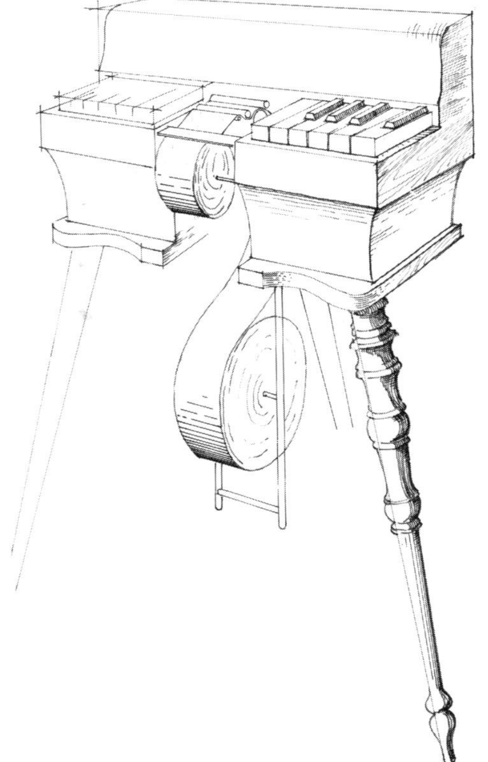

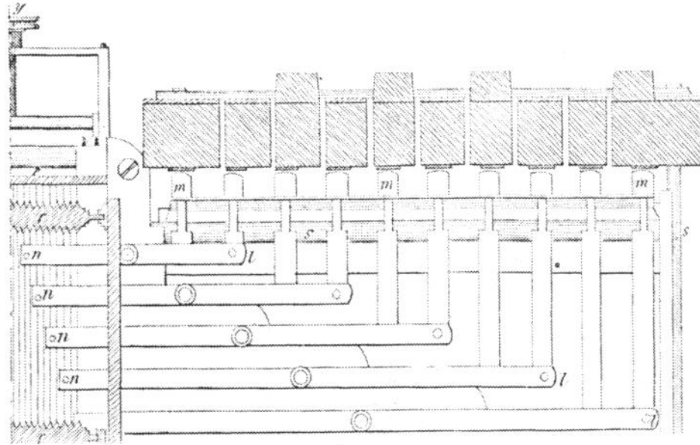

98. Cross-section of the right half of the Michela keyboard, showing keys and linkage. Printing was on the underside of the paper.   (IP)

97. The first manufactured shorthand machine was invented by an Italian, Antonio Michela, in 1862. It featured a separate piano keyboard for each hand, printing its own code on paper tape. It not only worked, but worked well.

the first model of which he completed in 1862 and presented to the public the following year. With the help of a carpenter he later produced it in considerable (for those days) quantities, so that his was the first shorthand invention ever to have been manufactured. The Michela Stenograph looked like a small harmonium —a handsome three-legged rectangular box with two sets of ten piano keys, one for each hand, separated by a roll of paper. Each set consisted of six white and four black keys so disposed that each finger operated one white and one black, except the thumbs, which operated two whites. The stenographer played 'chords', the individual 'notes' of which printed symbols for the sounds which the word contained. On the left, the first six keys printed initial consonants—the following four any subsequent consonants. On the right, the first four printed vowels and the rest the terminal consonant. A special shorthand alphabet was designed to be used with the machine and the paper was ruled into four columns corresponding to the keyboard divisions, to allow the relative position of the sign to be located. The paper roll was 300 metres long, enough for some seven hours of typing. The impression was by ribbon. Such was the instrument as it appeared when it reached finality, for more than a decade of development went into perfecting the first model. Patents were granted in 1876 and 1878 and the improved models won awards and a limited measure of international recognition. In 1880, the Italian senate bought six and put them to work; the French parliament also tested them, but no order materialized. Meanwhile, the Italian senate machines were operating at speeds up to 225 words per minute which, for those days, was approaching escape velocity!

The pressure was now on. More than twenty contestants crowded the field in the following five years; there were four in 1863 alone. All in all, the days of the lonely inventor, isolated from his fellow man, imprisoned in a mechanical and technical vacuum—those days were long past. It has often suited national or commercial interests to maintain the contrary, in the fear that admission of the inventor's familiarity with previous efforts might somehow reduce his worth. But as was pointed out previously, the general diffusion of information about writing machines began twenty or thirty years earlier; by now, the process was virtually complete.

This is particularly important in the case of John Pratt whose advent on the scene marks a kind of turning-point. Not only were his machines considered the best so far, but also they aroused a degree of interest and controversy hitherto unattained, with reports and descriptions in technical and other journals on both sides of the Atlantic. It has been rather facetiously contended[5] that the most important thing about the Pratt machine was that the later American inventor Christopher Latham Sholes read about it in the *Scientific American*. This is unjust, since the machines (there were several) had considerable intrinsic merits. Incidentally, Pratt himself was well familiar with other writing machine inventions, both earlier and contemporary, for he talked about them and criticized them freely. Foucauld, Hughes, Sweet, Bryois, Peeler, Hall . . . he publicly compared theirs unfavourably to his own.

Pratt built his first model in his home in Alabama, completing it in the spring of 1863 in the midst of the Civil War, at a time when he could not even patent his device, let alone promote it. So he packed his family and model off to England, where he was granted a provisional British patent the following year. The 1864 machine was encased in a box, the flat paper-carriage being attached to the lid.

99. John Pratt. The second model (1865) by this famous inventor featured a frustum of type-plungers controlled by a piano keyboard. Paper was attached to the flat carrier on the underside of the cover which had to be raised to reveal typing. (SML)

Twenty-seven piano keys controlled a horizontally mounted type-wheel and the impression was made through carbon paper by a hammer striking the paper against the type-wheel.

In 1865, a modified machine was complete, built with the collaboration of a Glasgow instrument maker. The type-wheel was replaced by a frustum of twenty-six plungers inclining upwards towards a common printing point. The lower end of each plunger was connected by linkage to a key on a piano keyboard, the type being at the upper end. Depression of a key brought the plunger up against a sheet of paper held flat in a carriage mounted on the underside of the lid of the instrument; this had to be closed during typing. Carbon paper produced the impression. Knobs were provided for carriage return and line spacing, with a universal bar for letter spacing.

Both the above machines were flops, by the inventor's own admission, and as a measure of his disgust with the whole design he abandoned it and departed radically from it in his next model. Piano keyboards, he said, were unpractical because they were too big and because they required too much effort. Furthermore, writing machines had to operate quickly, meaning (to him) at twice the speed of handwriting. Type-bar designs were doomed because there was no satisfactory way of preventing clashing at or near the printing-point (Sholes and other later inventors struggling with this problem bore eloquent testimony to Pratt's prophetic insight), and type-wheels were inefficient and slow because of the distance some of the characters had to travel (in a circumferential direction) in order to reach the printing-point. He later changed his mind on this score, but meanwhile the solution, he reasoned, was to have the type in even rows on a small square plate.

The keyboard could thus economize on keys by locating characters on the horizontal and vertical planes, two keys being depressed simultaneously. This concept was to be tried more than thirty years later in an instrument called the Saturn. In Pratt's case, a mere twelve keys (six for each hand) controlled a square type-index with a six-by-six configuration printing thirty-six characters. There were three extra keys for spacing etc. The character was brought over the printing-point by depressing a left-hand and right-hand key simultaneously, thereby locating the plate laterally and longitudinally, whereupon a hammer struck the paper against the corresponding type. Carbon paper was used for the impression.

This type-plate model was the basic design which he patented in 1866, the specifications of which offered the keyboard of twelve keys previously described or, alternatively, a separate key for each letter. A further model is elsewhere recorded as having sixteen keys controlling an eight-by-eight plate, obviously for upper

100. Pratt's final model was the type-wheel 'Pterotype' of 1866. The paper was held in a vertical frame and a hammer produced the impression. It was the seed which later grew into the great Hammond typewriter.   (CSM)

and lower case, but the principle is basically the same, being neatly described in the patent specification: the number of keys must be equal to the sum of the horizontal and vertical rows, 'that is, to twice the square root of the number of types'.

This was a great improvement over the original machine, but even this underwent numerous subsequent changes and improvements. Several examples were made in London, each one embodying modifications as they were developed. SML When it reached finality, the Pratt Pterotype (winged type), as he called it, was contained in a wooden case of L-shape, with the keyboard on the horizontal plane and the paper-carriage on the vertical. The keyboard now consisted of thirty-six cylindrical keys divided down the middle into two sections of four rows of

108

$5+4+5+4$ keys. The top row was of numbers and signs, the others of the letters in alphabetical order; the left hand therefore covered A to M and the right N to Z. The square type-index plate had now given way to a small vertically mounted type-wheel half an inch in diameter, on which there were thirty-six characters in three rows of twelve each. By keeping his type-wheel small, Pratt overcame his earlier objections to this system and proved the first to succeed in perfecting it. The mechanism was spring loaded so that depression of a key merely tripped the action, placing a stop in its track at the desired point, and releasing the hammer which made the impression through carbon paper. Letter spacing was automatic, word spacing was performed by partial depression of any key—a most ingenious arrangement indeed—and the clockwork was pump-wound at the completion of every line by the depression of a large knob on the right of the keyboard; carriage return and paper advance being accomplished automatically by the same motion.

'Pratt's machine was by far the most complete and practicable machine which had appeared to that date, and it is owing to its appearance, and the newspaper articles and discussions which it provoked, that we owe the typewriter of today.'[27] For Pratt introduced his invention (the type-plate model) to the Society of Arts in London on 1 May 1867. Two days later, the *Journal* of that Society gave him big coverage. *London Engineering* picked it up, as did many other publications, among which was the *Scientific American* of the same year. And theirs was the fateful article, prophesying fame and fortune for the inventor of a successful writing machine, which a man called Glidden is said to have shown a man called Sholes thereby providing him with the incentive to start work (see Chapter Six).

Meanwhile, Pratt himself was experiencing some difficulty in making that fame and fortune prediction come true. Since the machine was made almost entirely of wood, he told the Society of Arts, it could be marketed for a mere three guineas at great profit to the manufacturer and was almost impossible to wear out. But he could find no takers. More realistically, he promoted another feature of his design which was to prove historically important: the easy interchangeability of type-plates for different languages and characters at a cost of a mere ten-pence each.

Incidentally, there is much confusion regarding the order in which the various Pratt models appeared, and even the design of the machines themselves, and virtually every secondary source is inaccurate. Particularly, the conflict of priority concerns the type-plate and type-wheel versions. The type-plate model preceded the type-wheel instrument, both of which were developments of the first horizontal type-wheel and then the frustum design. However, Pratt appears to have continued displaying his type-plate machine quite some time after the other had already been completed.

Pratt returned to the United States in 1868 and patented his machine there that year. He found no backers for production and he himself lacked the funds to work on his own. Eventually he made a deal with J. B. Hammond who struggled with the design for over ten years and eventually produced some of the finest machines ever built, bearing, however, little resemblance to Pratt's except in the fundamental principle. Another pioneer, Crandall, took the Pratt type-wheel and stretched it out to make it into a sleeve which he himself was able to patent on his own, thereby by-passing Pratt's control. Perhaps Thomas's invention provided him with the idea, for by then his rolling pin type-sleeve was public property, the patent having expired.

All of this happened some twenty years after Pratt built his first model in 1863. In that year, F. A. De Mey (*not* May or Mey or de May) of New York virtually re-invented Thurber's Patent Printer, or at least, the De Mey machine was so

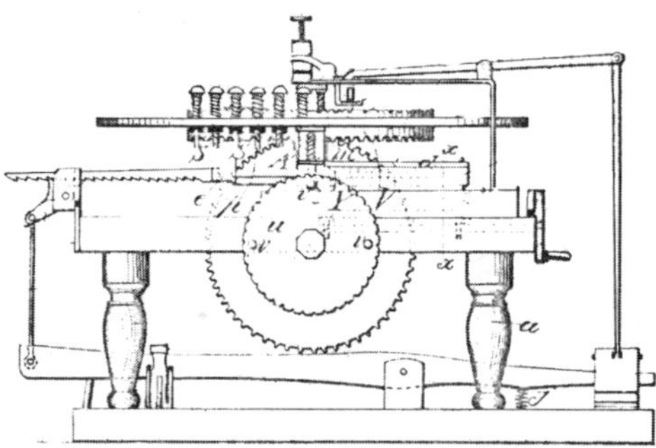

101. Thurber revisited . . . by F. A. De Mey in 1863. (USP)

similar in its essentials that it does not warrant a separate description. And Benjamin Livermore of Vermont patented an intriguing little instrument called a Hand Printing Device or Mechanical Typographer with only six keys, each

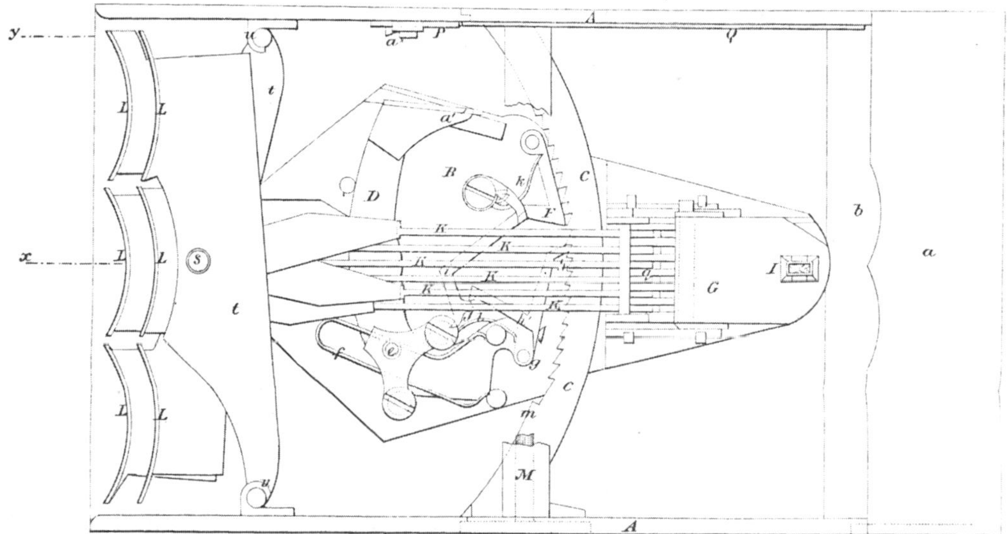

102. Benjamin Livermore used different combinations of up to six keys to build up the letters of the alphabet in his 1863 Hand Printing Device. (USP)

with a line or stroke on it, so that, when depressed singly or in combinations, symbols representing and resembling the letters of the alphabet were built up. The idea was not original, and met with no greater success than did Foucauld's experiment with a similar system twenty years earlier. It was, however, considerably more ingenious, not only because it used fewer keys but also because there was no need for backspacing to make additional overprints. The six keys were marked according to their corresponding function in completing the 'combination type'.

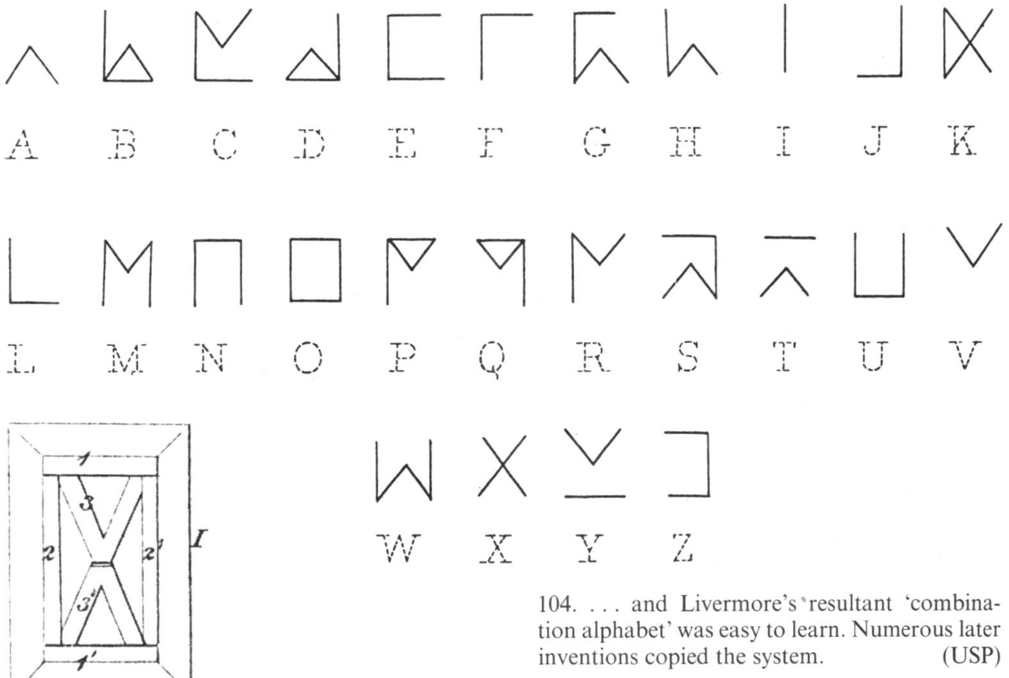

104. ... and Livermore's resultant 'combination alphabet' was easy to learn. Numerous later inventions copied the system. (USP)

103. The 'combination type', as he called it, looked like this at the printing point ... (USP)

A = part 3′, B = 3′ + 1′ + 2, and so on. Livermore's patent was for a 'portable device which may be held in the hand and operated with the greatest facility' and it deserved a better reception than it received.

And across the Atlantic, a Frenchman called Pierre Flamm had finished building a machine which printed syllables and even short words and which is usually included in reports on the writing machine although it was more of a mechanical compositor. A plate bearing 225 groups of consonants was fixed face down in a sliding frame, on the upper side of which was corresponding identification of the characters. A swinging sector hovered over this plate and vowels could be lowered at will into holes drilled in it. 'Un peu primitif'[15] is a generous evaluation of this

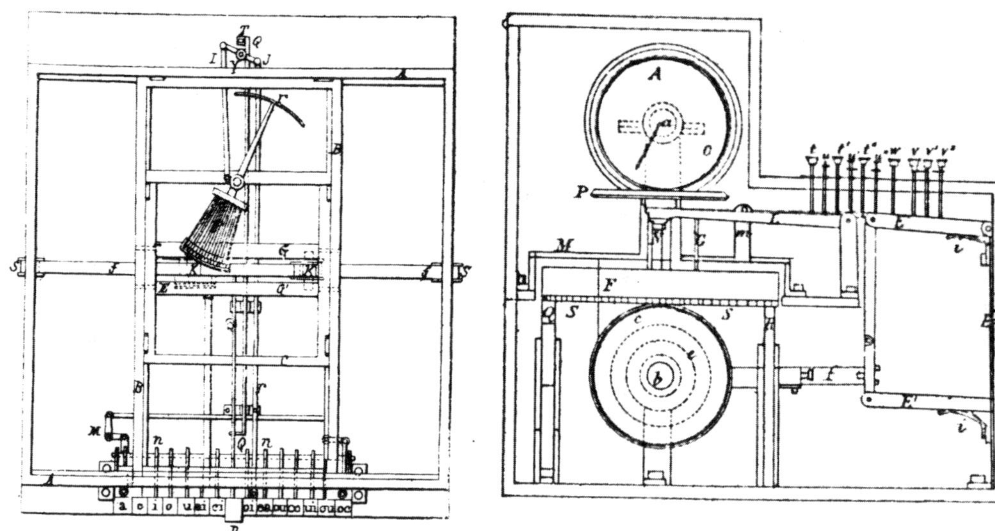

105. Flamm's compositor barely qualifies as a writing machine . . .　(37)

106. . . . although the specifications patented in England are considerably better.　(BP)

invention, the inclusion of which is justified by a fortuitous wording in a patent addition which extends the machine's use to printing on paper, using carbon to make the impression. Despite this, the device remains essentially a compositor, as does an improved specification patented in England during the following year. Entitled 'Machinery for Composing Type', it offered several alternatives, including one with carriages, and the other with keys, or stops, using 'fat ink' or inked paper for the impression.

TMV　　Machines mushroomed all over the world in 1864. In Austria, Peter Mitterhofer surfaced with the first of a number of models which will be described in detail in
SIW　Chapter Five, while across the Atlantic, Benton Halstead built a writing machine which he is reported[28] to have used in his office for eight years before finally patenting it in 1872. The specifications called for a stirrup hanging beneath the table, which at first sight might lead one to suspect that one rode this typewriter like a horse. The stirrup, however, was to assist in word spacing and makes the machine sound positively forbidding, the more so since the mechanism was to be built into a table 'substantially constructed to resist the vibrations of the operative parts'. The design was down-stroke, with type-bars set obliquely in a semi-circle behind the printing-point, in front of which was the piano keyboard. Paper was on a flat, weight-driven carriage, a ribbon was used for the impression, and a blocking device prevented jamming of the type-bars.

Meanwhile I. A. Peters of Copenhagen appears to have completed a vertical plunger machine similar in some respects to the later Writing Ball of his compatriot,

112

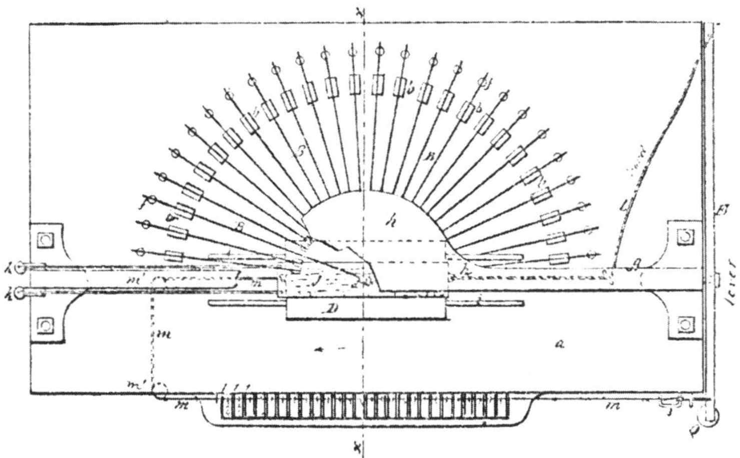

107. Halstead had a semi-circle of type at the back of the machine leaning over the printing point at 45°. The piano keyboard was at the front, and the whole affair was as big as a fair-sized table. (USP)

Malling Hansen. Depending upon one's source, the dates of Peters' invention are anywhere from 1862—when he is reported[28] to have begun work on his machine —to 1868, when he is recorded[55] as having 'invented' it. He unsuccessfully asked for public assistance for the production of the instrument for which in 1870 he filed a patent application, granted the following year; by that time however, he had forfeited his place in history to Malling Hansen who had already begun actual manufacture of his Writing Ball. A controversy as to the paternity of this design was bandied about by heirs and experts for over half a century, long after both participants had died. In fact, there are indeed similarities: both machines utilized the radial plunger concept originally devised by Galli and successfully applied by Foucauld decades earlier. But apart from this detail, Hansen's design

108. This is what remains of a type-plunger device invented by a Dane called Peters. A flat paper-carrier (missing) was inserted beneath the unit which was obviously intended to use the eyes on the lower right (one broken) for hinging upwards, permitting insertion of paper. (DTM)

113

presents a quite different aspect from that of Peters, a portion of which has survived.

DTM According to the director of the Museum, the machine occupied the space of a medium-sized round table and was like a baby grand piano in shape. Thirty-five vertical plungers in a five-by-seven format protruded from a flat horizontal plate; the plungers curved inwards and were guided through two additional parallel plates towards a common printing-point, being located vertically by coil springs between the upper two plates. It was a poor and inefficient design and one that would have stood hardly any strain—no provision appears to have been made to prevent the curved rods from twisting and thereby binding in their locating holes, and with a little wear this would certainly have occurred. Hansen's concept (and Foucauld's before him) was far more effective in that the plungers were straight, radial and direct.

There were two further inventions in 1864, both French and both remarkable for their originality, if for little else. One was a Sténographe Imprimeur built by a man called Bryois. It sported hundreds of keys, any number of which could be depressed simultaneously on the chord principle. All ten fingers were used and impression through a ribbon was on a paper tape advanced by clockwork. Each sound in the language was allotted its own sign or symbol, orthography being completely dispensed with. Two years later (1866), Bryois patented a second and

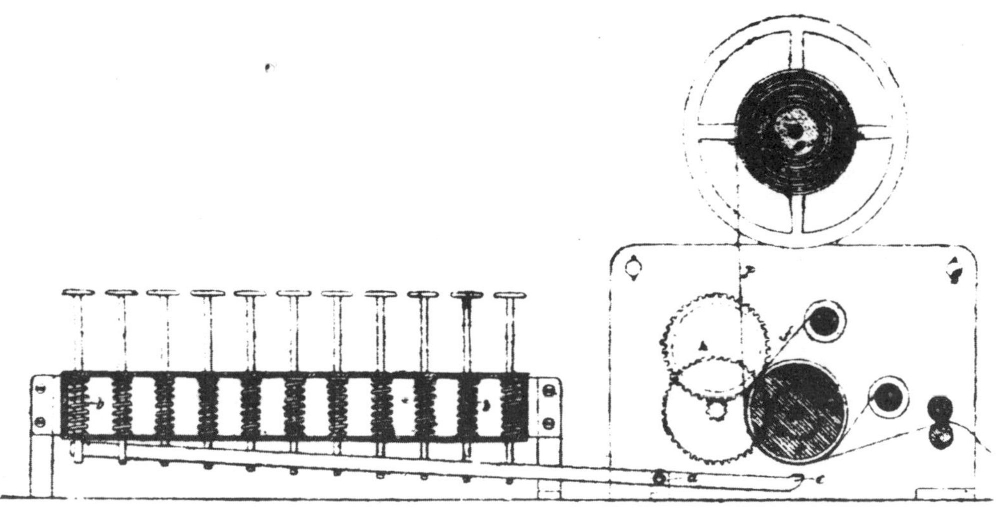

109. The Sténographe Imprimeur invented by a Frenchman called Bryois in 1864 featured hundreds of keys, any number of which could be depressed simultaneously. (16)

presumably improved model with 360 radial plungers and slightly reduced size, that printed whole syllables and short words. But he apparently recognized the impossibility of a machine with so many keys, for he resolved to build yet another

114

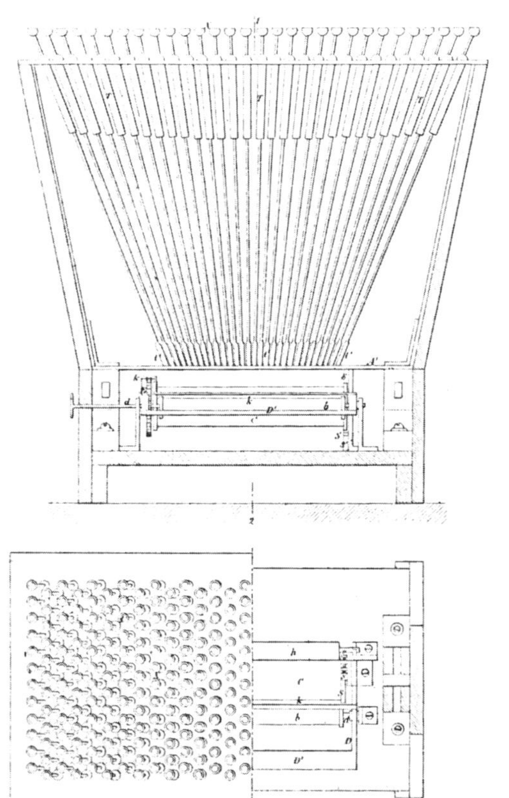

110. Bryois' second model used type-plungers. Above is a side view and, below, a bird's-eye view of half the keyboard of 360 keys. (FP)

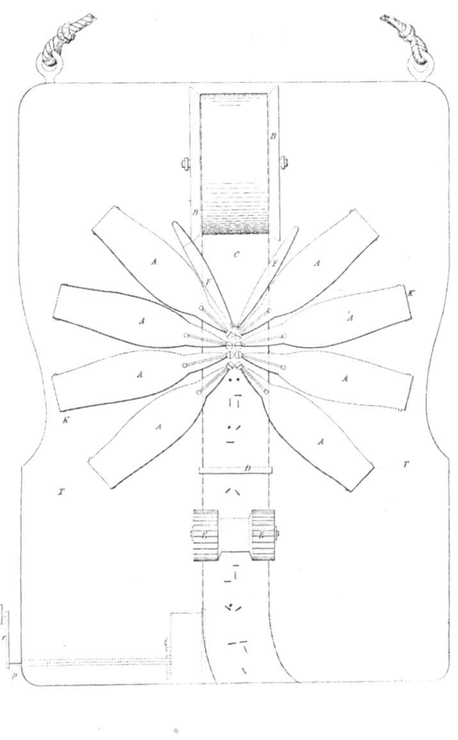

111. A shorthand machine designed to be worn like a bib was patented by Danel-Duplan in 1864. (FP)

with only twenty, although no indication remains as to whether he ever completed such a device.

The other stenographic apparatus from 1864 was designed to be worn around the neck. Danel-Duplan's novel machine had ten keys pivoted on a baseboard designed to be held vertically and operated by all fingers singly or in combinations. On the front of the board were eight keys radially disposed so that four were operated by the fingers of the right and four by the fingers of the left hand. The thumbs passed behind the board and each operated a key of its own which communicated with the paper on the other side. The ends of the levers had ink reservoirs with wicks for inking, like miniature modern magic markers. Thumbs printed dots, other fingers strokes, combinations of which produced a stenographic alphabet recorded on a thin paper tape advanced by clockwork. Not so bad . . . and, given the design, wearing it around the neck kind of made sense!

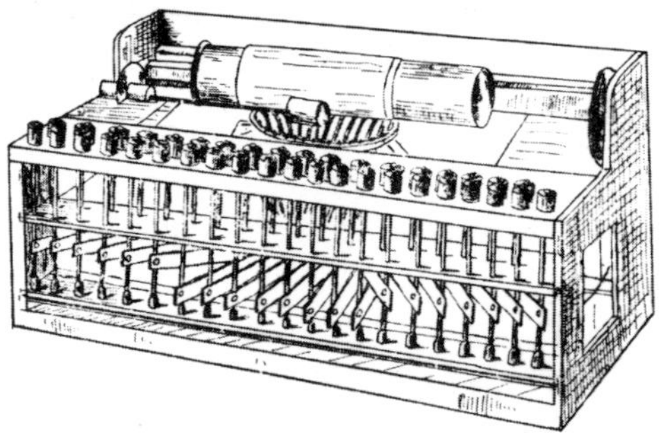

112. George House (1865) added an important element to previous up-stroke inventions: a cylindrical platen. Until then, all earlier machines of similar design had used a flat paper-carriage. (14)

So did a sophisticated machine of a different type, invented in 1865 by an American, George House. Of the up-stroke group, it shared important elements with a number of its predecessors, as well as with later Sholes models. On the House, type-bars were arranged in a circular basket striking up at a common printing-point on a cylindrical platen which revolved on its axis for letter spacing and moved linearly for line spacing—in essence, it was identical with some of Sholes's experimental models. Forty-one keys in two rows printed upper case only; inking was by roller. It was good but a bit of a brute, nearly three feet long.

PNY The same year Joel Smith, another American, is reported[57] to have invented the Daisy Point Writer for the blind, so-called because its keys were in a flower-petal formation. A small cube contained the six-key pin pricker with the 'petals' on top; these slid along lateral and longitudinal rails notched for letter and line spacing. The paper was held flat on the base of the wooden frame.

Finally in 1865, one of the most significant inventions to date. Pastor Malling Hansen of Denmark completed the first model of his 'Skrivekugle'—Writing Ball. More details on this man and his work will be found in Chapter Six.

And in 1866, back to France for two more stenographic machines. Henri Gensoul patented a Presse Sténographique which he claimed printed at the speed of speech. His system, brazenly borrowing from Livermore's, broke the alphabet down into a similar square, components of which formed the letters. Gensoul's machine was larger, however, and designed to print syllables and even words: the keyboard consisted of three identical blocks of keys, each block being composed of two rows containing four keys in each row. The first series printed the initial consonant, the second the vowel, and the third the terminal consonant in each word. Printing was on a paper tape advanced by clockwork. The inventor was apparently[16] commissioned by a newspaper to make a stenographic record of a court case but failed, showing that the machine was slower than he claimed. He later patented an application of the device for telegraphic purposes, in which he specified that, to keep components warm, gas jets must be placed under the ink reservoir, inside the inking roller and on the types themselves and that the

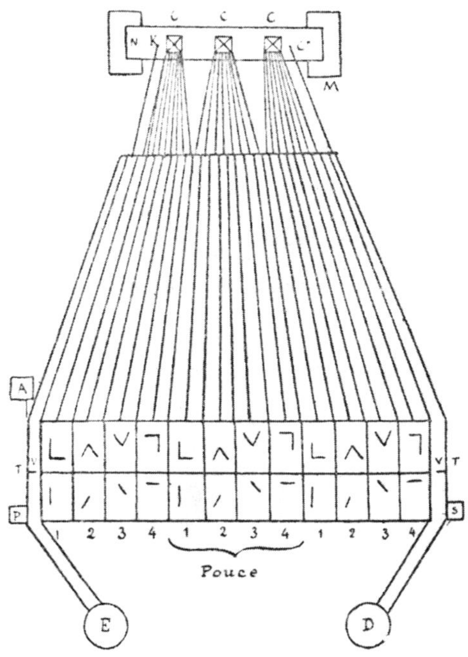

113. Gensoul's shorthand machine borrowed heavily from Livermore's 'combination' idea . . .
(16)

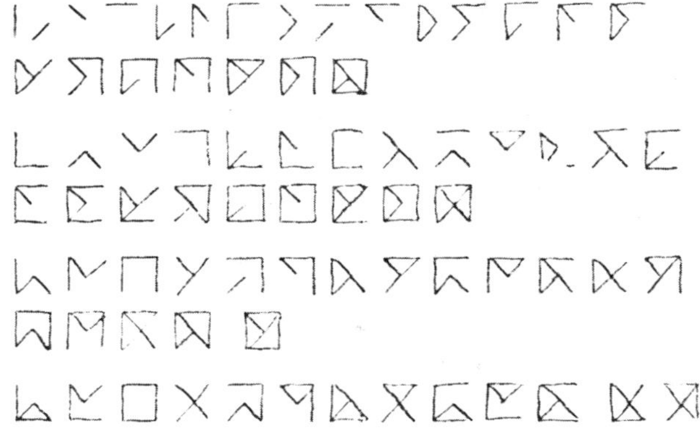

114. . . . as this sample of the printing clearly demonstrates. (16)

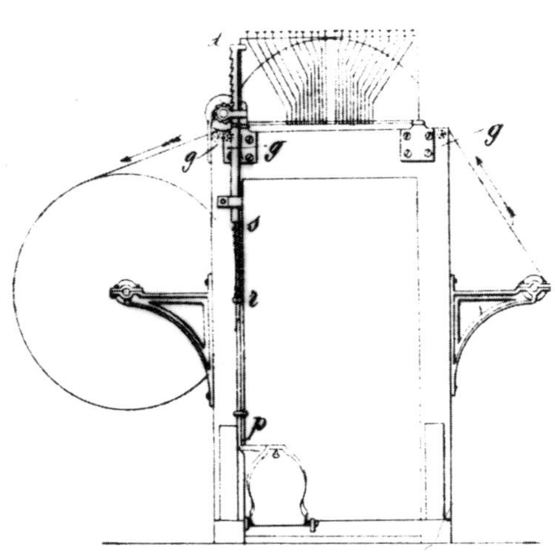

115. Dilliès, 1866, resorted to the well-tried plunger principle. (16)

gas must then be turned on and the whole lot lit. This is the sort of solution to an otherwise simple problem which one might normally expect only from a pyromaniac.

The other device was called a Phonotype or Phonographe by Henri Dilliès who economized on characters by the established practice of taking advantage of phonetic similarities (e.g. s=z). Twenty keys were the result: eleven on the right for the consonants and nine on the left for the vowels. In his patent application he proposed additional models with more keys to make the device adaptable for use with foreign languages and to improve orthographic fidelity.

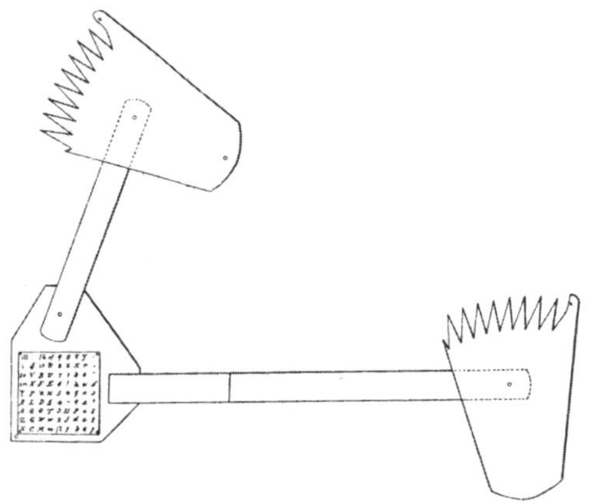

116. Peeler's type index, showing the attached arms, selectors and alignment fingers (or toes). (USP)

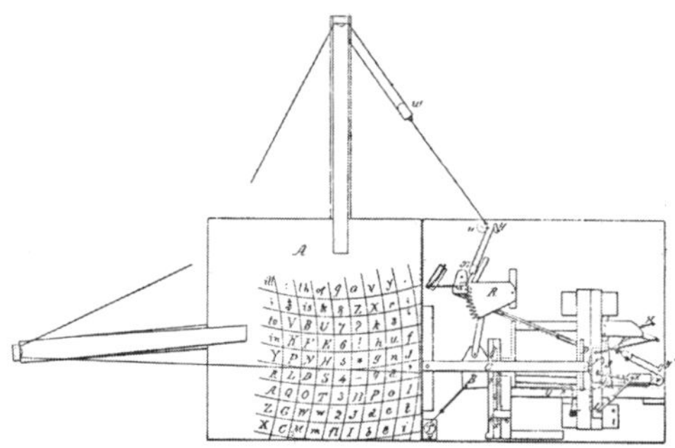

117. The complete Peeler machine, viewed from above. A character was selected by bringing a 'thimble' down over it, and cords attached to the thimble (hopefully) pulled the selectors, and hence the type index, into the correct position to print the desired character. (USP)

Meanwhile, back on the farm, Abner Peeler of Iowa invented his Machine for Writing and Printing, patented in 1866. It printed eighty-one characters from a square index of nine-by-nine configuration, controlled by one horizontal and one vertical selector with nine fingers each, one for every row and column on the index. To the side was a letter table on which the rows and columns were curved, as on the later Morris typewriters, and above it was suspended the indicator, as on the later Mignon, except that Peeler's patent referred to it as a thimble. So far so good, but now the operation became clumsy, for the thimble had to be brought down on to the desired letter, thereby exerting pull on the two cords suspending it which, by applying simple geometry to the selector fingers, were hopefully meant to locate the type-plate correctly.

The device was also interesting in that single characters printed the most frequent combinations of letters such as 'th', 'to', 'in', 'is', 'of', etc., thus making it one of the forerunners of later syllable typewriters, in which a single key printed a syllable at one depression, as against the 'chord' machines which performed the same operation by the simultaneous depression of more than one key. Baillet de

Sondalo, Flamm, Bryois and others had similar but even less practical intentions.

Rumour hath it that Sholes later plagiarized Peeler's machine, but this is clearly nonsense. Peeler himself never made this claim and in fact it only became an issue long after his death. It was obviously made by someone quite ignorant of the facts and yet Roby[32] became quite apoplectic in defending Sholes against charges which hardly seem worthwhile refuting—in a remarkable display of overkill, he dedicated the whole latter portion of his report to this.

That same year, however, a famous inventor, Thomas Hall of New York, who later gave his name to a small index machine that brought him fame and fortune, completed the prototype of a keyed instrument of progressive design, after six years of developmental work. Apparently his ideas had begun to germinate even earlier, in Francis's and Beach's time, although he did not become familiar with their efforts until much later. For some reason, his British patent predates

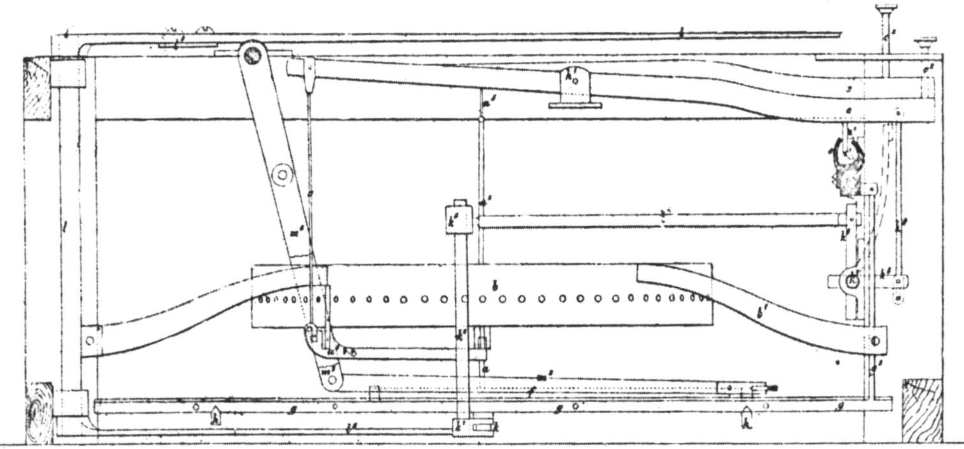

118. The first typewriter invented by Thomas Hall was this 1867 down-stroke. It was not a success, but his later index machine made him famous. The illustration shows one of the keys (upper right) pressed down, with the corresponding type-bar lowered to the printing point. (USP)

the US one. The former was submitted on 26 October 1866 and granted on 5 March 1867. The latter, however, is dated 18 June 1867. Hall's machine was considered[56] among the best up to that time and attracted considerable attention and admiration. It was of the down-stroke class, the keyboard controlling type-bars arranged in a circle above the printing-point. A flat paper-table was employed, the impression was by ribbon, and there was provision for 'spacing letters according to their thickness, giving the work a close appearance to letterpress printing', in other words, there was differential spacing. A cushioned ring suspended in the type circle offered alignment, and a special device prevented clashing of type-bars. It was a 'perfect success, so far as regards the variety and character of its work, and the amount it could perform', which was reportedly[56] 400 characters a minute—a most creditable performance indeed. The inventor demonstrated it to government departments in Washington and received orders; a manufacturing company was planned, but was abandoned as a result of arguments between the various interested parties.

Finally, an American called John Sweet is sometimes[15] listed as inventing a typewriter; his device, of hammer-operated type-wheel plunger design, was patented in 1866 as one for 'improvements in the moulds or matrices of stereotype plates, and in machinery for forming the same', and thus it falls outside the scope of this book.

1867 came, and five or six more inventions. Christopher Latham Sholes's were the most important—in this year, he began his tortuous experiments with his Type Writer, described in detail in Chapter Six.

Luigi Lamonica of Italy was ultimately less fortunate with his shorthand machines. They worked well, but the inventor was unable to launch them commercially. The first instrument appeared in 1867, followed by another, two years later; elegant in appearance, they were considered too big and complicated for practical purposes, consisting as they did of over 500 parts. Despite obvious difficulties in setting them up and adjusting them, the inventor nevertheless made successful stenographic records of sessions of local Municipal Councils as demonstrations first of earlier models, and, in 1874, of an improved and simplified third

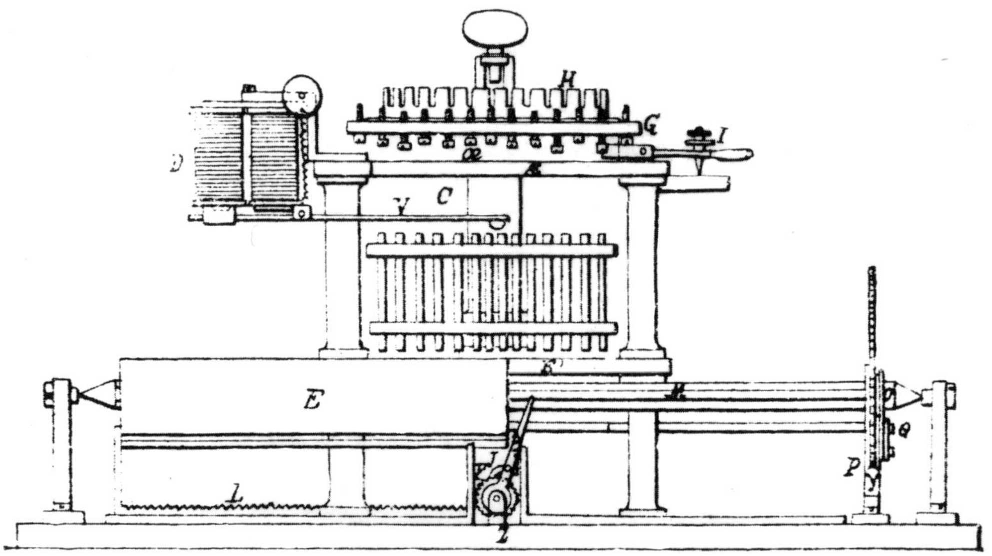

119. A most sophisticated piece of work was Fontaine's type-plunger design using a circular indicator. It featured provision for several different alphabets with means for the selection of the one desired, and full differential spacing for all the characters. It also offered a cylindrical platen, electric assistance, and lots more.   (37)

machine. He continued to work on his invention, allegedly finishing with one for telegraphic purposes called a Merografo. Neither authentication of this information nor details of the machine could be traced; it was attributed by secondary sources[16, 28] to an article in the 1 May 1881 issue of *Il Messaggero*, but no such article appeared in the paper on that date, or near it. The closest thing to confirmation is his 1879 Italian patent for a stenographic method which could be printed by any one of 'a thousand different mechanisms'.

Brilliant, however, was the machine patented in 1867 by a French lawyer called Louis Henri Fontaine and fully developed in a subsequent patent two years later. Both specifications refer to a type-wheel design, the more sophisticated later one using type-plungers disposed circumferentially around the wheel. This was mounted on a vertical shaft, above which was the type index with slots for alignment, and a handle for selection. Inking was by pad. Fontaine suggested that four or more alphabets be mounted simultaneously on the wheel and reproduced on the index: by moving the position of the alignment slots relative to that of the type-wheel, any alphabet could be selected at will, a small pointer indicating the position. For instance, if four alphabets were used, the slots on the selector would be spaced to correspond to every fifth plunger. Thus if Roman A was the first plunger, then B would be the fifth, C the ninth etc. Greek alpha might be second, beta sixth and so on. Shifting the position of the selector aligned the slots with successive alphabets. Ingenious, indeed. An electromagnetically-operated hammer triggered by the selector lever struck plungers against paper wound around a platen, which rotated automatically for letter spacing upon depression of the lever. And as if all of this were not enough, full differential spacing, for all alphabets, was achieved by regulating the vertical travel of this lever, thereby increasing or decreasing the number of teeth it picked up on the ratchet wheel. Screw stops corresponding to each character on the index offered fine adjustment: it was Harger's 1858 design brought to perfection.

And that virtually sums up the year 1867, except for a few scraps. Henry Worral of Hartford, in a letter to his sister,[28] reported having sold two typewriters of his invention—one to a scholar with undecipherable caligraphy and the other to a clergyman with writer's cramp. No other details. However, a man called Worrall (*sic*) was granted an 1886 patent for a machine (listed in Chapter Twelve) with keys attached to rows of horizontal levers pivoted at one end and with the type underneath at the other. These levers were contained in a frame which rode left and right above the flat paper-table to locate the desired character above the printing-point. Whether it is the same Worral is not known.

There is no doubt about a man called R. Allen who is erroneously listed[28, 40 etc.] as having invented a writing machine that same year. The date is incorrect (see R. T. P. Allen, Chapter Twelve); the only Allens who were granted patents in 1867 were one for a printing press and another for a smut machine. A *smut* machine? And a Brazilian called Jesuino Antonio Ferreira de Almeida was reported[46] to have been granted a patent for a typewriter which wrote very quickly and, im-

120. This beautiful type-plunger machine is languishing in ignominious anonymity in the Science Museum, London. Its identity is revealed at last: Cookson, 1885. (CSM)

mediately upon completion of the writing, produced hundreds of copies ... yes, they said 'hundreds'. Details are untraceable.

SML   All there is left to describe is an 'anonymous' machine of unknown maker and date, but believed to be from the decade 1860–70. It is, in fact, by F. N. Cookson of Wolverhampton and was covered by a British patent in 1885 and a US patent the following year. It does not, therefore, fall within the period covered by this chapter, and is dealt with more fully on page 261.

Chapter 5

On Claimants, Pretenders . . .

GIUSEPPE RAVIZZA

The year 1876 was one of the many milestones on the writing machine's long road to failure. That was the year of the Philadelphia Centennial Exhibition and the Type Writer was on display. Apart from buying the odd 25-cent typewritten message as a curiosity, people ignored the machine by the thousands. But not so the new-fangled telephone, which was also on show for the first time—this they thronged to admire, in equal numbers, thereby presenting us with the following ironic situation: the telephone, which the American, Alexander Graham Bell, had stolen from the Italian Antonio Meucci, was competing for attention with the typewriter, which the Americans Sholes and Glidden had stolen from the Italian Giuseppe Ravizza. According to the Italians, who usually believe that they invented it all anyway.

'Italy, the mother of knowledge and of the arts, makes the inventions: the foreigners try to snatch the merits' cried *Il Popolo Romano* as long ago as 1882. The cry has been heard ever since, and Aliprandi and others take it up with specific reference to the typewriter: 'The Italian invention emigrated and then returned to Italy "Made in U.S.A.".'[1] Ravizza has the honour of 'absolute priority in the invention of the typewriter' and this cannot be placed in doubt either by 'sceptics or by misoneists'. Evil intent was partly to blame for the loss of the glory, but so was ignorance. 'The Remington spreads itself also throughout Italy, favouring in this way the diffusion of the popular opinion that the invention of the typewriter was American.'[2]

Aliprandi was writing during the days of patriotic fervour in the 1930s and 1940s. But one would be mistaken in thinking that such nationalistic convictions are dated, for in the English translation of the booklet published in 1955 on the occasion of the typewriter's Centenary Commemoration (Italian version), the claim is repeated: 'The first to formulate and find the solution to all technical problems relating to it, and to manufacture an actually (*sic*) writing machine, was the Italian Giuseppe Ravizza (1855).'[65]

Other Italian writers are more level-headed. 'The education of our middle classes,' writes one[7] 'has a strictly anti-industrial bias. We still remain the sons of the Romans who left industrial tasks to slaves and freed men . . .' So much so, he adds, that history recalls the most mediocre preconsuls and poeticules but not even the names of the engineers who built the great roads, aqueducts and monuments of the Roman Empire. This anti-industrial mentality is responsible for the loss of 'many fruits of inventive Italian genius'.

How much truth is there behind all these assertions? Quite a lot. But ignoring the relative merits of other inventions and with reference strictly to the typewriter, many are those who detect that 'strange resemblance'[1] between the early Sholes and Glidden machine and Ravizza's Cembalo Scrivano, and frankly one needs only a minimal dose of national paranoia to see it. Completely impartial historians arrive at the same conclusion. The Frenchman Rousset, for instance, writing in 1911 (twenty years before Aliprandi) takes a decidedly pro-American stand throughout, and after (incorrectly) attributing the first manufacture of a typewriter to the Americans, he continues: 'We do not describe the Sholes and Densmore (*sic*) aparatus besides saying, by and large, the machine is similar to Ravizza's,'[36] which, an Italian would hasten to add, predated its American rival by well over a decade.

A considerable amount of 'borrowing' (not to use the word 'plagiarism') was going on all over the world. How else is the striking similarity between these

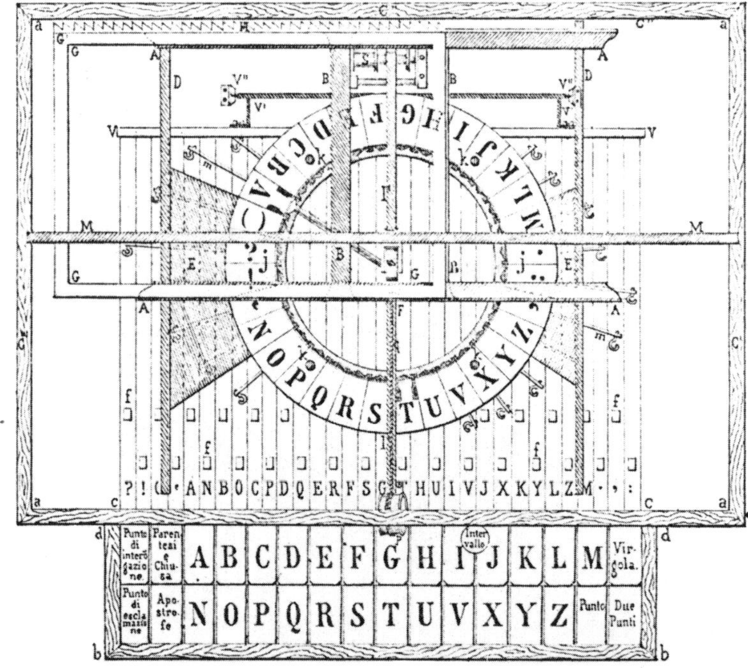

121. A diagram of Ravizza's Cembalo Scrivano, as it appeared in the 1855 patent. Italians are justified in believing that many later inventions bear a 'strange resemblance' to this machine, but they tend to minimize its own 'strange resemblance' to Conti's Tachigrafo of 1823.

(IP)

machines to be accounted for? Not only in appearance but also in mechanical conception the Ravizza, Francis, Mitterhofer, Sholes and Glidden, to mention but a few, were hardly isolated simultaneous developments. Between them all, they barely offered a single new component.

However, why the Italians should promote Ravizza's cause when he produced a machine that was virtually a copy of Conti's 1823 Tachigrafo and he himself admits in his diaries that he corresponded with its inventor and borrowed from him—why *he* should warrant the title of Father of the Typewriter and not Conti is another of those arbitrary decisions in which history abounds. Ravizza has been called[29] the 'first inventor and constructor of a truly functional typewriter' but neither was he first, nor was the Cembalo Scrivano, by the inventor's own admission, 'truly functional'; no more, say, than was the Sholes and Glidden or the Mitterhofer, as will be seen later in this and the following chapter. So why choose him and not Conti?

Giuseppe Ravizza was born in Novara, Northern Italy, in 1811, and he was in his early twenties when he began to delve into the problem of writing by machine. He finished up struggling with it for over fifty years, in the course of which he produced sixteen or seventeen different models, took out several patents, won a limited amount of recognition, but died in 1885 in frustration and despair at the futility of his efforts. His epitaph lists several of his achievements (he was also a lawyer, historian, archaeologist, and so on) but makes no mention of his Cembalo Scrivano, a circumstance which has proved no small irritant to later Italian typewriter historians. Once again, as with his contemporaries, the inventor's cause goes straight to their hearts: if it was not Mitterhofer the bard presenting himself as the ignorant peasant, it was Sholes losing control of his invention and dying of TB, or Ravizza pouring out his woes to his diary . . . all their lives ended sadly (as lives often have a way of doing if they are chronicled to the bitter end!) and served to fan the flames of patriotism for future generations.

The Ravizza diary, this valuable document, has left us a unique insight into the inventor's mind and activities. He began it fairly late—in 1856, in fact—but he reminisces frequently, allowing us to piece together his previous years. It seems he began corresponding with Conti in 1832 or 1833. How they came to meet or know of each other is uncertain, although Ravizza's mother-in-law came from the same town as Conti, which offers us a clue. Whether or not Ravizza ever actually saw the Tachigrafo is not revealed either; he almost certainly did, but it is of little consequence, for the specifications of Conti's machine were carefully documented and Ravizza's Cembalo Scrivano was almost identical, especially in the early stages of its development.

From Ravizza's diary, we learn that the first model of the Cembalo was begun in 1847. Just what happened in the fifteen intervening years between 1832 and 1847 is never fully explained: the inventor is often inaccurately painted as having worked on his machine consistently since the early thirties, although this is hardly likely, for there is nothing to show for these years of effort. From this

122. The operative parts of Ravizza's 1855 model. Originally, it was enclosed in a handsome wooden case to protect it from 'inquisitive fingers'. Note the use of a typewriter ribbon which is responsible for the popular misconception that Ravizza invented it. (STM)

point on, however, the diary provides a fair chronology of new models and improvements, although the numbering of machines is confusing for, as with Pratt and others, several different models were in circulation simultaneously. Others were broken up for parts, so that reference to them is restricted to a mere line in the diary, and then Ravizza would often rebuild the same model repeatedly, incorporating improvements and modifications, thereby further confusing the numerical order.

The most important period in Ravizza's career was in the middle 1850s. He was granted a patent in 1855, exhibited his Model Ten in the Novara Exhibition in 1856, where the Emperor Vittorio Emanuele II wrote his name on it, and in the same year his Model Four in the Milan Exhibition where the Austrian Kaiser saw it. These were important details at the time, precisely the sort of details that assisted in the diffusion of information on writing machines. Mitterhofer's unsolicited assertion that he was only a poor ignorant peasant (see page 132) and therefore knew of no other attempts at inventing the machine may well have been prompted precisely because he *had* such knowledge, the more so since he was travelling throughout Europe during these very years. Ravizza, who did not travel, nevertheless knew of the Hughes machine and of Devincenzi's patent, not to mention, of course, the Conti Tachigrafo—these three for certain, since he concedes it in his diary: how many others he knew of but failed to mention is open to conjecture.

The Cembalo Scrivano was a machine of the up-stroke group, the type-bars set in a circular basket underneath the printing point. A series of levers connected these type-bars to the keys on a piano keyboard, letter order originally alphabetical. There were thirty-two keys, with punctuation marks at either end and the twenty-five upper-case letters in the central section. The missing letter was the 'w', which is not used in Italian. Letter spacing was automatic. The paper was held in a flat horizontal carriage and the escapement was by a drum with a clock spring, a cord, and a wheel which was released one tooth at a time—in short, a sophisticated arrangement indeed. This was Ravizza's only major departure from the Tachigrafo, on which the paper remained stationary while the type-basket moved. Otherwise, the carriage was displaced progressively from rear to front for line spacing as it was on Conti's machine, printing of course being non-visible. A sheet of silk material treated with a mixture of fat and rust or graphite, and positioned between the type and the paper, made the impression. Later, this was modified into a strip soaked in dye and wound around a kind of spool, thereby providing his fans with their 'proof' that he invented the ribbon . . . except that Bain did the same thing a decade and a half earlier (see page 72). Two alternatives were offered to denote the end of a line: the warning bell, or, more poetically, a small window opening and revealing a sign announcing 'line is finished'. But in its more essential mechanical details, the machine was a masterful conception. 'The type-bars are placed in a circle in a bronze ring fixed over a wooden one which is itself attached to the

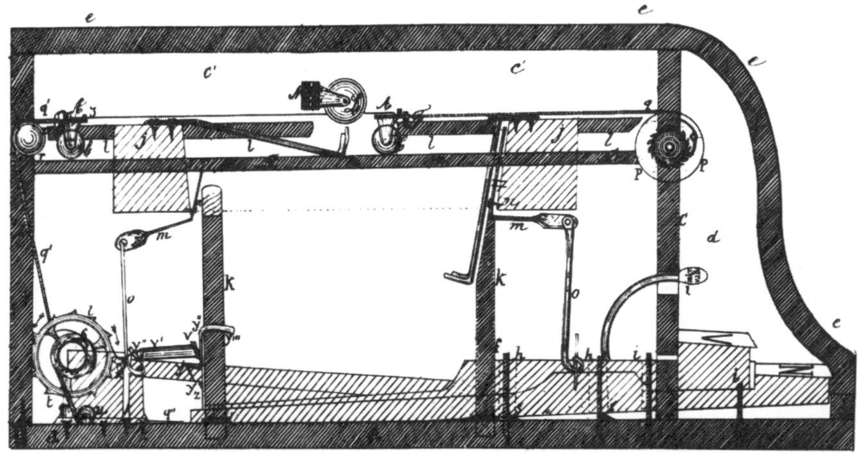

123. Type-bars controlled from a piano keyboard were suspended in a circular basket and struck up at the printing-point on the underside of the flat paper carriage. A cylindrical platen was suggested as an alternative.   (STM)

base of the machine by four legs'—these specifications, taken from Ravizza's patent, are virtually identical with those of later Sholes models, and, like the later American invention, this one appears to have suffered from the same problem of type-bars sticking and binding. Ravizza ingeniously solved it by swinging each bar on a small conical axle located in a steel bridge, thereby ensuring free vertical but limited lateral movement. A similar solution to the problem was ultimately applied to the American machine fifteen years later.

Ravizza's patent offered a platen as an alternative to the flat paper-table. It consisted of a cylinder of wood or hollow brass covered with leather and operating on the established principle of rotating for letter spacing and moving linearly for line spacing. Once again, this is an arrangement which was ultimately adopted by both Mitterhofer and Sholes, the former leaving it at that while the latter progressed to incorporate the 'continuous roll' system as we know it today, which was used intermittently since the 1840s.

This alternative carriage with cylindrical platen is one of the features listed in the first typewriter-advertising leaflet ever printed: a short eulogy on the Cembalo

Scrivano, composed in 1856 by a colleague of Ravizza's, the lawyer Costanzo Benzi. In praising the machine's superiority of impression, its speed and ease of operation, its portability for travellers, and so on, the pamphlet set a pattern for countless similar publications up to the present day.

The details of Ravizza's machines were widely diffused, and apart from the Novara, Milan and Torino Exhibitions in 1856 and 1857, where it was awarded medals, the Cembalo was prominent in the Florence Exhibition in 1861, and again in Milan in 1881. It has been further reported that a model won an award in London in the 1862 Exhibition, but no confirmation of this could be found and a confusion may well have caused the error: the Reports by the Juries[58] cited some unrelated Ravizza brothers for their Orvieto wines.

From the inventor's diaries,[3] the following information can be compiled about the different models he made:

Model Four: exhibited in 1856 in Milan. It was 'very small', and according to the inventor, was his best, his *non plus ultra*. After that machine, he wrote, it was all downhill. By 1876 (the last trace), it was on a wood-heap in Novara!

Models Seven, Eight and Nine: brief mention only. They had 'beautiful' characters.

Model Ten: this was Ravizza's favourite machine, the one he exhibited in Novara in 1856. 'Adorable', 'unequalled', 'insuperable' are some of the adjectives he uses to qualify it. He loaned the machine to Maggi (see page 220) in 1881, and it was returned to him in 1884.

Model Eleven: mentioned in an entry in 1876 as having coarse characters.

Model Twelve: an 1863 entry condemns this machine as being rejected in Florence because it was too big and because of its 'deafening noise'. It was later rebuilt to incorporate visible typing and exhibited in Milan in 1881.

Model Thirteen: 'the open machine works and visible writing exists', he wrote in 1879 of this model. The 'open' feature is one he tinkered with for many years in efforts at making the instruments operate more quietly: he correctly diagnosed that cases acted as resonators, but they were structurally indispensable on early models and were useful, furthermore, in protecting the works from inquisitive fingers.

Model Fourteen: started in 1879 and finished the following year. It wrote well but not beautifully, he said.

Model Fifteen: referred to briefly in 1879.

Model Sixteen: This was Ravizza's last heartbreak. With the participation of an industrialist called Carlo Fantoni, the Cembalo Scrivano was to be manufactured and marketed as a competitor to the Remington. Fantoni was determined to have Italy reclaim its invention from the Americans, Ravizza wrote. Consequently he took this sixteenth model to Fantoni in 1882 and left it with him; the following year, the industrialist, having clearly abandoned the project, sent it back without a word of explanation. Ravizza, already an old man, was pathetically demoralized, even though he conceded that the machine was truly ugly.

128

So much for the recorded models and information pertaining to them. A little ambiguity exists and there may have been another machine; in addition, there are numerous references to other important developments, mainly the inventor's obsession with visible printing. He began experimenting with this innovation as early as 1860, continuing to perfect it for over two decades. If the sewing machine had been built to sew non-visibly, he wrote, Singer would never have sold the quantities it did. And in later years, around 1880, after Remington had taken most of the wind out of his sails, he pinned his sole hopes on this visible machine. His judgement was not at fault, even if the device was never perfected. No one else was producing a visible type-bar instrument at the time and Remington actively opposed the principle until early this century. Like many other inventors, Ravizza swung back and forth from exaltation to despair. 'I have a visible Cembalo Scrivano' he proclaimed as far back as 1872 and 'finally, a Cembalo Scrivano with visible writing is finished' (1876). But it failed to fulfil the inventor's hopes and claims; it was, however, patented in 1883 in the names of Ravizza and Fantoni.

Other revelations in the diary include a recurrent preoccupation with lowering the noise level of his machines, leading first to cushioning under the keys and eventually to the 'open' framework. The keyboard and letter order concerned him: he lists four- and five-row possibilities, but prefers the 'small' keyboard. In 1872, the idea of letter order based on frequency occurred to him, but since he hoped to export the machine and offer it for foreign languages, he did nothing more than study the subject, and, in his last years, he experimented with a syllable typewriter, a model of which he apparently succeeded in building in 1882.

He appears to have received numerous orders for his machines, although whether or not these were honoured is uncertain. Some definitely were: three were given to the Milan Blind Institute, according to an 1860 entry, and from then until his death in 1885 five more were supplied to private individuals. This accounts, then, for eight machines, or roughly half of those he built.

Ravizza's achievements were many yet his love affair with the writing machine was by and large a succession of disappointments. His worst shock was the first time he saw the Type Writer—about 1880—and found it to incorporate virtually all his cherished ideas, except the visible writing on which he ultimately placed his hopes. He liked it, however, and professed to admire its solidity, beauty and construction, but he was unimpressed by the effort required to operate the keys and by their excessive drop—he was accustomed to the light touch of his own piano keyboard. Most of all, he was impressed by the carriage: 'the exact opposite of what I have done', he wrote, referring to the 'continuous roll' feature. And the speed! The advertising claimed that the machine was capable of thirty to sixty words a minute, he confessed to his diary, and a young lady was already known to have typed at thirty-five a minute. That speed is something to think about![3]

But even though he complains of seeing his ideas stolen or copied by Remington, he writes frankly about his Cembalo Scrivano: 'I am still far from the point of having a machine that writes well and fluently and that can be used daily, and sold . . .' (February 1880).

He never made it, and he died in 1885, embittered at seeing the Type Writer succeed where he had failed, using the principles which he had developed and patented thirty years earlier. Ultimately, however, he remains an also-ran in a cruel world that is callously geared to honouring the victor and ignoring the vanquished.

PETER MITTERHOFER

Mitterhofer got where he did, historically speaking, because there is only one of him. This does not make much sense and may be better expressed as follows: if Pratt, for example, had been an Austrian and Mitterhofer an American, then Pratt would have been their hero, and Mitterhofer would have remained a nobody. The thing is, he is the only hero the Austrians have got. And this applies more or less to the Germans as well. Of course, there have been some abortive attempts at making Drais the inventor of the writing machine, but to little avail. Interestingly enough, the Germans appear to have preferred Mitterhofer to Drais for much the same reasons that the Italians chose Ravizza instead of Conti—not because of chronological priority, but because of the similarity of their designs to those of later Sholes. All of this may sound most unfair to Mitterhofer, but the truth is there is no intrinsic reason for him to be in this chapter at all. He insinuated himself on the strength of those perverse national passions with which the history of the typewriter abounds.

Peter Mitterhofer was born in 1822 in a little village in South Tyrol called Partschins. It belonged to Austria at the time, but since the end of the First World War the area has belonged to Italy and the village is now called Parcines. Some Italians have annexed the man with the village: his house now bears a commemorative plaque in Italian and German (as is customary in that bilingual area) 'Peter Mitterhofer 1822–1893 inventò in questa casa nell'anno 1864 la macchina da scrivere—ERFAND IN DIESEM HAUSE IM JAHRE 1864 DIE SCHREIB-MASCHINE.' His tombstone is even more specific: 'Peter Mitterhofer, der erste Erfinder der Schreibmaschine' ('the first typewriter inventor'). The centenary of the invention of the machine was already observed by his followers in 1964, and in case there was still any doubt as to the origins of the invention—and clearly aimed at dispelling any possible disagreement—the following couplet is attached to his epitaph:

> Die Andern, die von ihm lernten,
> Durften die Früchte seines Talentes ernten.

(Others, who learnt from him, could reap the fruits of his talent.) The 'others' at whom this message is directed are, more specifically, Sholes and Glidden.

Not all Austrian and German writings on the subject plug this line. Some of them even insert scepticism into their reports, but such is the extent of nationalistic involvement that even the most level-headed and serious are prone to making the wildest claims and speculations. Dr Richard Current's[64] statement that the typewriter originated in Milwaukee is at least easier to swallow than Dr Rudolf Granichstaedten-Czerva's[20] claim that a 'mechanic' Carlos Glidden, while a student in Vienna, saw Mitterhofer's machine, copied it, and went back to the States to manufacture it with the help of Sholes and Co. Both Current and Granichstaedten-Czerva were historians, we are told, but while the former's statement is merely a partisan distortion of the truth, the latter's is sheer mischief. His authority? A 'confidence' entrusted to him by the son of an innkeeper in Mitterhofer's home town! One is hard-pressed to think of a more unreliable historical source than an innkeeper's gossip, let alone a generation late, and yet Dr G.'s assertion was picked up by subsequent writers in the German language, by text-books, journals and newspapers, and altered the typewriter's history for millions of German-speaking peoples who were clearly hungry for any new heroes they could find.

There are, of course, those writers who concede that Mitterhofer was not *the* inventor but just one of many; they are clearly irritated by their lack of a convincing alternative. For instance, in discussing Mill's patent of 1714 Martin[28] agrees that it is the first on record, but immediately adds that it would be wrong to consider the typewriter an 'English invention'. It had probably been invented independently in Germany at the time—or perhaps even earlier—but since a German invention could not be patented no records have survived. The absurdity of this *non sequitur* clearly did not strike the author: if nothing has survived, how can it be assumed to have existed?

Mitterhofer's claim to his exalted position dates from the publication of Dr G.'s pamphlet, which first appeared in 1924. Until then, the inventor had remained

124. Mitterhofer's machine was made entirely from whittled bits of wood, bent wire and leather patches. It had no platen, no inking device, no change of case—nevertheless, it has been proclaimed the 'first complete typewriter'.   (TMV)

virtually unknown. It suddenly transpired that the typewriter, that eminently American device, was German after all, except that everything had conspired to deny Mitterhofer the honours. It would be wrong, Martin says, to attribute credit for the invention to Sholes, because he probably would not have thought of building a typewriter at all if Glidden had not stimulated him and if he had not read the Pratt article. And if it hadn't been for Densmore—and if it hadn't been for Yost—And anyway, if Mitterhofer's invention had been taken up by German arms manufacturers they too would have made a first-class typewriter of it in time. Much the same thing was said of Ravizza,[1] as was pointed out previously. What Mitterhofer and Ravizza both lacked, according to their supporters, was the backing and assistance which Sholes managed to find. *If* they had had a team, such as Sholes's—*if* they had had financial backing—*if* they had had a Remington organization—if—if—if—

After all, Mitterhofer's machine was better than the Sholes model delivered to Remington, Martin says, but because of an unfavourable environment it remained unknown while the same thing was 're-invented' in America! Carl Müller of the Bayreuth Typewriter Museum is more level-headed, but, in an article he wrote on the occasion of the 1964 'centenary', he passes similar judgements.

Disappointingly enough, there is far less to report about the inventor himself than about the storm his followers kicked up over him. Son of a carpenter and cabinetmaker, Peter Mitterhofer grew into a man of many talents. Poet, singer, musician, ventriloquist—one has the impression of a latter-day wandering minstrel. Restless in his home town, he went on a walking-tour of Austria, France, Germany and the Balkans, presumably returning to Partschins at the age of forty with the secret determination to build a writing machine. The thought never occurred to Dr Granichstaedten-Czerva and others that their hero might have seen someone else's machine on his travels. They accept at face value Mitterhofer's unsolicited claim, contained in a letter to Emperor Franz Josef I, that his instrument was completely original. 'I can proudly say', he wrote, 'that the apparatus is my invention, since I, a simple Tyrol peasant from Partschins near Meran, have never seen anything that could serve even in the remotest way as a model for my invention.' Students more cynical than Dr G. might be inclined to wonder, the more so since the man was by no means a simple peasant but was in fact an educated and well-travelled person.

Much has been made of the similarities between Mitterhofer's machines and the later ones of Sholes and Glidden; these similarities undeniably exist and are quoted because they favour the Austrian's claims to precedence, but no mention is made (except by Italians) of the resemblance of them both to Ravizza's Cembalo Scrivano, which predated them by a decade or so. And what about Francis's machine? Is one to believe that the 'strange' similarities of such instruments as those of Conti, Ravizza, Francis, Mitterhofer, Sholes—to name but a few—are purely coincidental? One might as well reprint *Crime and Punishment* under a different title and swear that the resemblance is the product of chance alone!

Details of Mitterhofer's typewriters are as follows:

1864: Model One. This is the machine which has been called 'the first complete typewriter'.[26] Of the up-stroke class, its type-bars are contained in a circular basket striking at the common printing-point on paper held in a flat frame. There is a three-row keyboard, upper case only. The instrument is almost entirely made out of whittled bits of wood and bent wire. It has no inking device, since each

DTV

132

125. The first Mitterhofer model used needle points fixed in wooden plugs at the ends of the type-bars to perforate the paper, much as on machines for the blind. The inventor thereby avoided an inking system, also a mechanism whereby the carriage could be raised to reveal the printing. (DTU)

126. The Mitterhofer Model One, partially dissembled to reveal the type-bars with their needle points. (DTU)

character is composed of the tips of needles sunk into a wooden plug at the end of the type-bar: the impression is therefore perforated through the paper. This is not original—an apparatus exhibited by Gall at the 1851 Exhibition and listed in the Catalogue used the same system and of course so did other machines for the blind, as well. Hinges, where needed, are composed of leather patches. An utterly primitive device, the machine did however write—with difficulty— although how such an instrument without carriage or platen or inking device or change of case can be considered the 'first complete typewriter' is difficult to imagine. If anything, the only true sophistication in the machine is the use of differential spacing. But neither this nor any of the other components is Mitterhofer's invention. He was not, in fact, first with anything.

'MV 1865: Model Two. There seems to be some controversy surrounding this machine, which is almost identical to the one above but lacks any kind of paper-frame or carriage. Carl Müller believes this is a sub-model, used for experimental purposes by the inventor in the development of a cylindrical platen which first appears in:

1866: Model Three. A considerably improved machine, still basically following the same principles (up-stroke, etc.) but more professionally built: there is less whittled wood and more metal. It uses a cylindrical platen and regular type for the impressions; the inking mechanism is, however, unknown, since the machine was found in poor condition and this was missing. Such, anyway, is the explanation invariably offered—the unmentionable alternative of course is that there never was an inking mechanism to begin with and that carbon paper was used for the impression. This model had a total of thirty-nine keys printing seventy-two characters; upper and lower case were incorporated, with a primitive shift mechanism apparently designed to rely heavily on luck. Determined to present his invention to the authorities, Mitterhofer walked to Vienna, pushing the machine in a wheel-

127. 'He was the first typewriter inventor to have walked from Parcines to Vienna pushing a wheel barrow.'

barrow. This was quite a feat, and is invariably mentioned in accounts of his life. Rightly so, for the inventor chalked up a 'first' at last: he was the first typewriter inventor to have walked from Parcines to Vienna pushing a wheelbarrow. In his application to the Emperor for assistance, he presented an impressive list of advantages which his invention offered to diplomats, poets, sufferers from fatigue, invalids and so on. The reception was reported to have been hostile, but the Emperor nevertheless authorized a payment of 200 guilders, a considerable sum in those days. Thus, Mitterhofer may well have chalked up yet another 'first': which other typewriter inventor succeeded in getting a similar hand-out? Not only that, but he also received an important order for an improved model, which he went home to build. His reception in Vienna was hardly hostile!

TMV 1869: Model Four. This was the inventor's *magnum opus*. He took it to Vienna the following year and was paid 150 guilders for it. This model was believed lost until it was found recently in a crate thought to contain a writing machine of Austrian origin from 1842: it had been gathering dust in the Museum for some eighty years. It differs from its predecessors chiefly in having a full keyboard of eighty-two keys for upper and lower case, leading one to suspect that Mitterhofer was unable to find a solution to the technical problems of a shift mechanism and, rather than resort to the primitive device he used on his previous model, opted for the present alternative. Inking was by a unique combination of feathers and bristles. A letter, written on this machine by the inventor and still preserved, shows up some of the faults that plagued it. Among others, alignment and letter and word spacing were deficient. 'As you can see,' he wrote in that letter (1869) 'my writing machine still writes very badly but I have not yet despaired because I still know how to improve on some of it . . .'

So much, then, for another of the contenders, a man of undoubted merits and great talents, a few of which he used in the development of a writing machine. His image, like that of so many of his fellow inventors, has been enhanced by the passage of time and by the vested interests of nationalists and others for whom the tragic figure of the wanderer, the poet and the musician, the forgotten genius and the ignored hero lends passion to what should essentially be a coldly technical and historical appraisal. The centenary of the invention of this 'first' typewriter was celebrated by these men on 13 September 1964, and corresponding plaques were laid.

. . . and National Heroes

CHRISTOPHER LATHAM SHOLES

Meanwhile, on the other side of the Atlantic, members of the Herkimer County Historical Society, the National Shorthand Reporters' Association, and many other American organizations and individuals had already celebrated, in 1923, the half-centenary of 'the fifty-second man to invent the typewriter'.[5] And they will undoubtedly commemorate his centenary in 1973.

The fifty-second man?

Hardly! To his followers he was the first, of course, but if one accepts the alternative arithmetic and starts with Henry Mill, then he was, in fact, at least 76th and perhaps as much as 112th. But once people get their hands on a hero, they are reluctant to relinquish him. Roby, for instance,[32] could hardly have been more enraptured: 'A great want in the world will always summon the genius of the planet to its fulfillment, and the cry of "Eureka" goes up in many places.' He was writing of Sholes—not of Leonardo! And in their own eulogy[21] the Herkimer people allowed themselves to be fairly carried away by their adulation. After a brief and cursory dismissal of a few of the previous efforts at inventing the writing machine, including the remarkable statement that from 1850 onwards attempts were made only in America (thereby telling Mitterhofer, Ravizza, Malling Hansen and all the others just what members of the Society thought of them) they find themselves in the late 1860s.

'*The hour for the typewriter had struck*. And when, in the course of time, the appointed hour strikes, it seems written in the book of human destiny that it shall produce THE MAN.' (*sic*) And the book continues:

'*The time*—the winter of the year 1866–67.

'*The place*—a little machine shop in the outskirts of the city of Milwaukee.

'*The scene*—three men, all middle aged, thoughtful and studious . . .'

And so the curtain rises on a six-year drama which ends in the 1873 production of the first typewriter in the town of Ilion, which (few will be shocked to learn) is in Herkimer County.

Except that there are others who disagree. By 1925 the invention was being claimed by no less than 'seven wrangling cities'.[32] And thirty years later, the wrangle continued. 'The typewriter originated in Milwuakee,' wrote Dr Richard Current[64] on the occasion of a dedication by the Milwaukee County Historical Society,

19 May 1956 when Sholes was credited with having 'perfected the first practical typewriter in September, 1869'. Perfected? First? Practical? 1869? Clearly, as with Ravizza and Mitterhofer, there is much sick information to be examined.

Consider the following: 'In casting about for a suitable manufacturer for the new invention, the minds of the inventors turned naturally to the noted gunmakers who had already made the name Remington famous.'[21] 'Naturally'? Why 'naturally'? We are obviously meant to believe that Sholes and associates had determined *a priori* to call their invention a Remington! And why Milwaukee typewriter inventors should turn 'naturally' to a New York gunsmith is difficult to imagine, the more so since they had already exhausted all alternative manufacturing possibilities in the preceding years.

Or consider the word 'typewriter' itself. The term has usually been attributed to Sholes, who is credited not only with coining it but also with applying it to his machine. Weller,[43] whose partiality is extreme, is one of those who make this claim. The truth is that Sholes was not the first to use the word, and there is serious reason to doubt whether it was he who applied it to his invention. For a start, the *Scientific American* article on the Pratt machine, which gave Sholes the very idea of building a typewriter, used it. As to the naming of the instrument, an unpublished document (see Appendix C) corroborates information[12] to the effect that Densmore, Sholes's business associate, was responsible. This document is a ledger which Densmore opened in 1872 and in which he wrote: 'I have named the invention the "Type Writer" and the organization or company owning it, I think better be called the "Type Writer Company"; and accordingly, I have placed that name at the head of the statement . . .' This ledger statement is probably the earliest such reference and 1872 would thus appear to be the date on which the machine was formally christened. The Sholes patent of 23 June 1868 refers to a 'type writer' (Densmore filed this application) but the word was not specifically used at the time. Weller referred to it as a 'print writing machine' in an 1870 testimonial and, interestingly enough, Sholes himself persisted in referring to it as 'the machine' and not 'the type writer' in all his correspondence for years to come.

And so the controversy surrounding the American contender touches upon every aspect of the invention. Typewriter manufacturers for a hundred years have distorted the story to suit themselves and every eye-witness left a different version, heirs and relatives of the initial participants only adding to the confusion.

Depending upon one's source, Densmore was either a philanthropist or a scoundrel. He was either left penniless or made half a million dollars. He either invented an oil tank or he didn't because his brother did. He subsisted on a diet of raw apples and crackers either because he could not afford anything else or because he was an eccentric vegetarian. Glidden first invented the machine, or else he didn't. Maybe he merely suggested the idea. Or else one of the others did. He either participated from the start or entered later. Idem Roby. Idem Soulé. Sholes worked in a vacuum with no knowledge of previous patents, or else he

E*

studied them carefully and borrowed from them—and so on.

One is sorely tempted to dismiss this entire chapter in a single irrefutable paragraph:

'In 1867, Christopher Latham Sholes together with Carlos Glidden and Samuel Soulé invented a primitive writing machine which was patented the following year. A businessman called James Densmore bought into the invention and Soulé later withdrew. Over the years, Sholes produced twenty-five or thirty experimental models, most of them at Densmore's instigation, without succeeding in perfecting the instrument. In 1873, Densmore, with the help of a promoter called George Washington Yost, placed the manufacturing rights in the hands of E. Remington & Sons, who completely remodelled it and marketed it in 1874 as the defective and unsuccessful Sholes and Glidden Type Writer. It was not a mechanically or commercially successful machine until some years later, when a much improved model was marketed.'

There is, of course, much more to the story than this, and the account that follows is often contradictory and confusing. In these circumstances, all controversial information is attributed to its source.

Christopher Latham Sholes was born in 1819. A printer by trade, he enjoyed a modestly distinguished career in public and private life, finding himself employed as collector of customs for the Port of Milwaukee in 1866. By inclination, he was an amateur inventor who spent much of his time with other local tinkerers in a machine-shop run by a man called Kleinsteuber. Sholes was working on a numbering machine together with another printer[21] or machinist[43] called Samuel Soulé. A third man called Carlos Glidden was busy inventing a mechanical spader to replace the plough. And according to his own memoirs written fifty years later,[32] Henry W. Roby was at the same time inventing a magician's clock. These are the principal actors in a drama the script of which has never been conclusively written. Most versions concede that Glidden showed Sholes the article on Pratt's machine in the *Scientific American* (see page 109) and suggested that he rebuild the numbering machine to write letters instead of only numbers. The fact that most secondary sources agree on this point is not necessarily conclusive—it usually means that they all relied on the same original account. Variants include Weller's narrative[43] which gave Sholes sole credit for the early work, and Roby's[32] which agreed that Glidden made the suggestion, but after Sholes had already thought of the idea himself, and, an avid reader, had even studied earlier efforts. Another possibility[21] includes a General William LeDue who subsequently introduced the machine into government service in Washington: he claimed to have made the initial suggestion for converting the numbering machine.

By and large, however, all these reports agree that it was Sholes who played the chief part, and that the others either offered to help or were asked to participate. The one seriously divergent account was supplied by Glidden's daughter Jennie who, in later correspondence,[28] claimed that her father worked on the machine off and on from 1862, simultaneously with other inventions. When he was through

138

with his spader, he completed the writing machine and showed it to Sholes, who was immediately interested. Jennie pointed out that Sholes's obituary in the *Milwaukee Sentinel* (18 February 1890) corroborates her story: 'Mr Sholes is dead. He became interested in the efforts to perfect a typewriting machine.' She also quoted a lawyer who independently confirmed this version in his diary, describing the model Glidden showed him as a crude affair made of wood and string. Sholes and Glidden then worked on the machine together; according to Jennie, her father, who was fifteen years younger than his partner, permitted Sholes to put his name first out of deference to his age.

Jennie undeniably had a vested interest, but then so did most of Sholes's contemporaries. Others who recorded their reminiscences a half-century later were equally biased. Weller was one. He was chief operator for Western Union in Milwaukee, he said,[43] when Sholes came by his office asking for a sheet of carbon paper, a scarce item found only in telegraph offices in those days. Sholes then invited him over to his office to see something interesting, and when he arrived, he found Sholes explaining details of a little model to Glidden, implying that it was Glidden's first contact with the machine. This is in direct contrast, then, to Jennie's account.

The model consisted of a wooden ring with a circular piece of glass supported above it, standing on a wooden base with a telegraph key operating a single type-

128. To test the validity of his ideas, Sholes built a small model using a telegraph key and a single type-bar suspended in a wooden ring. Pressing the key kicked the type-bar against the piece of glass. (MPM)

bar by the up-stroke principle. A hard tap on the key kicked the type-bar up against the glass—by inserting a piece of paper and the carbon beneath the glass, the following message was recorded: WWWWWWWW. Or so Weller says—in fact, it probably looked more like this:

$$W^W{}^W \quad W \quad {}^{WW}W \quad WW^WW$$

because the paper was simply moved by hand, there being no provision for spacing or aligning the printing.

However, an impression was definitely made, and not only on the paper. Work was to begin immediately and Weller was promised the first machine for testing. Glidden, Soulé and Schwalbach, one of Kleinsteuber's machinists, were called in to help. According to Roby's account,[32] he himself was in on it too, but no one else appears to remember him or, at least, to consider him worthy of mention. Weller, on the other hand, attributed sole credit to his friend, but then 'throughout his long life, Mr Weller's devotion to the memory of Sholes has been unbounded'.[21]

The first primitive machine was finished in September 1867, and it worked, after a fashion. It was an enormous and ugly affair, made from a converted kitchen table with connecting wires running from protruding keys down to levers near the floor. A circular hole in the table-top supported the ring of type-bars which hung down into the space below. It was an up-stroke machine with ribbon inking, but with the ribbon above the paper so that the type pressed the paper against the ribbon. Only thin paper, held in a flat frame, could therefore be used. Escapement was weight-driven.

Everyone was bounding with enthusiasm. A patent application was filed the following month; Roby says they considered the thing ready for the market, and

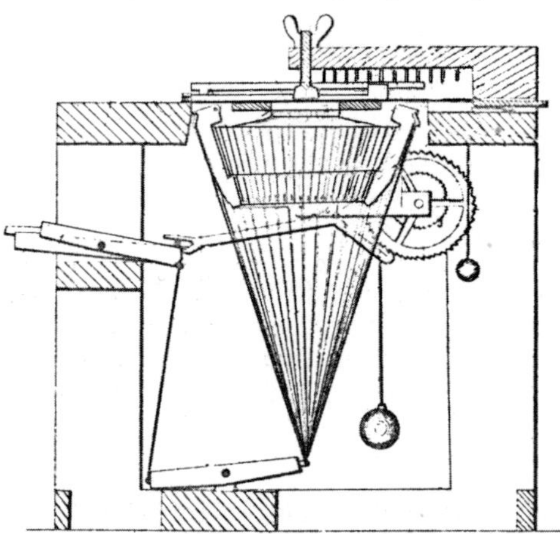

129. Sholes, Glidden and Soulé completed a machine, capable of printing the entire alphabet, in September, 1867 and immediately protected themselves by filing a patent application (October).   (USP)

the *Scientific American*'s prophecies of imminent fortune appeared on the point of materializing. Meanwhile, however, they were decidedly short of funds. Some[27] say that Glidden had supplied the money till then; Roby says he himself financed it. In truth, all of them probably chipped in, but their debts at Kleinsteuber's increased and they could raise no more money between them. They therefore decided to look around for investors, and meanwhile made some additional

SIW models. One had to be submitted to the patent office; another was sent to Weller who had moved to St Louis—he received it in January 1868. Soulé took yet another model to New York and Washington, but failed to interest an investor in it, even though it had already been considerably cleaned up. It was still as large as before, but a square wooden frame replaced the kitchen table, and so on. On the other hand, it was still 'far short of an acceptable, practicable writing machine', as a later Remington catalogue admits. Weller, however, was more nostalgic about it, finding it practicable only because he prepared copy for the printers and did not need the 'neat work that would have been demanded in depositions and transcripts'—in other words, ordinary typing. The fact that it wrote only upper case was equally unimportant for him. However, his description[43] reads like a mechanical nightmare: 'when one started out on a sentence commencing with the letter "T", in place of the sentence we would have a long row of T's . . . and the carriage would jump an inch or two, or perhaps half a line, stopping with a jerk . . .' For those amazed that a non-electric machine could print a whole row of identical letters, the explanation is simple: a type-bar, in this case the T, would stick at or near the printing-point and would not fall back in place. Every succeeding stroke would merely serve to hammer the T on to the paper. Since the machine was of up-stroke design, and hence non-visible, the typist would only discover the defect at the end of a paragraph, or whenever it was that he raised the carriage to examine what he was typing. The escapement also gave them trouble,

130. A model was built according to the patent specifications in the hope of attracting a financier. It was the size of a kitchen table; the plumb line at the rear was not for testing whether the machine was vertical but for supplying carriage tension. The enthusiastic Densmore invested in the project by proxy but pronounced it worthless when he eventually saw it.   (SIW)

Weller said, and, to increase its efficiency, they would hang additional weights from the cord, which would occasionally break, causing grievous bodily harm to any curious onlooker whose foot happened to be nearby. And the type-bars clashed, and the whole thing was considerably slower than the pen . . . but basically it worked, and Weller thought Sholes was marvellous and what did it matter anyway?

It mattered a great deal to James Densmore who pronounced the machine worthless when he finally saw it four months after buying a 25 per cent share in it, sight unseen!

In looking around for investors, the three partners wrote letters to a number of likely candidates. Sholes thought of Densmore, since the two of them had attempted to edit a newspaper together some fifteen years before. The project had failed and the two men had parted without shedding many tears, but reports that Densmore had subsequently made a killing in the petroleum boom must have filtered down to Sholes over the years. In truth, although reports disagree about Densmore's solvency at the time he received Sholes's letter, it appears fairly certain that he was almost down and out by then, or at least hard up and over-committed.

There is no disagreement, however, on Densmore's personality. Bullish, arrogant, domineering, overbearing, crude, eccentric—these are but some of the adjectives used to describe the man. Roby's treatment of him is generally sympathetic but nevertheless he describes him as follows: 'a great, ponderous, beefy-looking man of nearly three hundred pounds weight, with a florid complexion, a great shock of red hair, a shaggy beard, the eye of a hypnotist and the heavy jaw and animal force of a great Hyrcanian bull in *Quo Vadis* . . . restless as a tiger, a born bully with a fierce military spirit . . . he had an unfortunate personality that repelled many people instead of attracting them . . .'.[32]

For his part in the writing-machine venture he has variously been described as the hero and the villain. He remains, in fact, the most controversial figure of them all, but on one point, all are in agreement: without his intervention, the project would have come to nought. Not only did he raise the necessary capital, but also he supplied the driving force that kept the thing going when all others, including Sholes himself, declared it more dead than alive. He was never known to lose faith or determination and the image he projected to his associates was that of the irredeemable optimist.

When Densmore received Sholes's letter late in 1867 he was immediately interested. He replied by return mail and after some negotiation bought a quarter share in the enterprise for $600, this being the sum the three inventors had spent up to that time. (A 1939 Remington catalogue stated the figure to be $6,000 but this is almost certainly a typographic error.) He also glibly undertook to finance the manufacture of the machine, which did not strike him as an expensive pro-

131. The improved model ordered by Densmore was simpler, smaller . . . and very similar to some earlier inventions. It dispensed with the wires and substituted the original kick-up principle which had, however, been patented by Francis in 1857. The piano keyboard was a further refinement. (SIW)

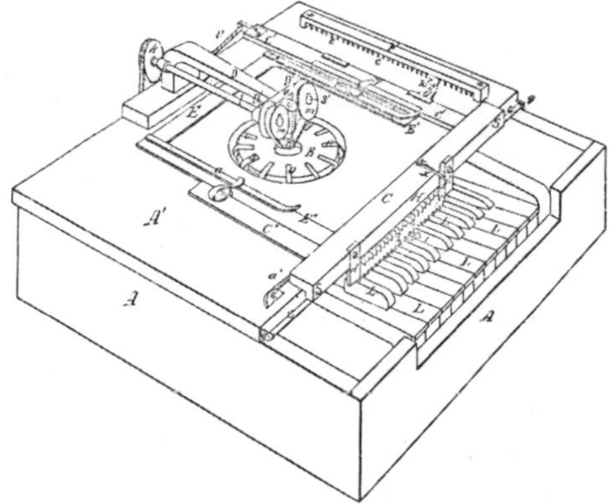

132. The patent covering this improved model was filed by Densmore in May, 1868. So far, not a single secondary source has this patent information straight: Densmore did *not* file the earlier patent on behalf of the inventors. However, the two patents were granted in inverse order to that in which they were filed, which is where much of the confusion originated. (USP)

position—obviously, for he scraped the bottom of the barrel to raise the initial $600.

However, when he finally saw the machine in March 1868 he was shocked. (Weller dates this 1870, but his memory failed him.) Immediately, he demanded changes—in fact, nothing short of a new machine would do and this was apparently[12] made largely by Soulé. The improved instrument was smaller and simpler, taking its inspiration from Sholes's original telegraph-key model. The long wires were gone and no longer was there any connection between the keys and the type-bars, which were merely kicked up against the paper whenever the keys were depressed, a feature of the earlier Francis machine (see page 99). A piano keyboard was adopted, flat paper-carriage and ribbon were retained, and the whole thing was now encased in a wooden box. It is pointless to deny the similarities between this and later Sholes models, and some earlier machines such as Ravizza's and Francis's. Whether one wishes to remain naïve enough to consider the resemblance coincidental is a matter of individual choice.

Densmore is reported[12] to have taken the two models (the old and the new) with him to Washington and to have patented them in the name of Sholes, Glidden and Soulé. The second machine bears an earlier patent date, which is attributed to the fact that Densmore liked it better. This is incorrect. The patent application on the first machine was filed on 11 October 1867, before Densmore even knew of the project; the inventors had wisely decided to protect themselves before showing their instrument to too many people. This patent was granted on 14 July 1868, having been submitted by the same attorneys who had previously handled the patent on Sholes's numbering device. Densmore did the patent on the second machine, filing it on 1 May 1868; it was granted on 23 June of the same year, a month before the first patent was sealed.

Densmore made his first abortive attempt at manufacturing the machine in Chicago later the same year, and a thousand dollars went down the drain with only fifteen machines to show before the project was abandoned and the instrument once again pronounced worthless.

The tone of the partnership was set: it was the promoter mercilessly hounding the inventor to improve the breed. One model after another was built and rejected. Experience was gained, but little else. Densmore sank more and more money into the enterprise, picking up Soulé's shares, as well as Roby's (according to Roby who never, however, specified just what sort of a holding he had). Densmore is also supposed to have bought out Glidden, although this remains a little unclear, for Glidden was back a short while later. Densmore's increasing share at the expense of the others has caused him to be severely vilified as a callous, ruthless and avaricious speculator preying on the ingenuity of his idealistic partners. This sort of thing makes for good reading, but is hardly historically accurate. It is true, of course, that his share increased, but it is equally true that all the money poured into the venture was his, or borrowed by him. But more important still is the fact that he, and he alone, kept faith, while the others sold out when it all looked like little more than a hopeless dream. Sholes himself begged Densmore to buy him out on more than one occasion, but this he refused to do because he could not have managed without him. The others were dispensable to Densmore, but not Sholes.

All in all, some twenty-five or thirty models were produced over the years (upwards of forty, according to some[12]), each one a slight improvement over its predecessor. Each one was pronounced perfect by its inventor—and each one was mercilessly torn apart by big bad Densmore. And not only he himself did the tearing, for he shipped a few machines out to trusted and knowledgeable friends, asking them to perform destruction tests on them. Roby claimed[32] that he was given a machine for testing, although the name most commonly quoted is that of James Ogilvie Clephane, a shorthand reporter, who 'faithfully and gladly tried out one model after another'[21] to which an old Remington publicity catalogue adds '. . . as fast as they could be made and sent him . . .'. The impression thus created is of an infinite number of sparkling machines dumped into Clephane's office; in fact there were two, and even one of these is in doubt. Destruction testing has always made good copy—these days it is more often conducted by publicity than by engineering departments. According to Densmore's ledger, Clephane was later charged $100 for one machine, and $125 each for two others, but it is hardly likely that a shorthand reporter would spend such large sums (in those days) for the sole altruistic purpose of performing destruction tests on another man's equipment. However, whether it was the intention or not, Clephane's

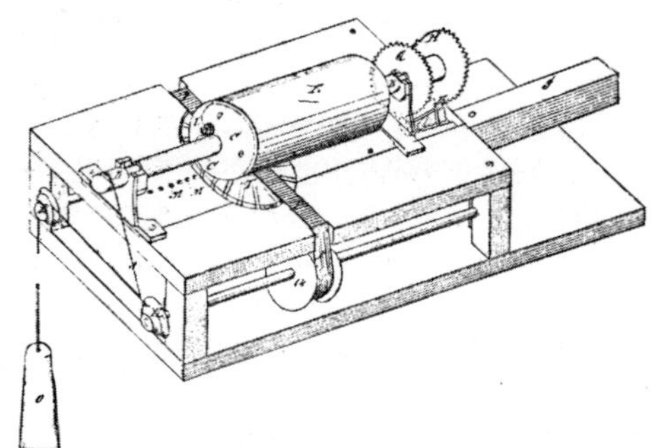

133. Densmore continued to demand improvements and this 1871 patent covered perhaps the most important of them all: the substitution of a cylindrical platen for the flat paper-table. However, the machine printed *around* the cylinder, which moved in a linear direction for line spacing. (USP)

machines clearly broke down and Densmore's entry in the ledger dated 4 June 1872 is perhaps the first typewriter service call in history: 'Gave Walter (Barron) to go to Washington to fix Clephane's machine . . . $10.'

Meanwhile Sholes was hard at work trying to produce something that would please his insatiable partner and, late in 1869, he was able to report to Densmore that he had perfected the machine at last. And indeed, out rolled a radical departure from his original designs. It followed his investigation into previous inventions—in a letter he wrote to Densmore at the time, he analyzed the reasons for failure of early efforts 'which I find, on research, have been many'.

On the latest model, Sholes borrowed a cylindrical platen of large diameter and abandoned the flat paper-table. Like almost all previous inventions, however, this worked exactly opposite to the modern arrangement: it rotated for letter spacing and moved linearly for line spacing. The machine now sported a keyboard of little brass buttons in place of the piano keys (maybe borrowed from Pratt or House, among others); the weight-driven carriage advance was better, and so on. 'I can think of no respect in which I can improve it,' he enthusiastically wrote to Weller in July 1870. 'The machine is done, and I want some more worlds to conquer.'

As Densmore saw it, Sholes had his hands full, conquering the one around him. It would not do. The machine was not good enough, and above all, the type-bars got stuck. Soon, Sholes was back at the grindstone. Three months later: 'I have made another most important change in the machine', he wrote to Weller. 'Everything now seems to me as perfect as it can be made, and I feel no inspiration to alter anything further.' And the machine was dispatched to Weller, who received it at the end of the year.

Sholes also wrote to Densmore, using his letter as a sample of what the new machine could do. Not only was it perfected, he said, but its speed was about twenty words per minute, although he believed it 'susceptible, I have no doubt, of being worked at the rate of sixty words a minute and, some are sanguine enough to believe that a hundred may be worked . . .'

But Densmore was unimpressed and, despite periodic crises of despair, Sholes continued to work on improvements. The design, as eventually patented in 1871, could now handle paper thick enough to satisfy the insatiable Densmore: Sholes facetiously typed on cardboard to prove his point. But things kept going wrong. In one component or another, Clephane's machine (and Roby's) kept breaking down. Densmore was dissatisfied. 'What more or different can be wanted?' Sholes cried. Plenty, was the reply. And more improvements followed, until Sholes once again declared himself done with the machine and done with inventing. He was fed up. He wanted to cash in his share of the invention. He wanted to get something out of it, he told Densmore, giving him 'full and free permission to do as you please with all the interest I have or may acquire in it'.

Poor Sholes! Time and again the elation he felt at his achievements turned to despair at his partner's dissatisfaction. Nothing appeared to please him, however hard Sholes tried. And yet Densmore was undeniably right—the machine was simply not good enough, and nothing could persuade him to lower his standards. He remained firm, but his optimism supplied the inspiration for renewed efforts. If he ever lost faith, he never showed it and many improvements followed in the course of the next few months.

'I have now a machine on which I am doing this work which is an entirely different thing', Sholes wrote to Weller in March, 1871. 'It has not the same appearance. The keyboard is not the same; the disk is not the same; very little similarity in any respect. I have been running this about two months, and it seems to get better . . . Since this machine has been running, I am getting more hope in the (i.e. Densmore's) promises . . .'

Promises? There were many. One of them was an unsuccessful attempt at selling the invention in 1870. Densmore had learnt from the mistakes of the Chicago manufacturing venture which, as Weller said, was like making a watch in a blacksmith's shop. (Ironically enough, this is precisely where the first watches were made, although Weller probably did not know it.) Rather than get involved in another similar deal, Densmore preferred to sell out, and a friend of Sholes's called Alanson Sweet assured him he could get his asking price: 'a round $100,000'.[12] Some[32 etc.] say $50,000. Armed with powers of attorney from Glidden and Soulé, who were partners in the original patents even if not in the machine itself, Densmore and Sholes went to New York to the Automatic Telegraph Co., having heard they were interested in an instrument for printing telegrams. But the price was too high and the company eventually backed out after one of its own mechanics, a young man called Thomas Edison, said he could build a better machine for less money.

What exactly happened during that encounter is anybody's guess, for there are almost as many 'authentic' versions as there are people who record them. Edison himself maintained in his biography[17] that he worked on the Sholes machine even before its presentation to the Automatic Telegraph Co. He recalled that a Mr Craig, one of the founders of Associated Press, brought Sholes to him

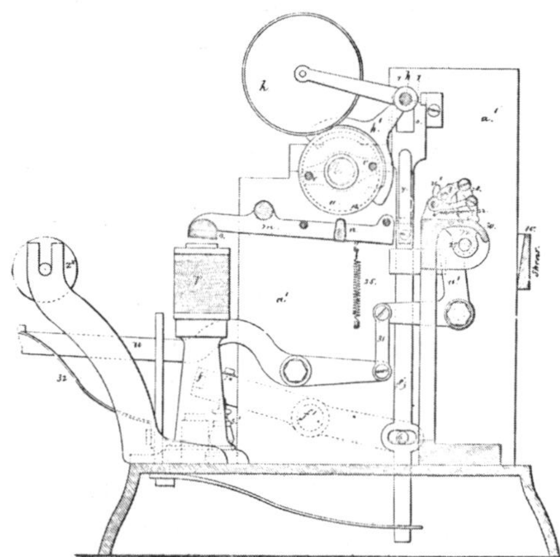

134. Thomas Edison spoiled Densmore's chances of selling the invention to the telegraph people by designing a 'better machine for less money'. His electric type-wheel device, using comb and pin barrel, was patented in 1872. However, Edison did help Sholes with the final major modification: the 'continuous roll' platen.

(USP)

early in 1870, asking him to perfect his machine. He had a hard time with it, he said, but eventually succeeded in getting fair results, whereupon several models were made and used in the telegraph company's offices. 'The typewriter I got into commercial shape is now known as the Remington,' Edison bluntly stated.

According to some reports[21], Sholes later admitted that Edison gave him valuable suggestions for improving the device. Others disagree. One of them[12] even maintains

146

that Edison's memory failed him and that he confused his own electric type-wheel machine, which he patented in 1872, with the later Remington non-visible type-bar instrument. This Edison 'Type Writing-machine' resulted from the inventor's promise to the telegraph company to supply them with a cheaper and better alternative to the Sholes, and its essential features were borrowed from Brett, Hughes, etc. It had a constantly rotating type-wheel with pins protruding spirally from its spindle. This traversed the machine the length of the two-row keyboard so that each key had its corresponding pin on the spindle with which to arrest the movement. The type-wheel rode along the shaft for letter spacing, the paper-carriage being stationary; a padded bar operated by an electromagnet served as a hammer and secured the impression. Paper in a 'continuous roll' was used.

This same 'continuous-roll' feature appeared on the Sholes machine soon after the inventor's encounter with the telegraph people, and it may well have been one of the 'valuable suggestions' Sholes received from Edison. What it meant, in fact, is that the carriage was modified by being turned through 90° and made to travel linearly for typing and rotate for line spacing, just as on modern machines. And it is just the sort of thing a telegraph man would have specified: he would have wanted his machine to print on paper unwinding off a roll, so that he could cut it off at will according to the length of the telegram.

This explanation of Edison's involvement in the Sholes machine is far more plausible than the theory based on his failing memory. He exaggerated grossly, it is true, but it is almost impossible to believe that a man of his mechanical calibre could have confused the up-stroke type-bar machine with his own electric type-wheel design which was 'an instrument . . . very different from the Sholes type-writer'.[12] Furthermore, if Edison had in fact confused the two machines, he would surely have referred to it as the machine he *invented* and not as the one he 'got into commercial shape'.

In any event, the attempt to sell out to the telegraph people failed, and the partners were soon back in Milwaukee, squabbling as before. Densmore pressed for more improvements; Sholes hovered on the brink of despair. But once again, Densmore's tenacity came through loud and clear, and another improved model, the 'final' machine, made its appearance. It had the 'continuous roll' feature mentioned above, plus a number of other refinements. But best of all, Densmore turned up with orders for ten units!

Manufacture was once again undertaken. This time, it was by Sholes, Schwalbach, Glidden (who either returned or else had never wholly left the project) and Walter Barron (Densmore's step-son), with Densmore supervizing. Twenty-five machines were built in 1871 before the project ended in disaster, an uncharacteristically desperate Densmore casting the blame far and wide. Sholes was a printer, Glidden a gentleman of leisure—neither of them had any knowledge of tools. As for Schwalbach, he was dismissed as a 'blacksmith . . . quite deficient in sight and . . . a very bungling, coarse workman'. It was because of him that the work was always botched, Densmore exclaimed.

It was said in anger; predictably enough, Schwalbach himself disagreed and later made some extravagant claims as to his contributions to the invention. Densmore himself treats the man far more respectfully in the ledger statement, referring to him as an 'ingenius' (*sic*) mechanic. Furthermore, a patent filed by Densmore in 1872 (granted 1876) in the names of Sholes and Schwalbach covered certain improvements common to later Type Writers and Remingtons (such as the guide frame for key levers and the transverse spacing bar) and is eloquent recognition of Schwalbach's contributions.

Densmore quickly recovered from his despair and became aggressive again. He was determined to keep trying. Sholes, as always, had lost heart. Densmore failed once again to sell out the entire property, this time for a flat $1,000 plus a $1 royalty on every typewriter manufactured,[12] and several other attempts ended abortively. Frantic for more money, he offered to sell his brothers a 10 per cent share each for $1,000. This is reported in the ledger and in correspondence in which he warned his brother Amos that the whole project might be risky and not to buy into it unless he could spare the cash. He himself had spent $10,000 so far, and the sum was growing, especially around the middle of 1872, when yet another attempt at manufacture failed. This time, the partners were so short of money that they could not even afford proper tools, and yet the machines they produced are somehow of better quality than those from the previous batch. 'We shall be in a position to furnish good machines' Sholes wrote to Barron, 'provided any person is in a position to want them after they are furnished.' Sholes was still afraid that the typewriter was a passing novelty but 'Densmore laughs at the idea when I suggest it.' That was surely the only laughing Densmore was doing! By the time they dropped this latest attempt at manufacture, he had sunk a total of over $13,000 into the venture.

There is now considerable confusion as to just who owned what part of the invention, and the arrival on the scene of George Washington Newton Yost only makes matters worse. Yost turned out to be the right man in the right place at the right time. A super-salesman, he had, and in abundance, all those qualities which the aggressive and irritating Densmore lacked: magnetism, royal bearing, diplomacy and so on. Some sources in fact now assign Yost the important role in the enterprise: 'He did more than all others combined in financing and placing the machine on the market.'[32]

Just how Yost got interested in the first place is not clear. According to one version,[12] he owed Densmore large sums of money and agreed to help his creditor out of a sense of obligation. Others[5] have Yost buying in: Densmore, having swallowed up everybody else, is supposed to have sold part of his holdings to Yost and then proceeded to divest Sholes of his remaining shares for an unknown figure, 'probably around $6,000'. According to this version, the Sholes and Glidden typewriter now belonged to Densmore and Yost, who proceeded to sign it over to Remington.

This makes for good copy, but is historically inaccurate. For a start, it is almost certain that Glidden was still involved in the ownership. The fact that the manufactured product was later called the 'Sholes and Glidden' bears this out. After all, it was now five years since Densmore was supposed to have bought Glidden out; so what obligation was there to retain his name? It is true that early patents were partly his but Soulé also shared them, and yet his name had been dropped long before. Furthermore, a manufacturer is in no way obliged to use an inventor's name merely because it appears in a patent. So if it is true that Densmore had bought out Glidden, how is the use of his name to be explained?

Jennie Glidden maintained[28] that her father never completely relinquished his interests as Soulé did. Others[12] have Glidden selling out and returning later. As for Yost, he initially participated only in the marketing company which he and Densmore formed to sell the machines. He did buy shares from others later on, but the astute Densmore is hardly likely to have sold his own holdings in the machine itself to a man who already owed him a large sum of money which he was unable to repay. The ownership of the machine, however, is complex and confused and is discussed more fully in Chapter Eight.

In any case, it was Yost who suggested that they approach E. Remington and Sons and try to interest them in the machine. The arms manufacturers were reportedly looking around for something to help fill excess industrial capacity—they had made their fortune on guns during the Civil War and when it ended the demand for their product dropped off drastically. Sewing machines were already being made by them.

So with their latest and best machine under their arms—the 'spring-motor' model in which the carriage-return weight was replaced by clockwork—the trio went off to visit Remington. Sholes stayed well in the background, Densmore succeeded in holding his tongue, and Yost did the smoothest selling job in his life. The result was the negotiation of a manufacturing contract which was eventually signed on 1 March 1873. Everybody was to get a slice of the thin pie as soon as production got under way, and royalties were dealt out like cards in a poker game. Remington's mechanics, Jefferson Clough and William Jenne, demanded and were granted a flat fifty cents per machine for their work in re-modelling it. Their improvements were later embodied in a patent filed in 1875 and issued in 1878. Other royalties were distributed to the original patent holders, including some to a Charles Washburn who held an 1870 patent covering carriage design. Sholes and Densmore obviously expected to get a return on their five-year investment of time and money in the development of the machine, Remington had to make their profit on manufacture and, of course, Densmore and Yost were to get their cut at the retail end. There was going to be more than enough to go around, and trade was expected to be brisk.

It turned out differently. Bills began to pile up as quickly as unsold machines, and debts as high as defective typewriters returned for repair. The commodity most readily available was ill-will, culminating in fights and squabbles between all parties. Remington, itself in serious financial straits, progressively took control.

After Glidden's death in 1877, his widow permitted the name of the machine to be changed (Glidden had consistently refused during his lifetime), whereupon Remington stopped paying her husband's royalties. It was only after she sued that they settled out of court for half the previous figure.[28] Densmore and Sholes had a falling out, as did Yost—and so on. But this is covered in a later chapter.

Back in 1873, things looked much more rosy. The contract called for the manufacture of 1,000 machines, plus a blissfully unrealistic option for 24,000 more. Remington were understandably cautious and determined to run as few risks as possible. An advance payment of $10,000 was stipulated, and Densmore, already deeply in debt, had to sell off some franchises to help raise the money. He and Yost opened the first American typewriter shop in Manhattan and awaited the arrival of the machines. Remington, meanwhile, having handed the Sholes spring-motor model to their sewing machine technicians Jenne and Clough, succeeded in ironing out a few of its bugs; others, unfortunately, stayed on to plague them all for years to come. It had not been a practical machine when Remington got hold of it, as no less a Sholes enthusiast than Weller[43] readily concedes. And it was only moderately more practical after Remington got through with it by September 1873, when manufacture actually began.

The first machines were shipped out of the factory early the following year. The 'Sholes and Glidden Type Writer manufactured by E. Remington and Sons, Ilion, N.Y.', as the label read, was a large enclosed up-stroke instrument on a sewing-machine stand with a foot-treadle carriage-return. It printed upper case only, and non-visibly. But it worked—after a fashion. At least, when it was new it did. And it printed almost as fast as some people could write, and even faster than a few. And it looked pretty, all painted black, with scrollwork and stencilled flowers and hinged covers to hide all the works. 'It is graceful and ornamental', said the blurb. 'A beautiful piece of furniture for office, study or parlor.'

And it was.

HANS JOHAN RASMUS MALLING HANSEN

Jennie Glidden was not the only dutiful daughter to defend her late father's place in history. The year was 1925, and a fierce public controversy was raging in Denmark over who was the original inventor of the 'Skrivekugle', the Writing Ball. This remarkable design had been attributed to Pastor Malling Hansen; suddenly, a Professor Hannover of the Danish Technical College and others began insisting that Peters be given priority for his primitive instrument described on page 113. And Hansen's daughter, Johanne Agerskov, would have none of it: she even published a book in which she unequivocally staked her father's claim.

In truth, the ancestry of the Writing Ball, on examination, goes back even further than Peters. Essentially, the machine is of the plunger group, with the keys on the ends of spring-loaded rods disposed radially in porcupine fashion out of a semi-sphere: the characters are at the bottom of the rods which are all directed

135. A closer view of the crude Peters type-
plunger invention which Malling Hansen has
been accused of plagiarizing.   (DTM)

towards a common printing-point, while the keys are at the top. It is obvious,
therefore, that neither Malling Hansen nor Peters can claim absolute priority:
Galli and Foucauld came out with essentially that principle some thirty years
earlier. Not that this fact in any way detracts from the importance of Malling
Hansen's 'epoc-making'[55] device. After all, the inventor, with typical nineteenth-
century extravagance, set himself no mean goal to achieve. He wanted his instrument to print not one but up to ten copies, using ordinary letters but at the speed
of shorthand. All this, he said, had to be possible under one or more of the following
conditions: in the darkness of the night, or while being pitched and tossed on the
waves of the high seas, or while driving in a carriage over corrugated roads or
lying in a bed . . .

He didn't quite make it. He did, however, invent a fine machine which was
already finished before Glidden had even suggested the idea of a typewriter to
Sholes. Hansen's Writing Ball was manufactured from 1870 onwards, went
through a number of improvements and model-changes, was exported to many
countries, sired several imitations, and continued in production for several decades.
And in use for many more. More than forty years after it was introduced, the
machine was 'still to be found in many offices on the Continent'[27] at a date when

151

conventional four-row front-stroke machines were already being mass-produced and sold relatively cheaply; for the Writing Ball to have survived in widespread use as late as 1909 attests to its excellent design and construction.

The machine is therefore the first regular typewriter to have been produced and sold in any sort of quantity. As was seen in Chapter Four, both the Hughes and the Foucauld machines for the blind were the first to be manufactured (discounting printing telegraphs) but their use and distribution was limited. The Jones instruments were all destroyed by fire and never actually made it to the market place. Then came the Writing Ball and next the Sholes and Glidden.

Pastor Malling Hansen was born in Denmark in 1835. After a brief period of uncertainty, he began studying for the ministry and took a job as a teacher at the Copenhagen Deaf and Dumb Institute, of which he was appointed director in 1865 after a three-year term in a similar capacity in the provinces. It was in this position that he first began concerning himself with the construction of a writing machine, the first model of which he is reported to have completed in 1865. This date is confirmed by his daughter in the book she wrote about her father's invention—there might be reason to doubt the authenticity of such a clearly biased source except that a contemporary of Malling Hansen's, another cleric called I. A. Heiberg, kept a diary in which there was an entry for the year 1865 stating that he had seen Hansen's machine. It apparently developed out of experiments which the inventor had made with a porcelain semi-sphere on which he painted the letters of the alphabet, in an experiment aimed at helping the deaf and dumb inmates of his institute to communicate.

It is of little importance to anyone except the heirs involved whether Peters or Hansen came first, and whether or not there was plagiarism, or whether they even knew of each other's experiments. This kind of argument is as unproductive as its American counterpart as to whether or not Sholes had studied previous patents and borrowed ideas from them. However, anyone claiming that Peters should be credited with inventing Hansen's machine merely because of the radial-plunger similarities might be reminded of Galli's and Foucauld's previous inventions.

The 1865 Skrivekugle was clearly a long way from the final form the machine was to assume. The principle had been established, the design conceptualized, but much carpentry was still to be done before the instrument was perfected. In fact, over the next quarter century or so that the progress of the Writing Ball can be traced, albeit sketchily, it underwent numerous model-changes and face-lifts. Documentation is unfortunately incomplete and many pieces of the jigsaw puzzle are still missing. Contemporary newspaper and magazine articles, however, provide considerable information, for Hansen's machine stirred up a storm almost from the moment it was completed. Still more has been compiled by the Danish Technical Museum, whose director, Mr K. O. B. Jorgensen, has been kind enough to make it available. Despite this, the history of the Writing Ball is still fraught with contradictions from conflicting sources.

136. The first model of the Hansen Writing Ball (1867) was enclosed in a wooden case. It was by far the best design of its time: simple, direct and virtually foolproof.   (DTM)

137. Spring-driven clockwork kept the cylindrical platen under tension, the escapement controlled by an electromagnet. The 'ball' was screwed to the cover, with the plungers striking directly at the printing point.   (DTM)

There remains little doubt that the first model can be dated 1865. What happened for the next few years, however, is uncertain. Presumably, developmental work continued, for the earliest model in the Danish Museum, dated 1867, has the movement enclosed in a wooden box, so that only the 'keyboard' is visible. Inside the box is the cylindrical platen which is partly mechanically and partly electrically operated. This feature was retained for many years, until the inventor considered that a purely mechanical escapement had been perfected to the point where the electrical assistance could be dispensed with altogether. On the combined arrangement, carriage tension was supplied by clockwork, while an escapement permitting release of the carriage a tooth at a time was controlled electromagnetically as the keys were depressed. Impression was by carbon paper.

Further developmental work followed, and the first production model was completed. One might well draw a parallel here with the later Sholes and Glidden machine on the other side of the Atlantic. The model which Densmore and Yost delivered to Remington in 1873 was reworked by Remington's technicians and the improved machine was manufactured the following year; in Copenhagen, Hansen presumably delivered the 1867 model to Jürgens Mekaniske Establissement some time between that year and 1869, for in October 1870 manufacture began.[45] The Jürgens firm had been founded in 1852 and was responsible for the manufacture of the Writing Ball throughout its lifetime; in 1887, the company incorporated and became 'C. P. Jürgensens Mekaniske Establissement'.

As it now stood, the production machine had shed its wooden case and had

138. Full-scale manufacture of the Writing Ball began in 1870 after some further developmental work. Precision engineered in brass and steel, it had shed its wooden case and a flat carriage was substituted for the platen, although this was still available as an option. More keys had also been added, but the essential design remained unchanged. (CSM)

SML become, in fact, a massive piece of exquisite precision engineering. Weighing in at no less than 165 lb (75 kg), the brass and steel mechanism remained essentially unchanged: the sphere had sprouted fifty-two plungers by this time, but otherwise the same carriage escapement was retained and the whole exposed machine stood on its massive brass base. A cylindrical platen was available as optional to the flat paper-carriage. The machine was offered for sale in London in 1872 for £100, the price dropping to as low as £17 six years later, when improved versions became smaller and more manageable. A telegraphic model that printed on paper tape was produced and as early as 1872 Hansen was offering delivery of a cryptographic model.[55] Later, when it had reached finality, the Writing Ball shed its flat carriage and offered only a conventional one with a cylindrical platen; an interim model also appeared on which the paper was clipped to a curved frame that swung in an arc beneath the printing point.

139. The interim model used a curved paper-frame which swung in an arc beneath the plungers. The electromagnetic escapement was replaced by a mechanical one. The machine had been stream-lined, lightened and further simplified.   (DTM)

140. Details of the interim model. The sheet of paper was simply fastened in the frame without any backing, printing taking place against the vertical anvil in the centre.   (DTM)

142. The Writing Ball was modified for various special applications, such as this ingenious telegraph version: the Morse code was cut into each plunger (as on the original Morse 'type rule') so that as the plunger was depressed to print the character on paper tape, it inter-mittently closed a circuit, thereby automatically transmitting the Morse equivalent.   (DTM)

141. And this is how the Hansen Skrivekugle looked when it reached finality.   (DTM)

155

Patent dates are deceptive, for they were not granted in Denmark at that time. The first Hansen patent, therefore, coincides not with the date of invention but with the date of manufacture. This 1870 British patent covers the cylindrical platen model, with the flat paper-table as an alternative. All of it is enclosed in a box and depression of a key closes an electrical circuit, actuating an electromagnet which rotates the platen one tooth. A purely mechanical escapement is also suggested. The device is then ingeniously offered as a Morse telegraphic transmitter by cutting teeth corresponding to the code on the type-plungers and causing those teeth to break and close the circuit as the plunger is depressed.

The 1872 patent covers a departure from the invention with which Hansen is commonly associated. It is listed as a machine for reporting and it retains the ball only as one of the alternatives for the superstructure; otherwise, the plungers are modified so that they stand in a straight line or lines, and strike not at a common point but at a row of them, as long as the sum of all the characters required. These plungers make dents in a strip of conductive material which is later passed through another machine so that the dents close the circuits of electromagnets, one for each character, which do the printing. Thus, the first unit can be used separately from the second, making it portable for reporting, since twenty-three electromagnets—to say nothing of the clockwork etc.—are more than the average reporter can be expected to take on an assignment!

The third specifications, patented in 1875 (British) cover the later improved Writing Ball with the curved paper-carriage. The machine has been streamlined, the body hinging above the platen; the escapement is mechanical and swings the curved surface for printing.

That the Hansen Writing Ball enjoyed considerable success and a healthy following is amply documented. With the exception of fanatical local Sholes supporters, it is generally recognized as the first machine to have been consistently produced and sold to the public in any sort of quantity.[28, 40 etc.] It stirred up a great deal of interest, was often exhibited at national and international expositions (Copenhagen 1872, Vienna 1873, Paris 1878, and so on) where it accumulated an impressive number of medals and awards, and was the first machine to have had a truly world-wide distribution. It was regularly sold throughout Europe and, as previously noted, was in use until at least the first decade of this century. It was manufactured under licence in Austria by Albert von Szabel and sold there; the Austrian models were almost identical with their Danish counterparts. But then, it also reached some of the most unlikely places.[45] Orders for forty machines placed at the Paris Exposition at a price of 300 Kr. each were distributed as follows: 9 for Denmark, 7 for France, 2 for Sweden, 2 for Norway, 3 for England, 2 for America, 1 for Italy, 2 for Bohemia, 1 for Hungary, 1 for Egypt, and 10 for Peru. 10 for Peru? Well, there it is!

One cannot help feeling that Sholes, Densmore and Yost would have celebrated for a solid week had they succeeded in selling forty machines at the Philadelphia Exhibition in 1876, where they did not, in fact, succeed in selling a single one.

But then, the Writing Ball was a better typewriter than the one they were offering at that time. It printed better and more quickly and its service record was impeccable. In fact, the Writing Ball may well be considered the most nearly perfect piece of precision engineering in typewriter history, past or present. This fact can hardly be appreciated from photographs, but is readily apparent on direct examination.

The progression of this direct radial-plunger design from Galli through Foucauld to Hansen has already been considered; the lineage did not end with the Writing Ball, however, which begot several off-spring. One of these was the Cryptographe of Alexis Kohl, circa 1885, which was virtually identical with the Hansen telegraph model: a monalphabetic substitution machine, it was never manufactured. And despite Rudolf Schade's vociferous protestations to the contrary, his instrument, of which a few were made in Germany in 1896, was a copy of the cylindrical-platen Hansen model: virtually the only difference between the two was that Schade's machine clamped on to the edge of a table, whereas Hansen's stood on its own base. Not only the inventor himself, but also compatriots of his,[30] maintained that Schade arrived at his design without ever having heard of Hansen's machine—this may be dismissed as sheer nonsense. The battle, however, raged until as late as the 1920s, when everyone seemed anxious to score as much national kudos as possible.

Of all nations, the Danes are characteristically reticent when it comes to proposing the candidacy of Pastor Malling Hansen as 'the' inventor of the typewriter. They satisfy themselves with discussing the facts surrounding the Writing Ball rather than involve themselves too deeply in conflicting national passions. They make less noise than the Germans or Italians or Americans, but not because they lack justification. Perhaps in spite of it.

Chapter 7

Some Critical Evaluations

Having now considered the salient facts, we should be able to proceed to a critical appraisal of the contributions made by each of the major inventors. And the result of such an impartial assessment should be the designation of *one* man as the inventor of the typewriter. In deference to the sanctity of this terrain, however, kindly remove nationalistic and patriotic passions before entering!

The history of the writing machine begins with Henry Mill. This statement is axiomatic, at least for the time being. Efforts prior to his have been rumoured, but no corroborating evidence has yet been located and, in its absence, these reports must be considered purely speculative.

However, Henry Mill did not invent the writing machine. He did not invent it any more than Daedalus and Icarus invented the aeroplane or Flash Gordon invented the spaceship. An inventor must be more than a dreamer, more than a theorist or philosopher. A good idea, a brainwave, is not enough: the inventor must translate this personal intimate language into practical reality. Mill *could* have built his machine but there is no proof that he actually *did* so.

It has, of course, been argued that since he was an engineer it is to be assumed that he did just this. That he did, in fact, build his machine, that it worked, and that it was on the strength of this that he applied for and was granted a patent, and one, furthermore, that even specified the type of impression. But once again, this is pure hypothesis and as such is unacceptable. Something or other of a more concrete nature is required, some proof—even minor. For in 1714, when Mill's patent was granted, it was not necessary for an inventor to present either a model or even a drawing of his device, and he was in no way required to prove that his invention was practical. Thus, until some evidence is presented to the contrary, his patent must be considered an interesting historical curiosity but nothing else.

Other eighteenth-century efforts are similarly unacceptable. Unger, for all his long descriptive tract, left the mechanical problems relating to his invention unresolved. He was the only inventor who might otherwise have qualified for the honours, even though his was a musical writing machine. And towards the turn of the century, Bramah designed his interesting device which is, however, a printing rather than a writing machine.

Which brings us to the nineteenth century, when indisputable evidence of the existence of early typewriters begins to be found. National controversies now result purely from the arbitrary evaluations of impassioned patriots.

Consider, first of all, the man whom American historians[12, 21, 32, 42 etc.] consider 'the Father of the Typewriter':[43] Christopher Latham Sholes. In order to justify this choice, a qualification must always be appended, for Sholes was clearly not the first man to invent a writing machine. When he is not simply called *the* inventor, then his is said to be the first machine acceptable as 'fast' or 'practical' or 'successful' or 'efficient', or any other arbitrary qualification which is never defined.

Before considering the validity of these claims, it is perhaps in order to decide whether there is even any justification for their very consideration. Take speed, for instance; why is it such a vital issue? And anyway, how fast is fast? For time after time, one reads[21, among others] of rival machines being dismissed as 'slow', as if this were the sole feature of any importance. Speed, surely, is a vital aspect only of those machines devised purely for the purpose of equalling or breaking an absolute or established velocity. If one were to design a vehicle for the express purpose of breaking the sound barrier, then the performance of the machine relative to the speed of sound is readily apparent and it either succeeds or fails. In the case of typewriters, the issue has remained arbitrary, as was obviously the intention of those who raised it, and the very application of the qualification itself has never been justified.

There are probably few fields in which the assignation of such a purely arbitrary set of performance standards is made a prerequisite of the concession of priority in an invention. One might as well disqualify the inventor of the aeroplane because his machine did not fly faster than a bird (however fast that is), or the inventor of the automobile because his 'horseless carriage' did not out-gallop a horse (whatever speed that may entail). And applying this kind of reasoning fully, one might add that Marconi's radio looked nothing like a pocket transistor, nor did Edison's home recording outfits in any way resemble the tape deck. The first telescope magnified a ludicrous two or three diameters; the first telephone barely produced a squawk in an adjoining room, let alone a thousand miles away; and the first watches were such atrociously erratic timekeepers that many had built-in sundials for double-checking and resetting.

None of these inventions is disqualified on the grounds stated. Yet are we still to believe that a typewriter which was slow was not a typewriter? Or that it failed to qualify if it did not pass arbitrary tests of practicality or efficiency or commercial success?

But if there is no justification for even applying the criterion of speed to pre-Sholes typewriter inventions, then even less justification can be found for contending that Sholes passed this test where his predecessors failed. The crude, defective version of the Sholes and Glidden is billed as the 'first practical typewriter'[64], yet in its remodelled and perfected form it was barely capable of thirty words a minute. For some reason or other, never explained, thirty w.p.m. is supposed to be dandy, even though typing speed was inevitably compared with that of handwriting, which exceeds this figure (see page 42). And further to complicate the issue, no comparative speed tests were ever performed on pre-Sholes typewriters, some of which were undoubtedly fast by standards in those days.

As for the other claims, these are equally false. To distinguish it from its ancestors by calling it a 'practical' machine is quite out of the question. One has only to recall the verdict of Charles Weller[43], one of Sholes's greatest champions: his account of carriages jumping, type-bars jamming, weights breaking and falling, and so on, is mildly terrifying. In this respect the Sholes and Glidden scores lower than some earlier inventions which were masterpieces of engineering. Furthermore, it was not until the 1880s that the machine can honestly be considered practical: Sholes and his associates laboured through thirty or forty models over six years without ironing it out and Remington made one improvement after another for a further five years, without getting rid of the bugs even then. By this time the inventor, Sholes himself, dismissed and disowned the instrument, repudiating it in the severest terms. He refused to own it, or even to use it or recommend it, and he was convinced that it and Remington would go bankrupt, a prediction that proved at least partially correct. He begged Densmore to buy him out because the machine was worthless, and his correspondence leaves no doubt as to what he thought of it. All the Ilion typewriters he ever heard of soon got out of order, he wrote, and were consigned to the trash heap; the machine contained defects so basic that Remington mechanics could not possibly 'palliate and bridge over them'.[12]

Was the 1869 model of this instrument the first 'practical' typewriter, then? An instrument of the up-stroke class which typed non-visibly and in upper case only, which was the size of a kitchen table, which had a flat paper-carriage and weight-driven carriage tension, which constantly broke down, jumped and jammed, which was slower than handwriting. Practical? Or was the perfected 1874 model the first 'successful' machine, perhaps? In fact, commercially it was an abject failure. Part of the fault lay with the instrument itself, and part with the times, but the fact remains that the machine was a commercial failure. Densmore and Yost gave up on it and Sholes himself disowned it. Remington came perilously close to washing their hands of it; Wyckoff, Seamans and Benedict, who inherited the machine from a line of unsuccessful sales agents, admitted in 1878 that the Type Writer had been defective and that its pathetic sales picture had been due to its inherent imperfections. Meanwhile, Yost had been obliged to pick up broken machines and return them in batches to the factory for repair, implying that dangerously few Type Writers continued in service, for sales were very slow indeed in those days. The exact figure has never been accurately established and every source quotes a conflicting number. The nearest one can get is to say that, until the 1880s, annual sales remained in the three-figure bracket, and in the low three-figures during the early years.

This would appear to dispose of the 'success' argument, leaving the only other qualification—that the Sholes and Glidden was the first machine to have been commercially manufactured; but this has already been proved inaccurate in previous chapters. The Hughes and Foucauld were produced in small quantities in the 1850s and the first batch of 130 Jones machines (1852) was destroyed by fire, an Act of God which negatively affected the sales but certainly not the fact of its manufacture. And, finally, the Hansen Writing Ball (1865) was not only manufactured and sold for a quarter of a century or so, but was also exported and adapted to a number of different uses.

So the Sholes and Glidden was not the first typewriter, and Sholes was not 'the' inventor of the machine. On every score, one or more inventors can be found who achieved the same goal many years earlier than he.

Who else is there, then? For a start, Mitterhofer, but he hardly stands a chance. The man who most vociferously launched his candidacy, Dr Granichstaedten-Czerva, was either ignorant of previous efforts or else it suited him to ignore them. Without wishing to repeat the ground covered in Chapter Five, it is clear that Mitterhofer did not in fact add a single new element to the writing machine's history or development and the claim that his was the first 'complete'[20] machine is unfounded. Of all the major candidates, Mitterhofer has the least hope of acceptance: no original contributions (even if this is not necessarily a charge of plagiarism), rough and unimpressive workmanship (when compared to some of the earlier machines which were gems of precision craftsmanship), and a date far too late to be even considered seriously.

Ravizza stands a better chance. His machine was also toted as the first 'complete', also as the first 'practical'[1] machine, and the first one that 'actually'[65] wrote. Once again, none of these claims is accurate, but of all the candidates so far listed, he is the closest to success. After all, he predated Mitterhofer and Sholes by more than ten years, and his machine contained all the elements of theirs, plus others, such as ribbon inking. But, once again, the claims are wide of the mark: why, for instance, is a machine which prints only upper case, and non-visibly; which has a flat paper-table in place of a platen; and so on—why is this machine more 'complete' than other earlier ones? And why is it more 'practical' than its predecessors? Because it worked? So did all the others. It was slow, but so were they. It seems fairly certain, furthermore, that the machine was modelled after Conti's Tachigrafo which preceeded it by over thirty years, and most of the important elements of the Cembalo Scrivano were 'borrowed' from Conti. And the Tachigrafo worked; there is no doubt of this for it was demonstrated to and tested by independent and, in those days, invariably sceptical and negatively predisposed authorities.

That leaves Malling Hansen, whose claim is more than valid, given the qualifications invariably applied by Sholes advocates. For the Writing Ball was invented earlier than the Type Writer and it was also manufactured some four years prior to it. It not only worked; it also worked better than the early models of its American competitor and, while the Sholes and Glidden was rather roughly and loosely made, the Writing Ball was exquisitely engineered. But regrettably for Hansen, as for Sholes and the others, none of this is good enough. They appeared too late on the scene; choosing one of them would be the equivalent of stating that the world's first motor-car was the Ford Model T.

161

Since none of the so-called 'fathers' stands the test of critical and objective inspection, it is necessary to establish the framework within which the subject is to be considered, and then to go back to the earliest machine which fulfils those conditions. And, in fact, stripped of superficial nonsense, the problem becomes a simple one indeed: the 'inventor' of the typewriter is that man who first built a machine that can be proved to have worked. That is all. The two critical factors, the *only* two, are firstly that the device be mechanical and secondly that it fulfil its purpose and actually *type*. In this context, therefore, it is completely immaterial whether the machine was fast or slow, since no satisfactory terms of reference have ever been established, nor can any absolute measure be devised nor, even, is the issue a valid one. Nor is it of any moment whether the instrument was big or small, whether it looked like a modern machine or not, whether it was manufactured, mass-produced or a 'one-off' special, whether it was 'successful' or 'practical' or 'complete' or 'true' or any other arbitrary epithet designed to conceal reality. For the purpose of assigning priority of invention, a typewriter is a machine that types, and that's that.

A perusal of the list of inventors prior to Ravizza, Mitterhofer, Hansen, or Sholes reveals scores of names of men who built machines that worked, and of which proof exists. One goes back immediately to 1833, to the French candidate Progin. His was a fine machine of an excellently conceived design; it worked and, clearly, it worked only marginally more slowly than those of later generations, if at all.

But let us leaf back even further. Burt? Gonod? Conti? It is now 1823 and all of them existed and worked. And one can legitimately go back even further.

The typewriter would appear to have been invented by Pellegrino Turri, in 1808. His device was definitely mechanical, and has been called 'the first invention which can be accorded the designation machine'.[40] And the present author's research has revealed more proof that the machine 'worked' than that which has survived for any other instrument for more than sixty years to come.

As is so often the case, a stack of correspondence typed on the Turri machine was brought to light largely by accident, through an article printed in a general interest magazine called *La Lettura*, in March 1908. The magazine was a kind of unsophisticated, mass-circulation publication—a kind of turn-of-the-century *Readers Digest*, with articles on everything from the use of real African insects in the manufacture of costume jewellery to the customs of Albania.

And then there was an article by Umberto Dallari entitled 'Old Letters Written by Machine' which begins 'Nihil sub sole novum'. There is nothing new under the sun! Dallari was not interested in the typewriter at all. He was, in fact, a civil servant, and as director of the State Archives in the Italian town of Reggio Emilia he discovered a file containing the correspondence in question while involved in classifying and cataloguing the contents of the archives. The correspondence was from precisely a hundred years prior to that date, and Dallari obviously considered

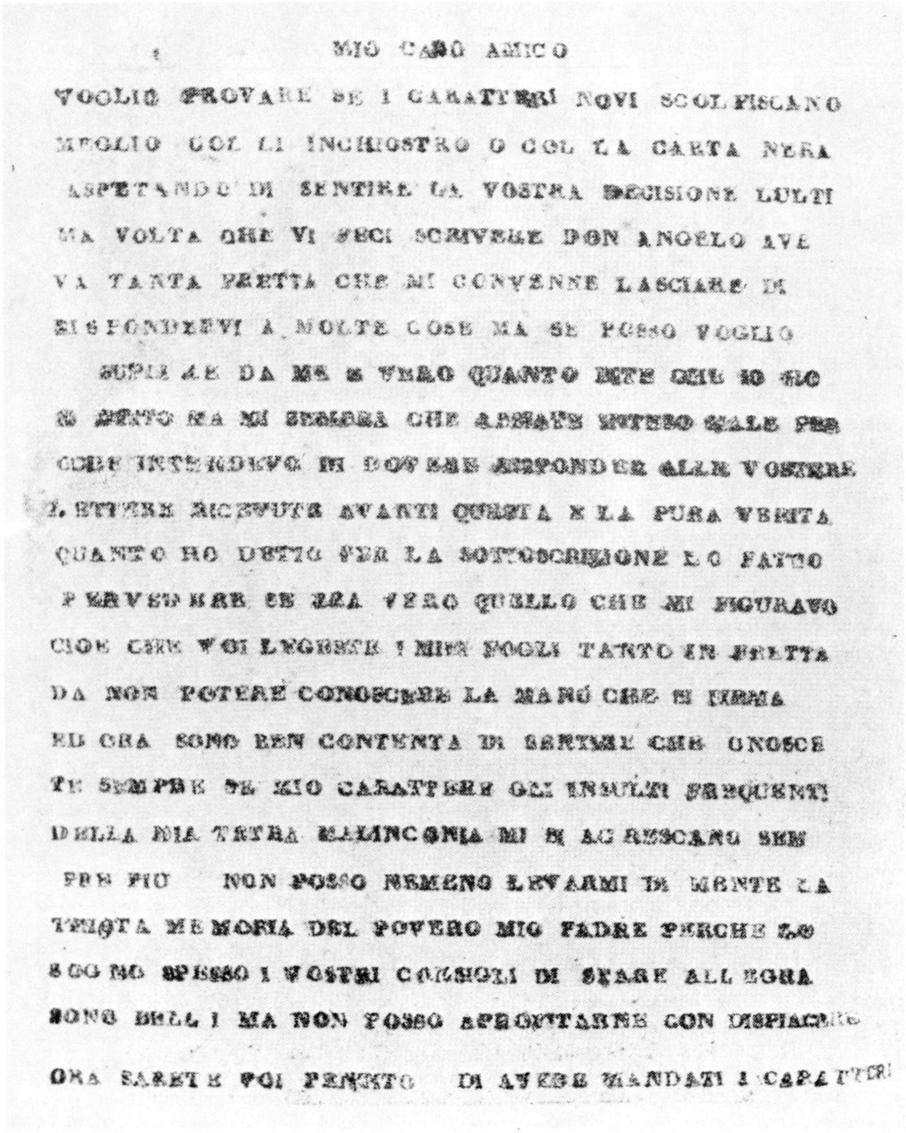

MIO CARO AMICO

VOGLIO PROVARE SE I CARATTERI NOVI SCOLPISCANO
MEGLIO COL LI INCHIOSTRO O COL LA CARTA NERA
ASPETANDO DI SENTIRE LA VOSTRA DECISIONE L'ULTI
MA VOLTA CHE VI FECI SCRIVELE DON ANGELO AVE
VA TANTA FRETTA CHE MI CONVENNE LASCIARE DI
RISPONDERVI A MOLTE COSE MA SE POSSO VOGLIO
SUPLIRE DA ME E VERO QUANTO DITE CHE IO HO
O DETTO MA MI SEMBRA CHE ASSAI INTERO QUALE PER
COSI INTENDEVO DI DOVERE RISPONDER ALLE VOSTRE
LETTERE RICEVUTE AVANTI QUESTA E LA PURA VERITA
QUANTO HO DETTO PER LA SOTTOSCRIZIONE LO FATTO
PER VEDERE SE ERA VERO QUELLO CHE MI FIGURAVO
CIOE CHE VOI LEGERETE I MIEI FOGLI TANTO IN FRETTA
DA NON POTERE CONOSCERE LA MANO CHE SI FIRMA
ED ORA SONO BEN CONTENTA DI SENTIRE CHE CONOSCE
TE SEMPRE IL MIO CARATTERE GLI INSULTI FREQUENTI
DELLA MIA TETRA MALINCONIA MI SI ACRESCANO SEM
PRE PIU   NON POSSO NEMENO LEVARMI DI MENTE LA
TRISTA MEMORIA DEL POVERO MIO PADRE PERCHE LO
SOGNO SPESSO I VOSTRI CONSOLI DI STARE ALLEGRA
SONO BELLI MA NON POSSO APROFITARNE CON DISPIACERE
ORA SARETE VOI PENTITO   DI AVERE MANDATI I CARATTERI

143. 'Nihil sub sole novum' is how Umberto Dallari began the article he published in *La Lettura* in 1908. The occasion was the centenary of some typewritten letters sent by Carolina Fantoni to Pellegrino Turri. The one reproduced in the Dallari article was selected because it was of good quality—it is now missing from the State Archives in Reggio Emilia. (60)

his discovery a curiosity worthy of publication. Nowhere, however, did he make any claims of priority for the correspondence; he was merely surprised that typing out letters, which became popular during his own lifetime, was in fact a practice at least a hundred years old.

Pellegrino Turri came from a prominent provincial family whose records go back for centuries. On his death in 1828, he left all his possessions to his son Giuseppe, who in turn bequeathed the family's papers to the Archives when he himself died in 1879. And among these documents, which Dallari sorted out, were the typewritten letters which Turri's lady-friend Countess Carolina Fantoni wrote to him on the machine he built for her.

Turri's personal file opens with a certificate dated 1806, announcing his admission by unanimous vote to the Reggio branch of the Società d'Arti Meccaniche, a fact which may well be related to the typewriter that appeared two years later. Pellegrino was, in fact, well known for his mechanical aptitude. Dallari, quoting the Turri family biographer Dr Prospero Fantuzzi, paints a picture of the inventor as a short stumpy man with an ugly face but a winning personality and many fine qualities, including sufficient money to allow him to lead a life of leisure. Born in 1765, he married a rich woman at the age of thirty, entered public life in 1800 as a councillor, among other posts, and led an active life until his death in 1828. His family appears to have been friendly with the Fantonis, and Pellegrino probably met Carolina while the latter was staying at Reggio. Born in 1781, the Countess became blind 'in the flower of her youth and beauty'[60] and Pellegrino was so overcome with grief at her affliction that he resolved to build her a machine to enable her to correspond in private with her friends. The Fantoni family left Reggio in 1808 and from then until Pellegrino's death, he and Carolina conducted an irregular correspondence, mostly during the early years.

A separate file contains the typewritten correspondence, a total of sixteen letters plus an original essay, numbering in all thirty-one typewritten sheets. The file is incomplete; there are references to typewritten letters and to sonnets which are now missing. Furthermore, the letter which was reproduced in Dallari's article[60] and which was selected because it was the clearest and the best preserved, is also missing: it was among those used by Fantuzzi in his biography. Almost all the pages are $18\frac{1}{2}$ by $24\frac{1}{2}$ cm (some are $20\frac{1}{2}$ by $25\frac{1}{2}$), but they were originally sheets double that length, folded down the middle. Our Countess was either exceedingly mean or else suffered from a chronic paper shortage, for she often used both sides of the page and, when the letter was only short, she cut the sheet down the middle and apparently saved the blank half.

However, we must be kind to the Countess, for fate treated her cruelly. For a start, she was blind and—as if that were not enough—she was crippled by arthritis and paralysed by migraine headaches. 'If you saw with what sick hands I am printing, it seems impossible that I can move them at all,' she wrote on 6 November 1808, and there is hardly a letter in which she did not bewail her fate in heart-breaking terms. 'Have mercy on an unhappy woman' she exhorted him on 8

8 settembre (1809?) A. C.

OTTO SETEMBRE

IO HO PASSATI DEI POTENTI DISPIACERI MA UNA AG
TRISTI RU[?] TORMENTOSE DI ADESSO MAI HO PROVATI
IL DESTINO MI E SEMPRE AVUERSO INSINO LE COSE
CHE PER REFLESIONI HO DESIDERATO ORA MI FANNO
TREMARE VOI CHE AVETE CUORE E CHE MI SIETE VERO
AMICO NON MI ABANDONATE COL LA VOSTRA ASISTENZA
ABBIATE PIETA' DI UNA INFELICE CHE SE MI VEDESTE
OGGI VI FAREI VERA COMPASIONE, VI LASCIO PERCHE
NON HO PIU FORZA DI STAMPARE SONO VOSTRA AMICA

CAROLINA

144. One of the many letters in the Turri file is this one from 1809 in which Carolina exhorts her friend to 'have pity on an unhappy woman'. Note the overtyping in the second word of the second line: it should read 'piú'. In the original, the accent appears after the letter. (SAR)

September 1809, 'for if you saw me today you would really pity me. I leave you because I don't have any more strength to print . . .' So prevalent are these complaints and so heart-rending are the terms in which they are expressed, that a cynic might be tempted to suspect the Countess of using and even perhaps exaggerating her afflictions as a means of binding her 'friend' to her. There is, indeed, little new under the sun. On the other hand, there is no doubt that she was severely handicapped both by the blindness and by the arthritic hands, and it is to Turri's credit that he built a machine she was capable of using at all. In fact, there were times when she was so sick that she was unable to type and was obliged to dictate her letters, a practice she preferred to avoid, which she did when she was able: 'Just to show you I have been busy with my printer I send you two sonnets, one of which you have had, and the description of winter in prose done by me' (9 October 1808).

The dating of these letters was done largely by Dallari, basing himself on references and comparing them with known dates. Sometimes the Countess put no date on the letter at all; usually, however, she started with the day and month but the year was not included. However, since she came from a prominent aristocratic family, the year can easily be supplied. For example, in one letter she informs her friend of her imminent marriage, which is known to have taken place in 1809. All the letters are from the period 1808–10; what happened to later correspondence is unknown. There was certainly more, for the machine provided her only means of private communication. 'I will never forget that it is a precious gift made by you,' she wrote, and she preserved it carefully until her death in 1841, when it was returned to Pellegrino's son who clearly did not consider it as 'precious' as did the person for whom it was built, for it has not been heard of since.

Were it not for the fact that this correspondence has survived, Turri might have been destined to play no greater historical role than so many of his colleagues who invented machines of which details are lacking. In the present case, however, we can reconstruct many essential elements. We are, of course, dealing with a writing machine and not a printing press. Carolina refers to 'printing' (*stampare*) on numerous occasions but the word is not to be taken in its modern sense. There was, after all, no other word with which to describe the action of the machine, although on one occasion she used *scolpire*, meaning to engrave or carve or impress clearly or deeply, when referring to the action of the machine. The instrument was not some form of printing press, however, because

(a) there are examples of overprinting of isolated letters which could not have been done if the word were composed in a press;

(b) on occasion, she needed a single letter to finish a word at the end of a line and this letter was printed a half-space above, as if it implied the use of a caret, and a blind person could not have composed this on a press;

(c) letter spacing was slightly irregular, but erratically so, not following any fixed or set pattern;

(d) the left-hand margin was similarly irregular;

L'INVERNO

DI BREVI ORE SONO I GIORNI E , LUNGHISSIME SONO QUELLE
DELLE NOTTI , E MINACCIATE L' UNE , E L', ALTRE DA ORRIDE
TEMPESTE . TARDI SI VEDE SPUNTAR L' AURORA , E DOPOSCAT-
RIR POI DAL MONTE LO SPLENDIDO , MA FIACCO SOLE, SI SCOR-
GE , E IL SUO CALORE E'SI POCO ARDENTE, CHE FORZA PIU
NON HA DI DISTRUGGERE LE DIACCIATE ACQUE , SOLO LE ROMPE

APPENA NEL PUNTO DEL SUO PIU FORTE CALORE, E LE LA-
SCIA A DIACCIARE DI NUOVO LENTAMENTE ANCOR PIU FORTE
AL SUO SPARIRE. TUTTI I MONTI POI, CHE CI CIRCONDANO RICO-
PERTI SI SCORGANO DI BIANCHE NEVI , CHE SEMBRANO ANNUN-
CIARCI CHE PRESTO ANCHE NOI L'AVREMO SOTTO I PIEDI. LA
CAMPAGNA ORRIDA ALTRO NON CI MOSTRA ORDINARIAMENTE,
CHE ALBERI SPOGLIATI , A CUI SOLO SI VEDE AL CALCIO LE PRO-
PRIE INGIALLITE FOGLIE NATURALMENTE CADUTE, POCHISSIME
SONO LE PIANTE, CHE MANTENGANO LA LORO VERDURA , E
QUESTE NELLA MASSIMA PARTE INFRUTTIFERE QUALE SAREBBE L'
ALLORO L'ORBACO, IL BUSSOLO, IL LEGGIO. ABBIAMO POI LA PRE-
ZIOSA PIANTA DI ULIVO QUALE SEMPRE CONSERVA LE SUE VER-
SCURE FOGLIE NON MENO CHE IL FRUTTO CHE SPESSO DASIPERTO
SI VENTI DISTACCATO VIENE DA SUOI PICCOLI RAMI , E SPESSO

145. '. . . I am sending you . . . the description of winter in prose done by me . . .' she wrote on 9 October 1808. And here is the opening page: entitled 'Winter', it is the earliest original type-written manuscript. The five-page story was typed with the 'new characters' and ink, instead of with the 'black paper', and the impression soaked through the page like a blotter. 'The hours of the day are short, and those of the night are very long . . .' (SAR)

(e)   the impression was by 'black (that is, carbon) paper', requiring some form of blow; and

(f)   above all, a blind operator with arthritic hands could not possibly have manipulated three-millimetre-high letters in a printing press, or even distinguished between them.

The machine, then, was a typewriter. It printed the full upper-case alphabet in 3 mm. letters, plus four punctuation marks. Considering the afflictions of the operator, it was provided with a highly effective system of line and letter spacing; line spacing was foolproof while letter spacing was good but was not perfect, indicating either that the Countess had to do it manually and at times forgot, or else that the escapement was somewhat defective. But that was not the only problem. 'My printer has broken down,' she cried. 'The *tavoletta* does not close any more.' The *tavoletta* was probably a cover or board which was hinged in some way, obviously for storage or portability, since despite the fact that it would not close, she continued to type on the machine as before. One can safely assume that, in common with most early machines, this one was large, the more so since it was made for an untrained blind arthritic person whose digital dexterity must have been close to zero.

Alignment was excellent—so good, in fact, that it would seem to preclude the use of type-bars, at least at that time. From this, as well as from the preceding, one may infer that the machine was of the plunger class, and this assumption is supported by the fact that the characters were once changed in an attempt at improving the impression. Carolina made several references to the new characters ('I want to see whether the new characters print better with ink or with the black paper'). The first set of characters was sharp and made deep impressions, indicating that the bed was of a relatively soft material (leather?); the second set of characters was blunt and the printing was more superficial.

The impression was first accomplished by 'black paper', later by some form of ink applicator. The black paper was prepared by Turri, who was the Countess's only source of supply. 'I am desperate because I find myself almost without black paper,' she wrote (6 November 1808) and she had good reason to be concerned, for the impression was indeed weak. Otherwise, however, Turri's carbon paper was remarkably effective, considering that the impression has survived for over 160 years and has received anything but gentle treatment. The very fact of this 'black paper' is interesting, for its invention is generally attributed to the Englishman Ralph Wedgwood who was granted a patent for 'carbonated paper' in 1806; Turri may well have made it independently at virtually the same time.

The substitution of ink for 'black paper' proved less successful. Presumably, the sharp characters were replaced in order to obviate the need for carbon paper, for Turri must have tired of having to make it. The new blunt characters were designed for use with ink; this was messy, since the quality of the paper was not good and the infuriating Countess often insisted on using both sides of the page, so that those sheets can become virtually illegible: the impression goes straight

X   A . C .   *(1809) 30 luglio.*

TRENTA LUGLIO

QUAL SORPRESA VI RECHERA IN LEGGERE QUESTO
FOGLIO, E PIU GRANDE E LA MIA NEL CASO IN QUAN-
TO MI TROVO ESSENDO VICINO IL PASSO DA VERIFI
CARE I MIEI PRELIMINARI GRADO DI AVICINARMI A DO
VERE FARE IL PREAMBLE PASSO IL GRANDO, UNA
COSA CHE HO DA ME DESIDERATA ORA MI SPAVENTA
SAPPIATE CHE E O IO ACCETTATO LOFERTA DI MATRI
MONIO CHE MI HA FATTO AVANTI DA SE E PER MEZ
ZO DI MIO FRATELLO E MININI CON IL CONSENSO DI
TUTTI I MIEI DI CASA, MA VOI CHE CONOSCETE QUAN
TO IO SONO RIFLESIVA POTETE IMMAGINARVI QUAL SIA
LA MIA AGITAZIONE, FRA GIORNI DEVANO FARE LA
SCRITTA IN FORME PER COMPIRE IL RESTO HO BISO
GNO DELLA VOSTRA ASSISTENZA E MI LUSINGO CIE
AVENDO VOI AVUTO SEMPRE TANTA BONTA PER ME
NON MI ABANDONERETE IN UNA CIRCOSTANZA TANTO
IMPORTANTE SENTO IL PIU VIVO DISPIACERE DI NON

146. '. . . I have accepted the offer of marriage . . .' Details such as this helped Dallari to date the correspondence, for the lives of the two aristocrats are amply documented. Later correspondence (4 April 1810) already refers to 'my husband'. For the moment, however, Carolina confesses that the 'terrible step' she has so much desired now terrifies her, and she pleads with her friend not to abandon her, which is curious indeed, considering it was someone else she was marrying! (SAR)

through them as it would through blotting paper. However, the new characters were tried with carbon paper and worked very well indeed. In one letter using carbon, Carolina complained 'I cannot use the ink because I am alone like a dog,' indicating that the use of ink was a complicated affair for a blind person. It required periodic replenishing or attention, as would an inking roller or pad, because one can see on a page a few messy opening lines where it had just been inked, followed by lines becoming progressively dimmer as the supply exhausted itself. One inking, however, was sufficient for a page. With the carbon paper, this phenomenon of the printing becoming fainter does not occur, indicating that fatigue or other factors relating to the operator were not responsible.

There is little to be added to this description, although in time more information may perhaps come to light. One immediate source which unfortunately proved fruitless was an examination of the Società d'Arti Meccaniche's papers—one would have expected Turri to have presented his machine to his colleagues and fellow inventors; but if he did, the evidence has not survived.

Whether one accepts Turri as the inventor or not, the fact nevertheless remains that his is the first instance of an invention that fulfils the two requirements of a typewriter. He can hardly be disqualified by the fact that the machine itself has not survived, for its recorded existence is beyond doubt. This is exactly analogous to the case of Johann Gutenberg, who is generally recognized as the inventor of printing: bibles produced by him are in existence, even though the press on which they were printed has long since disappeared. This can hardly be considered Gutenberg's fault, and recognition of his invention is not withdrawn on account of it.

In the case of the typewriter, more than enough is known for Turri to be conceded the honour of having invented it. But unfortunately for the many patriots all over the globe whose loyalties will be offended by this report, a denial of Turri's rights makes it no easier for either Mitterhofer or Sholes or Ravizza or Malling Hansen or Progin or Drais or Alissoff. There were too many others before them!

Chapter 8

Pioneers and Others, from 1874 Onwards

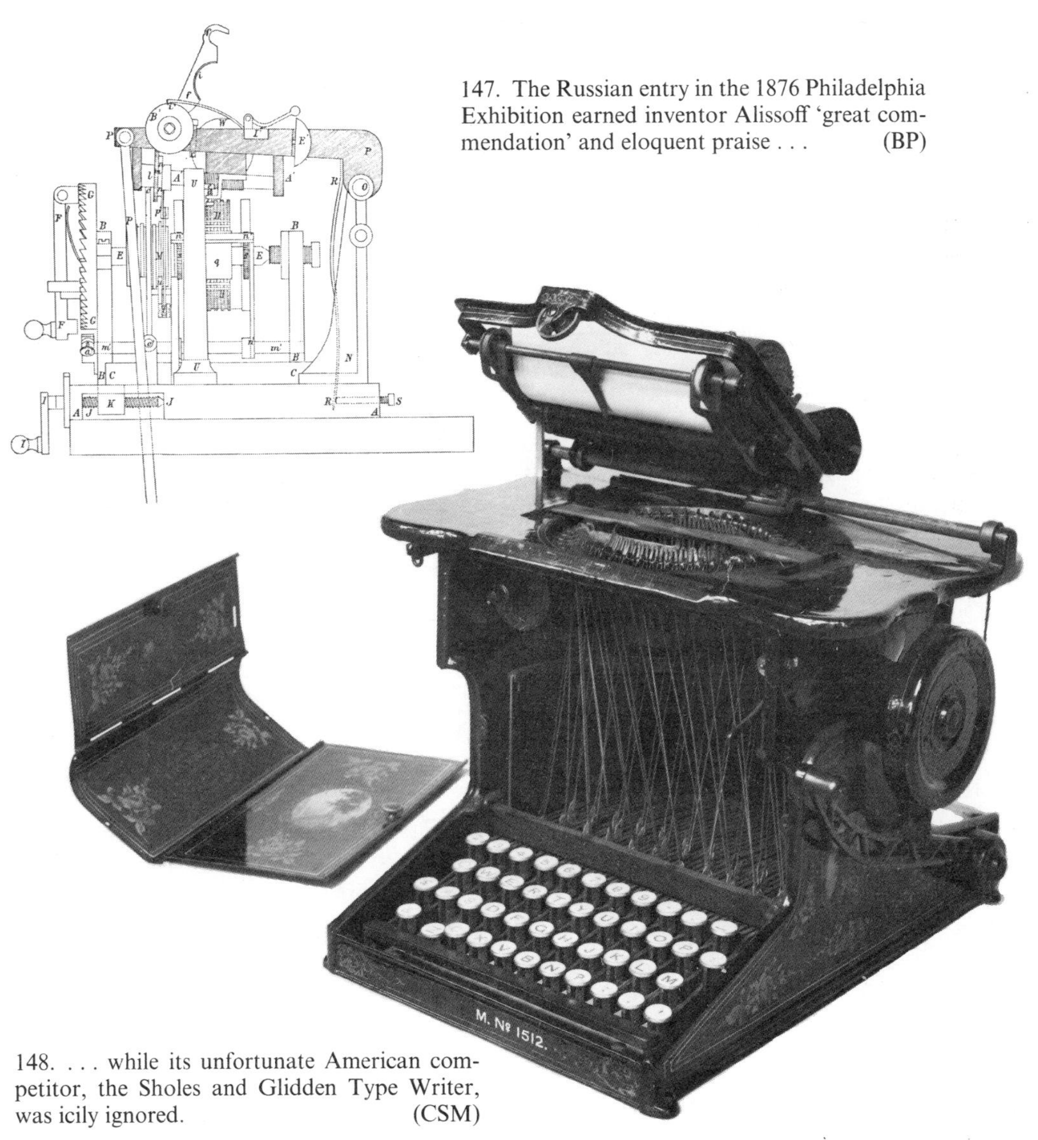

147. The Russian entry in the 1876 Philadelphia Exhibition earned inventor Alissoff 'great commendation' and eloquent praise . . .   (BP)

148. . . . while its unfortunate American competitor, the Sholes and Glidden Type Writer, was icily ignored.   (CSM)

The 1876 Philadelphia Exhibition may have offended supporters of the Italians Meucci and Ravizza, but that was nothing compared to what it did to Sholes, Densmore, Yost and Remington. The Alissoff typewriter (the one the Russians claim was the first) had completed its maiden voyage across the seas and was on display. It received eloquent praise in the official report as a beautiful and carefully constructed machine deserving 'great commendation'. The Sholes and Glidden was not even given a mention!

In fact, things could hardly have been worse for the infant Type Writer. Launched in 1874, it had to contend not only with a sceptical and apathetic public but also with a general economic depression which was destined to become even more acute before it finally let up. Furthermore, the machine, as we have seen, left much to be desired, so that in part it had only itself to blame for its unenthusiastic reception.

According to the terms of the agreement with the Remington factory, one thousand machines were to be manufactured, plus an excessively optimistic option for a further 24,000. The indomitable Densmore and the super-salesman Yost then formed a partnership to sell the machines, and judging from the financial commitments which they glibly accepted, they expected to do a brisk trade. Firstly, they borrowed heavily to pay Remington an advance of $10,000 to get production started. This was one of the cautious arms manufacturer's terms, and the fact that the two men agreed to it at all is proof of their confidence in the eventual success of the enterprise. And then, they had royalties to pay on all sides: they had Sholes to take care of, plus a few other minor contributors, plus the fee per machine to Remington's mechanics for improvements they had made, plus royalties to a Charles Washburn for a feature on which he held a prior (1870) patent (the longitudinal movement of the platen for letter spacing) plus principle and interests on past debts.

All of which might have been possible had they sold machines hand over fist, as they expected. But they didn't, and however thinly they spread the profits, there were just not enough to go around. Recriminations were plentiful, that was all, and the financial history of this period is still largely controversial. If we discount the biased reports of those[43] attempting to portray Sholes as the altruist, the benefactor of mankind, the victim who got nothing for his troubles but tuberculosis, an account of the ownership of the invention does little credit to anyone. It is a study of that universal aspect of human nature which makes a man desperate to abandon a sinking ship only to want to clamber back on at all cost after it has shown that it can stay afloat. Man is never at his best under these circumstances.

Throughout its early history, Densmore was the captain of this ship, apparently determined to go down with it if need be. The Sholes and Glidden, and hence the Remington, was Densmore's achievement more than anyone else's, and its ultimate success is a credit to him alone. Without him, Sholes would have made one or two models and then packed it up. History would have heard no more of him or his machine. It was Densmore who changed that sequence of events, although his devotion was motivated by nothing other than materialism: he was a businessman and a promoter who saw in the typewriter his chance to make a fortune. He was something of a visionary, perhaps, for he was the only one who kept faith throughout, and as a consequence picked up an ever larger share of the ownership as one man after another lost hope and sold out at moments when the invention appeared to be worthless. Later, when it began to acquire some value, everyone had second thoughts, but by then, of course, it was too late, and the only trading done was in insults.

Briefly, the history of the ownership is a confusing collection of contradictory claims, but after sifting through them all, we find that what actually happened is probably as follows: back in 1867 Sholes, Glidden and Soulé divided the machine between them and then brought Densmore in for an equal share, for which he paid them each $200. Glidden then backed out for $250 and Soulé, later, for $500, leaving the invention in Sholes's and Densmore's hands. Glidden was the first to regret his action, reclaiming a part of the credit for the invention (and getting Densmore to agree to call the machine the Sholes *and Glidden*) and a part of the ownership as well[12]; Sholes, believing the invention to be worthless anyway, dished him out a part of his own holdings, but Densmore refused.

Glidden, still unappeased, demanded something more concrete, in writing, and in this Densmore obliged. But the document was broadened to include both of Densmore's brothers, who had bought into the enterprise when funds had been desperately short. As it now stood, James Densmore owned four-tenths, Sholes three-tenths, having ceded a tenth to Glidden, Amos Densmore and Emmett Densmore one-tenth each. Sholes and James Densmore were to manage the property in trust for the other shareholders. (This agreement superseded the earlier ledger statement, which fails to mention Glidden's share.)

For the moment, everyone seemed appeased—quite possibly because it looked like worthless property anyway, and one-tenth of nothing is every bit as much as two-tenths of nothing! But then Sholes began rocking the boat by unloading his tenths: one, in fractions, was sold in 1873 to Amos and others for $5,350 after the Remington contract was signed. He wanted to sell the rest, as well, and disregarded pressure from James Densmore that he hold on to his shares. Densmore's concern was not for Sholes's eventual welfare but for his own: while Sholes held on to his shares, Densmore felt he could control the enterprise, but he was not so confident if outsiders came in. If anyone were to buy Sholes out it ought to be him, he felt, but what with the debts he had accumulated in launching the invention, he could not even afford to eat.

Then Yost appeared on the scene, and he and James Densmore made a concerted effort at buying up all interests in the enterprise. In 1874 they bought one of Sholes's remaining two-tenths for $10,000 in promissory notes, plus Emmett's tenth and the other fractions that had been dispersed. James and Amos Densmore and Yost now formed a new company in which James held the reins. But the

Gliddens would not sell their tenth, and even agitated for more share in the goodies, while Emmett, according to a previous agreement, retained control of all the overseas rights.

The company was out on a limb, financially, and James was more heavily committed than the others. One wonders how he kept track of all his debts and the interest on them without the help of a computer. Yost, on the other hand, was also liable, to a lesser extent, but he appears to have taken his responsibilities lightly—or, at least, more lightly than James. For apart from all the debts Densmore had accumulated in getting the enterprise on its feet, including the $10,000 paid to Remington, he also had to contend with a veritable pack of wolves snapping at his pocket-book all the time. There were the progressively more fantastic claims his brothers made for a share of it all; there was the fast-spending Yost who accumulated debts and expenses faster than Densmore could sign for them; and, last but by no means least, there were all the royalties that had to be paid for every machine manufactured. If they had sold thousands upon thousands of them it might have been different, but as it was, after paying out to Sholes and Glidden and Washburn and Clough and Jenne and Emmett, the company showed a tidy loss without anything in James's pocket for his troubles, and without even repaying current notes to their creditors.

Clearly, something had to be done, and the uneasy truce between Densmore and Yost ended. The former wanted more austerity, the latter a more lavish image. There was nothing for it but to split. And so a change was wrought: the Type Writer Company was incorporated at the end of 1874 with James Densmore and Yost each owning 40 per cent of the stock and Amos the remainder. The first act of this new holding company was to lease manufacture and selling rights to Yost and a new partner of his called Luxton. These two went into business under the name Densmore, Yost and Company, even though Densmore had no part in this retail undertaking and in fact protested against the use of his name.[12]

Free of restraint from Densmore, Yost was now in a position to be lavish all he pleased. And lavish he was, firstly with more promises to all and sundry (including $1,000 per month each to James and Amos in return for the Type Writer Company's franchise) and secondly with the additional credit his smooth tongue succeeded in raising, including $20,000 worth of machines from Remington! And as if that were not enough, he even managed to borrow money from Remington to buy half of Amos's stock in the Type Writer Company, thereby becoming the largest single stockholder.

Things were now even worse than before. Densmore had lost control, Sholes was getting next to nothing but promises in which he no longer believed, and creditors were beginning to sue. And they sued not only Densmore, Yost & Co. but also its individual members as well, including James Densmore who had committed the folly of allowing Yost to use his name. In desperation, he convinced Yost that it was time to pack it up and renegotiate with Remington; the new agreement which they all hammered out in November 1875 transferred

149. The characteristic profile of the Sholes and Glidden conceals key levers that extend the length of the base from the keyboard on the left. The small horizontal hook is for holding the keyboard cover open and the vertical lever is for carriage return and raising. (AC)

manufacturing and selling rights to Remington in return for a fee to the Type Writer Company for use of its patents. Now it looked a little better, on paper, but by the time Remington had deducted a part to offset Yost's debt and the patentees had received their royalties, there was still nothing left, or next to nothing. Sales continued to be poor, and the new retail agents—Locke, Yost and Bates—were themselves replaced in 1878 by Fairbanks and Company, a well-known scales manufacturer, who fared little better. The Type Writer Company remained financially sick; Sholes and Densmore started to break and their relationship never fully healed; the Densmore brothers themselves split in strife; Remington began tottering on the brink of bankruptcy. There was no money in the thing. Densmore could not keep afloat, Sholes wanted to unload his shares and start manufacturing a new machine, and all seemed lost.

It seemed that way to Remington as well. They eventually decided to sell out their entire typewriter division in order to raise the money to pay off their own

150. Success at last! The double-case Improved Model Two was introduced in 1878 but it was not until the 1880s when the bugs were finally ironed out that the machine began to sell well and its name was changed to Remington. (AC)

175

creditors and stave off bankruptcy. The sales agent at that time was a dynamic organization composed of Wyckoff, who was Yost's former star salesman; Seamans, who was the head of Fairbanks' old typewriter division; and a long-time Remington official called Benedict. Between them, these three men finally made a success of the typewriter and went on to buy up the Remington operation in 1886, having established themselves as the principal contenders in the rapidly growing field. The boom had begun.

As far back as 1878, however, the seeds of change were being planted. Yost, parting company with the Type Writer, allied himself with Walter Barron and a man called Franz X. Wagner (who was later to go on to even greater things) and began work on a machine of his own. He needed the use of existing patents, however, and he negotiated for them with Densmore. The upshot was a machine called the Caligraph which Yost began marketing in 1881. Densmore was out of it all by then: he was left sitting on the fence, selling rights to both Remington and

151. George Washington Newton Yost was responsible for the Caligraph which became Remington's chief rival until they joined forces (with others) in a Typewriter Trust. Note the full keyboard and the characteristic 'bird cage' type basket, with one of the type-bars on its up stoke. The illustration is of Model Four.   (AC)

Caligraph for use of the same patents. But even so, things were far from bright for him, for the backlog of old debts and interests rested fairly and squarely on his shoulders. Furthermore, Remington promptly refused to pay the Type Writer Company any more fees, resenting the competition of the Caligraph in a field which had been exclusively theirs till then. They claimed all sorts of things, plus a bit of arbitrary rough stuff such as demanding Sholes's new machine in compensation, intending not to promote it but to shelve it. This machine was to be Densmore's and Sholes's last individual attempt at entering the field and was a hot item which the Type Writer Company had picked up. Densmore was in a desperate position, however, and was obliged to make all kinds of concessions; as part of the final agreement signed in 1881, Remington got hold of the new machine and snuffed it out, despite Densmore's later efforts at having it returned. But at least the Type Writer Company was to receive fees again, even though initially most of them went to repaying a $59,000 debt to Remington (of which $50,000 was Yost's) plus amounts payable to Sholes, Glidden and the other creditors.

Despite this, Densmore was beginning to smile. The Improved Model Two was now selling briskly in the hands of its capable agents, and Yost was in the market with his competitive machine. This was of the up-stroke class and the first model printed only upper case, even though Remington was already offering its shift-key model with both cases. But they themselves expected to sell ten upper-case machines for each one with both cases, and so Yost may be forgiven for his temporary retrogression, as it now appears to be. A second model Caligraph soon followed, however; this printed both cases and departed from the Remington shift-key principle by providing a 'full' keyboard, with a separate key for each character. In other words, 'A' and 'a' were not placed one above the other on the same type-bar, but each had its own. It looked clumsy, perhaps, but for several decades it was a popular alternative to the shift key. And the Caligraph sold well although more slowly than the Type Writer; but what did this matter to Densmore, who was receiving royalties from them both?

While it was still progressing, Yost severed his relationship with the Caligraph organization (under circumstances which have never been fully explained, but which may easily be conjectured, given Yost's well-documented personality) and together with a new set of associates he placed on the market the first of a long series of machines which bore his name. The non-visible, type-bar Yost departed radically from established construction by replacing the ribbon with an inking

152. Breaking away from Caligraph interests, Yost gave his name to a double keyboard, non-visible machine with a unique 'grasshopper' action. In this illustration of the first model, the platen is in its raised position and a type-bar approaches the guide.　　　　(AC)

pad. It too had the large keyboard for which Yost showed a particular weakness and although he died in 1895, some eight years after it was introduced, his type-writer continued to enjoy brisk sales for decades, mainly in Europe.

James Densmore, on the other hand, never actually succeeded in producing other typewriters. With Franz Wagner he looked into the possibility of re-entering the market, but their projected machine came to naught, as did the Sholes portable which Remington had obliged him to surrender years before and upon which he had placed much hope. He therefore spent the last years of his life collecting royalties from all sides and fighting fierce battles for the rights to a number of properties of which no one seemed able to decide the ownership. Oden[31] maintains that some of them were 'sham' battles, designed to drum up public interest in typewriters in general—these suits were invariably settled out of court. However, when Densmore died in 1889 he was reputedly worth a half million dollars, so the machine treated him well in his old age.

Meanwhile, his brothers Emmett and Amos, together with Walter Barron and the ubiquitous Franz Wagner, designed and began production of a machine of their own, which they rightly named the Densmore. It, too, was of non-visible

153. James Densmore was out of it, busy collecting royalties, but his brothers Amos and Emmett were responsible for the marketing of a rather conventional up-stroke machine that bore the family name. (AC)

up-stroke design and boasted a unique type-bar arrangement plus other innovations such as platen release, ball bearings for the type-bar pivots and so on. The machine enjoyed some success, was later absorbed into the Union Typewriter Company, and its demise has already been recorded in Chapter One. The two Densmore brothers became large stockholders of this typewriter trust, and, what with one thing and another, Emmett became a millionaire by the time of his death in 1911.

Sholes was not so generously rewarded and it has been long argued by his supporters that he was gypped. To hear them tell it, one might easily believe that he got not a cent out of the whole business, but this is not quite so. The greatest ultimate fortunes, undoubtedly, were made by those who stuck their necks out

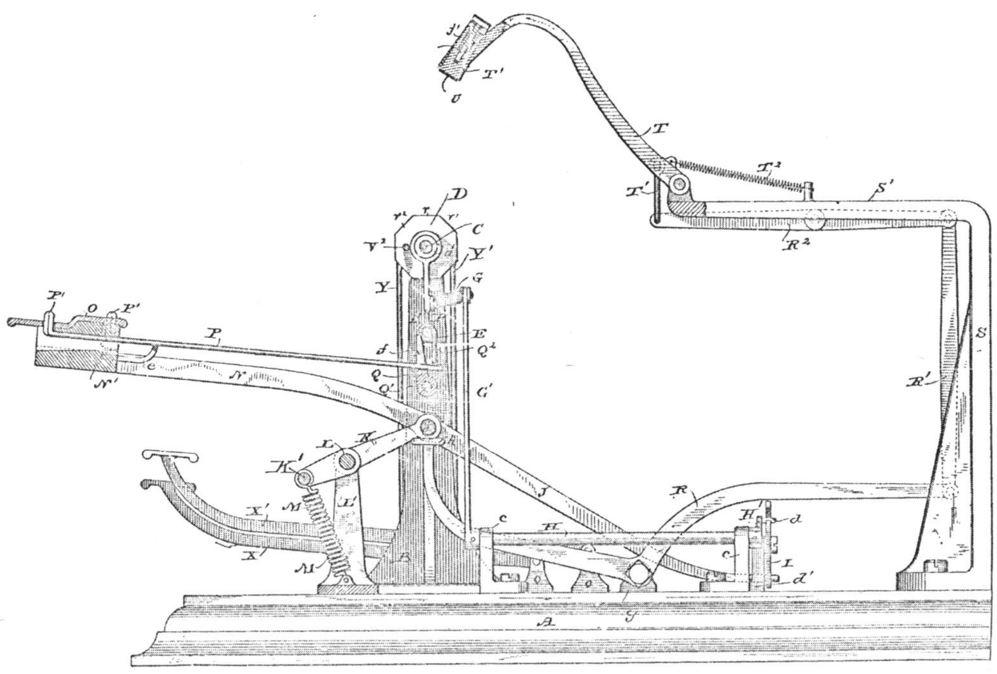

154. Christopher Latham Sholes, on the other hand, repudiated the principles of his invention —and the practice as well—and busied himself with less orthodox instruments such as one patented in 1889. It was never manufactured, and in fact his only design to leave the drawing board, apart from his first, was his last: (USP)

155. . . . the Sholes Visible, a novel machine in which the type-bars moved down a corridor to the printing point, thereby overcoming alignment problems inherent in contemporary type-bar instruments. His son Louis was responsible for manufacture, but the project was not a success. (LC)

furthest; considering the extent of Densmore's commitment and the reluctance of Sholes to take any risk at all, the outcome is hardly unjust. Sholes not only refused to sink his own money into the enterprise but was also desperate to rid himself of his holdings and liquidate his interests, convinced that the thing was doomed to failure. Like Carlos Glidden, he later regretted his actions and blamed Densmore for his misfortunes. Just how much he did get out of the invention is disputed—some[21] say he sold out in 1877 for $12,000 while others[12] more realistically quote Sholes himself to the effect that he was paid $25,000 from 1872 till 1882, with $5,000 still owing. The amount was probably even higher, but in any case till then it was more than that received by Densmore, who took all the risks. This position was, of course, reversed by the time Sholes died in 1890, a year after his former partner.

Despite his total rejection of the original machine, Sholes continued to experiment and perfect other typewriters which he was convinced would outperform and outsell the Remington. On and off, between bouts of illness and even during them, he continued inventing. One product was the small portable already mentioned; another was a type-wheel machine designed after he despaired of ever achieving correct alignment with type-bars. An octagonal type-sleeve instrument was patented in 1889, and there were more, but they never reached production stage. Only one of his designs was actually manufactured, other than the original one. The intriguing Sholes Visible tried to solve the alignment problem by a unique front-stroke action and by forging the type on to the type-bars, but it was slow and never a serious contender in an already highly competitive field: it came too late, after its time. Sholes's son Louis, who failed with it the first time around, attempted to revive the machine even later, in 1909, but the project never got off the ground. Much money was lost on it.

In fact, several of Sholes's sons basked in their father's glory. Their intrinsic human and commercial worth was seriously placed in doubt not only by Densmore but even, on occasion, by their father, and their contributions to the history of the typewriter would not be worth recording individually were it not for the fame of the man who sired them. To begin with, there was Louis, as already noted, who tried to make it with his father's type-bar machine. A further model attributed

MPM

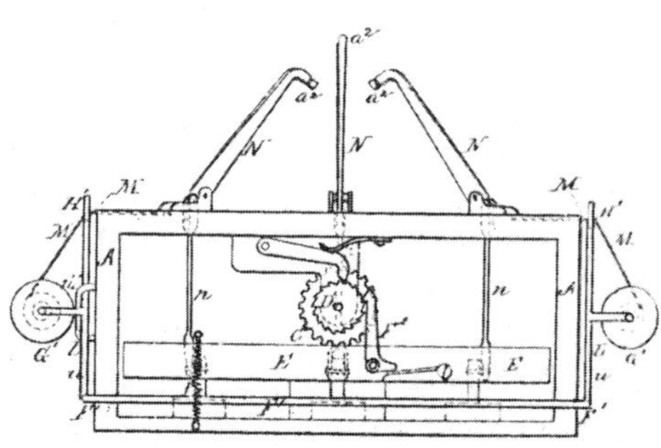

156. Fred, another of Sholes's sons, did a little work on the sidelines but nothing of any account, not even his down-stroke invention patented in 1880. (USP)

to him was never produced, almost certainly because it infringed too many existing patents. And there was Fred, who flitted around the typewriter scene for a number of years. He is said[12] to have persuaded his vindictive father to sign some patents which the dying man had refused to do, since he no longer directly benefited from them. That was in 1890, and to the last the old man resented the wealth others had made out of the machine. Fred's affairs, however, remain largely uncertain. Together with a William Miller, he was granted an 1879 patent on an awkward horizontal type-bar machine and the following year for a down-stroke design with type-bars behind the platen. Neither was manufactured. He is also recorded[28] as being involved with the Remington Junior; as one of the organizers of the Rex enterprise; as a liquidator of the Pittsburg Visible in 1913—but nothing creative.

Zalmon, on the other hand, turned out to be much more prolific. Together with Franklin, the son of one of the original Remington brothers, he began

157. Most successful of Sholes's sons was Zalmon who teamed up with one of the Remington boys to manufacture the Rem-Sho, characterized chiefly by a shift linkage which moved the type-basket and by the fact that early models were beautifully cast in bronze.   (AC)

production of a machine which they called the Remington-Sholes or Rem-Sho and which bore far more than a passing resemblance to the Remington Standard. Although at that time the Rem-Sho was beautifully sculptured and cast in bronze, the essential difference between it and the older competitor was merely a shift linkage which moved the entire type-basket back and forth for change of case as against the conventional one which moved the carriage on the Remington. Although the Remington typewriter was no longer connected in any way with the Remington family, their name continued to be used on it and the appearance of a competitive machine bearing a similar label sparked a fierce legal battle which Zalmon S. and Franklin R. lost in 1901. The Rem-Sho was then renamed the Fay-Sho after the company's president, Charles Norman Fay, and production continued while appeals ran their course. They went all the way to the Supreme Court, which eventually reversed the lower court decision, and the name Rem-Sho was restored. The machine was manufactured till 1909, by which time it had gone visible. In that year, however, the company went into receivership and its assets were sold to the French clock manufacturer Japy Frères & Co., who transferred them all to France and brought out essentially the same machine calling it the Japy. For the many who have made the mistake in the past, this instrument is NOT Japanese!

Undaunted by this failure, Zalmon then formed the Sholes Typewriter Co. of Waterbury, Conn. which was unsuccessful in 1911 with its Waterbury and Acme machines. Two years later he was in London and trying for the third time, with a machine variously called the Zalsho and the Z. G. Sholes. It rapidly met the fate of its predecessors. Back in the States, he was still at it when he died in 1917.

So much, then, for all those who were initially connected with the Sholes and Glidden and the first Remington, plus their numerous heirs and relations. There were other pioneers, however, who influenced the typewriter independently and played at least as important a role as many of those already mentioned. Lucien Stephen Crandall, for instance, made several contributions, the main one being the commercial application of the type-sleeve. He also incorporated differential spacing on some of his machines.

Crandall's first patent, granted in 1875, was for an impractical up-stroke device with a curved keyboard of eight keys controlling type-bars with six characters on each. This was the first of a number of departures from the design most commonly

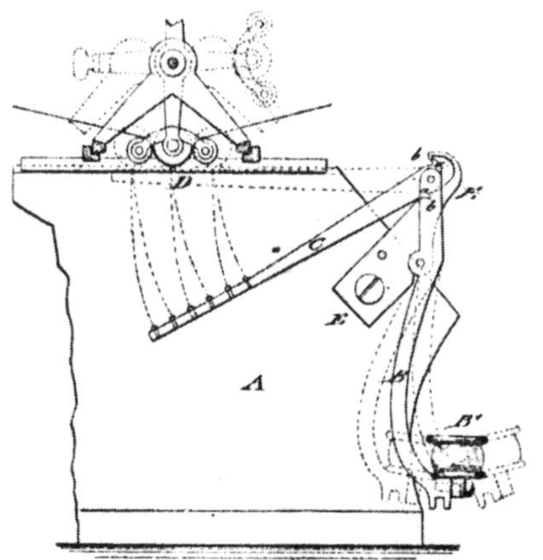
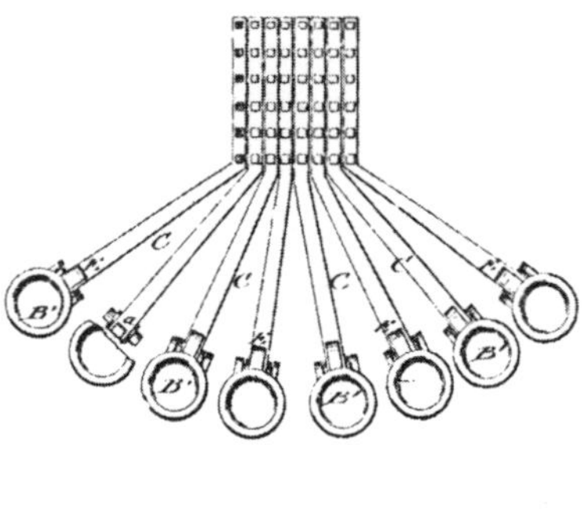

158. Lucien Stephen Crandall started off with a loser in this unusual 1875 patent for an up-stroke which was never manufactured.   (USP)

associated with him, in the development of which he and his principal competitor, James Bartlett Hammond, borrowed heavily from their compatriot Pratt whose inventions have been discussed in detail in Chapter Four. After a certain amount of juggling, Hammond secured Pratt's patents, or rather the use of them, while Crandall preferred to depart sufficiently from the design to make his own version patentable and thus avoid paying royalties. He placed the type around the surface of a sleeve which was made to revolve to the required letter and come down on to the printing-point, locking itself in place for correct alignment. The US patent was filed in 1879 and the Crandall was the second make to appear on the market.

159. Crandall then changed his approach and came up with a type-sleeve design which was manufactured for several decades in a number of models. (LC)

It is said that Sholes and Densmore were at one time interested in it and that Crandall, while an employee of Remington, got his ideas from developmental work carried out by that firm on Hammond's prototypes. Whether this is true or false, the first Crandall was a failure. Subsequent improvements were made: the machine passed through a marginally better Model Two until it appeared in a successful Model Three, labelled the Universal Crandall, in 1893. A further model appeared as late as 1906, so that this make enjoyed a long life indeed.

Meanwhile, however, Crandall was responsible for two design departures. One was an 1889 up-stroke non-visible machine called International, similar in many respects to the Remington, and the other was an Improved Crandall of 1895, a novel type-bar roller-inking machine sporting an overhead hammer. Both these typewriters were short-lived and Crandall, for all his efforts, barely succeeded in eking out an existence.

Hammond was more fortunate, but he deserved to be, because he came up with a better machine. What with clashing type-bars, haphazard alignment and defective printing pressure, Hammond's approach was fundamentally sound: he eliminated the type-bars altogether by using a swinging type-sector (two sectors on his early

160. One of the greatest makes of all time was the Hammond, which began its illustrious career in an attractive wooden case with thick ebony keys in two curved rows. The swinging type-shuttle was its major mechanical contribution.
(AC)

machine) thus automatically obtaining perfect alignment. Regularity of impression was achieved by having a hammer strike the paper against the type-sector from the rear, thereby ensuring that, no matter how hard the key was struck, the impression would always be uniform, since the key did nothing more than trigger the hammer mechanism, apart from selecting the correct character, of course. Much of this was taken from Pratt's Pterotype: Hammond was not lumbered with Crandall's handicap of having to produce something different enough from Pratt's design to make his own patentable.

For all that, developmental work on the Hammond took well over a decade, during which time virtually everyone gave up all hope of the machine ever achieving its purpose—except its maker. Even Remington tried their hand at it without success. However, from 1880 when the patent model was completed and the following year when the first units hit the market, the Hammond carved an illustrious niche for itself in the history of the machine. It also made the inventor a millionaire, proving that not all typewriter manufacturers met an inclement fate! The design was one of the best ever, and when it first appeared it far excelled its type-bar competitor. In fact, it was never essentially changed, and for sheer versatility it outstripped anything ever built: literally hundreds of different type-shuttles easily adapted the machine to all languages and scientific purposes. Only the contemporary Blick was in any way comparable.

James Hammond died in 1913 at the peak of his typewriter's popularity, but his death did not signal its demise, as happened only too frequently in the case of its competitors. The factory continued to prosper until 1927 when the machine acquired its final form as the Varityper which is standard equipment in the printing trade to this day.

Thomas Hall, whose earlier abortive efforts were recorded in Chapter Four,

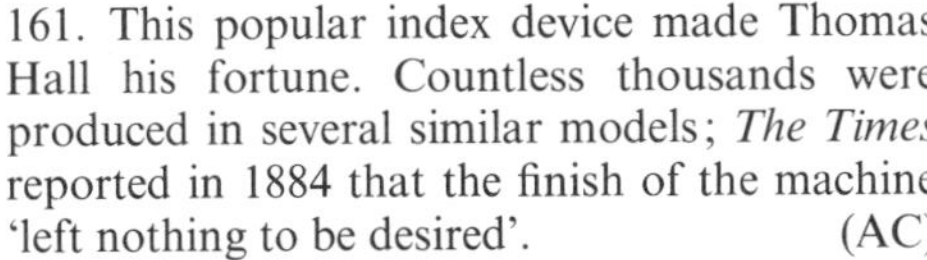

161. This popular index device made Thomas Hall his fortune. Countless thousands were produced in several similar models; *The Times* reported in 1884 that the finish of the machine 'left nothing to be desired'.　　　　(AC)

162. The type-wheel Commercial Visible was one of the most popular of the fifty or so inventions of the prolific Richard Uhlig.　　(LC)

patented the first commercially successful index machine in 1881. It was a curious instrument with a square rubber index plate bearing the type. Non-visible, rather awkward, it was nevertheless enormously popular and clearly filled an important need in the market. How many thousands of these machines were sold is not known, but there were enough to make Hall his fortune. It was a 'beautiful little machine', according to an 1884 review in the *Times*, and its finish left 'nothing to be desired'. It continued to enjoy brisk sales for a couple of decades and spawned numerous progeny in different countries; these were little more than barely disguised chips off the old block: Kneist (1893), Graphic (1895), Eureka (1898), and so on. The inventor himself later built a modified version called Century, similar in appearance but using a type-sleeve in place of the index plate.

And there were others, many others, whose legendary involvement with the typewriter is worth recording. There was the prolific Richard Uhlig, who sired no less than fifty different designs of which thirty were still-born, but of the rest, a good baker's dozen reached maturity. These included the Allen, Arlington, Atlas, Commercial Visible, Delta, Emerson, Model, Modern, Perfection, Uhlig and so on. His patents date from 1897 and while none of them was a great machine, several met with some success, notably the Emerson and the Commercial Visible. The latter was more successful with the public than with the critics: 'No machine has ever, to our mind, possessed more graceful lines, very few have had such ambitious aims, fewer have had more good points, but certainly no machine ever proved more disappointing to us in actual use.'[27]

Or take Charles Spiro, who entered the picture with an 1885 patent for a delightful little typewriter called Columbia and who went on to invent a further seven machines of which several made headlines. The Columbia assured its place

185

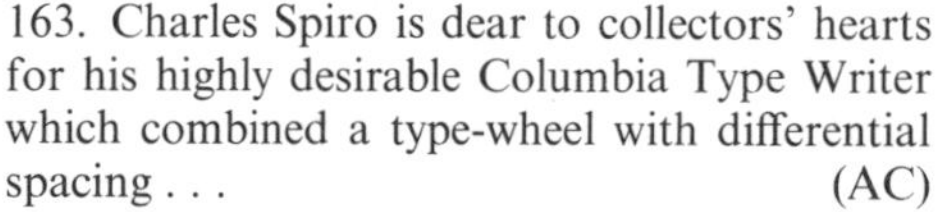

163. Charles Spiro is dear to collectors' hearts for his highly desirable Columbia Type Writer which combined a type-wheel with differential spacing . . .   (AC)

164. . . . but it was the down-stroke Bar-Lock which brought him fame and fortune.   (GC)

in history by being one of the first on the market with differential spacing, and it enjoyed widespread sales for many years despite its relatively slow operating speed. However, it was little more than a by-product of Spiro's major inventive efforts which resulted in the Bar-Lock, a down-stroke machine of the highest quality. It was claimed to be the first visible type-bar design and while this is not true (Prouty and Grundy patents preceded it, as did the Horton, which was actually manufactured; furthermore, it is questionable whether the Bar-Lock is, in fact, 'visible' unless the operator stands, straddles the machine, or is endowed with a telescopic neck), it was nevertheless the best for almost a decade after its introduction in 1889. There were other machines to Spiro's credit, too, but they hardly did him justice after the heights he attained with the Bar-Lock. He patented a linear index machine with keyboard control in 1885; four variations of type-wheel designs the following year; and so on. He also invented the Visigraph of 1910 and the Gourland of 1920, both of which were manufactured, but it is with his earlier machines—the Columbia and the Bar-Lock—that his name is mainly connected.

Wellington Parker Kidder was another pioneer. His first effort appears to have been a circular index machine for printing large showcards and signs, for which he and an associate were granted a joint patent in 1887. He then invented a meritorious down-stroke instrument called Franklin with a simple and effective movement, for which a patent was filed in 1889 and granted in 1891. But in the following year he patented a different design altogether: a machine with a thrust action which he called after his Christian name. The Wellington, later known as the

165. Reliable statistics are lacking, but it is probably safe to state that the thrust-action Wellington in its various guises was one of the very best sellers of all times. Even today it crops up with monotonous frequency all over the world. (AC)

Empire and in Germany as the Adler, enjoyed a brisk and virile longevity and although Kidder borrowed its basic principle from the earlier Rapid, he is nevertheless responsible for one of the great machines of that era. The same concept was later developed into the more sophisticated noiseless typewriters described on page 233.

Or consider the prolific and versatile Franz Xaver Wagner, whose best design provoked a revolution in typewriter history. He was ubiquitous in the early days, a man whose skill placed him at the top of his profession. He first collaborated on the development of the Sholes and Glidden in the Remington factory; later he paired with Yost on the first Caligraph. He joined the Densmores and Walter Barron to devise the first Densmore machine and participated in the development of the two succeeding models. There are many, many patents in his name, covering all kinds of principles—too many to permit individual treatment. His first was granted in 1880 for a swinging sector machine, another in 1884 for a swinging segmental comb (cf. Wheatstone *et al.*), and in 1885 for an up-stroke in which the type-bars rocked to select upper and lower case; in 1887 for a type-wheel with swinging indicator, and the following year for one with a linear index; in 1889 for a keyboard machine with type on linear combs, and so on.

But then he changed course, and together with his son he invented a visible front-stroke machine with a four-row keyboard for which he filed a patent in 1893. Further development by the Wagner Typewriter Company resulted in a machine bearing the inventor's name; the entire operation was then bought up by the typewriter-supply company of John Underwood, the resulting instrument becoming the prototype for all later upright front-stroke instruments. Wagner cannot be given credit for the invention of the front-stroke principle, but he perfected its application and thereby ushered in the age of standardization in the typewriter industry.

An equally versatile genius was Byron A. Brooks whose first invention was also his greatest single contribution to typewriter history: while employed in the

Remington factory on the development of the Sholes and Glidden, he devised the shift principle, using lateral movement of the platen to select upper and lower case from the same type-bar. He filed his patent covering this improvement in 1875 and it was granted three years later. (Yost witnessed it: a historical curiosity in the light of his declared aversion to the shift principle.) Brooks has many other patents to his credit, including three in 1883, one in 1885 for his successful down-stroke design, a further type-wheel patent in 1891, two more the following year,

166. Byron Brooks is best remembered for a down-stroke machine which bears his name, yet his greatest single contribution was the invention of the shift principle of placing upper- and lower-case characters on the same type-bars.

(GC)

and so on. His down-stroke Brooks and type-wheel Crown were both manufactured, the former being absorbed into the Union Typewriter Co., and eventually phased out.

No less fertile an inventive genius was Lee S. Burridge of New York who,

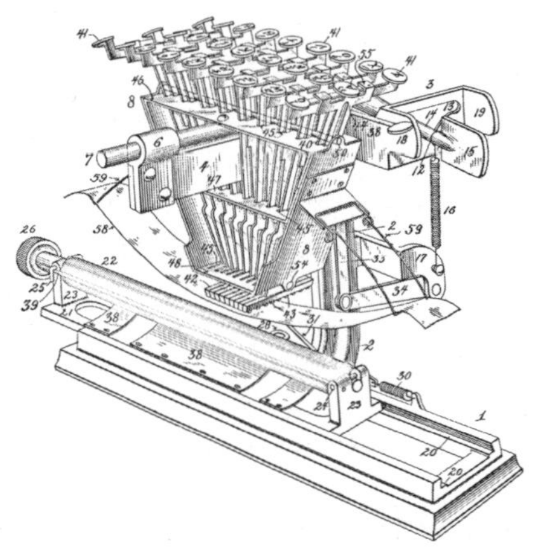

167. A novel design which has the unmistakable Burridge touch: almost everything he invented was unorthodox yet perfectly feasible. Regrettably for us all, this 1897 device was never manufactured. (USP)

both alone and in collaboration with others, held literally dozens of patents. Together with Newman Marshman, he designed the Sun and the type-bar American, the two machines for which he is best known. But there were many others, and all of them share a certain intangible quality despite the fact that the designs are totally different. It is the Burridge imprint and there is no mistaking it: the 1897 Century has it; so do the American Double Wheel and the Burridge, all of which are described in Chapter Twelve.

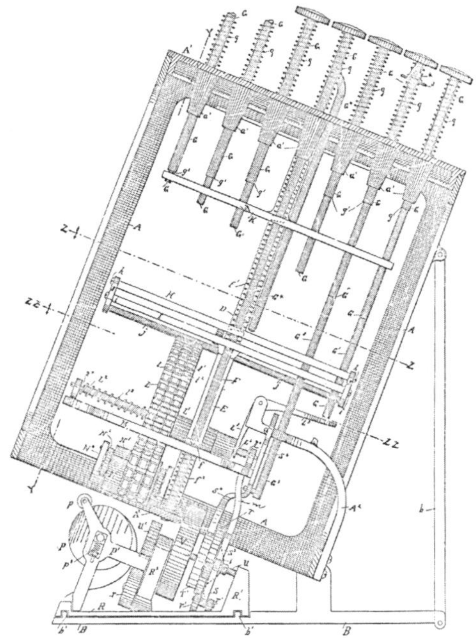

168. It all started for George Blickensderfer with an unusual 1889 type-wheel patent—and it finished with one of the finest and most versatile machines ever made. (USP)

George C. Blickensderfer, on the other hand, can make no claim to versatility, yet he produced one of the finest machines of all times. He hit on a good thing right from the start, and stuck to it throughout his active career. His first 1889 patent was for a curious type-wheel device which contained the germ of an idea, further developed in a more compact design patented two years later. However, it was the Model Five, introduced in 1893, which established the famous profile of the make which was to flood the world with many hundreds of thousands of units during the following quarter of a century until the death of the inventor in 1917.

Finally, a unique individual in typewriter history ought to be singled out—a man with the unimposing but not inappropriately anonymous name of Harry A. Smith. He was the scavenger who cleaned up the mess everyone left around, and he holds the dubious distinction of having marketed many different machines of which not one was original—the vulture who hovered over the dying remains of typewriter companies.

Smith's activities are often clouded in uncertainty. It is reported, for instance, that he bought up the Blickensderfer company after its demise in 1917, and the last four-row front-stroke Blick-Bar reappeared soon afterwards as the Harry A. Smith. Before that, however, the Emerson suddenly showed up sporting the name Smith: this was around 1914, after the original machine had failed and the factory, owned since 1910 by Roebuck Typewriter Co., was moved to Woodstock, Ill., where a well-known front-stroke machine named after the town was subsequently manufactured. How that use of Smith's name on the Emerson came about is an interesting mystery, since Roebuck did not relinquish control of his company in the interim. To complicate the issue still further, the 1914–15 Woodstock Models Three and Four reappeared in 1922 as the Annell (Smith's middle name), marketed by the Annell Typewriter Company. In all probability, Smith bought up lots of these machines after they were discontinued and merely relabelled them, a practice which is considerably easier than manufacture but hardly likely to be more successful.

In passing judgement on these early pioneers there is little that needs to be said. Most of their work was completed by the turn of the century; much later, in 1923, on the occasion of their own version of the typewriter's half-centenary, the Herkimer County Historical Society prophesied for the following fifty years: 'The future of the typewriter will be wonderful, more wonderful than anything we have yet known.'[21] If this prediction is to materialize, someone had better spring something on us quickly, in the short while left, for time is running out and, so far, we can only sum up this half-century in words which Dallari (and others) used many years before: *Nihil sub sole novum.*

Chapter 9

Technical Classification of Early Machines

Of the many alternatives, the most efficient method of classifying old typewriters is according to the means by which the impression is made on the paper. This criterion was the one used by Tilghman Richards[40] and applied to the London Science Museum collection. Other methods based on carriage location or design are confusing and unsatisfactory because of the numbers of exceptions to the rules. Even the present classification is at times arbitrary, where a machine fits equally into more than one category.

However, if one proceeds strictly according to the standard set, namely that the impression is the determining factor, then at times the 'feel' of a machine is all there is to tip the scales in borderline cases. To assign the Lambert to the radial plunger group[40] is less convincing than to consider it an index machine. Furthermore, there is always the danger that an instrument may be accidentally misclassified because it looks, superficially, as if it belongs in the wrong place. The same source inserts the Mignon in the index class, for instance, when in fact it is quite clearly a type-sleeve design.

Several important modifications have been made to previous attempts at classification in order to accommodate designs which belong on their own. For instance, there is a small group of plunger machines which cannot justifiably be lumped together in other groups simply because there are only a few of them. A prime example of this is the Edison Mimeograph, which might easily have slipped in among the type-wheel or index machines, but which belongs, with the Hughes and others, in a class of its own, for there is an essential difference between its action and that of, say, the Simplex.

169. This early Williams (serial no. 1471) illustrates the 'grasshopper' movement which made it famous. One of the type-bars can be seen midway in its leap. Note the baskets on both sides of the platen: a blank sheet of paper is rolled up into the one in front and passes to the one behind as typing progresses.   (AC)

191

Furthermore, there is a subdivision of the type-bar movement which has come to be called the 'grasshopper'. Following the reasoning outlined previously, this has been expanded to include other designs which share its vital hopping quality, for it is this which clearly gave it its name. The term is usually applied only to the Williams, in which the type-bars hop from a pad to the printing point and then hop back, as, essentially, they do on the Yosts and Maskelyne, of course, although this is generally overlooked.

1 TYPE-BAR CLASS

This is the most important, not only because of the large number of instruments which belong in it, but also because it developed into the conventional design which monopolized the industry for half a century. In fact, it was only on the re-introduction of the 'golf-ball' that the type-bar's stranglehold was broken.

To this class belong all machines which utilize pivoted bars (occasionally non-pivoted, as on the Improved Crandall) to carry the type to the paper. The type is fitted or moulded to the end of the bars, sometimes singly as in early upper-case

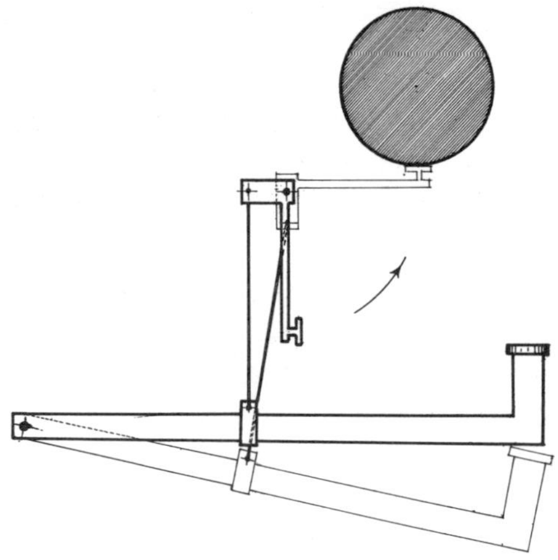

170. Schematic representation of the up-stroke principle.

or full keyboard machines, but usually in pairs as with the conventional shift machines, often in triples with three-row keyboards, and occasionally in multiples.

This type-bar movement offered the inventor considerable flexibility in choosing the position of the printing-point. On some of the earliest machines (Sholes and Glidden etc.) this point was underneath the platen, which had to be lifted or swung upwards for the written word to be seen. These were the non-visible or 'blind' machines of the up-stroke group, so called because the type-bars were suspended downwards and swung up to the printing-point. There are no 'visible' machines in this group, which also includes several models with grasshopper actions, such as the early Yosts.

The down-stroke design, on the other hand, was primarily devised to permit

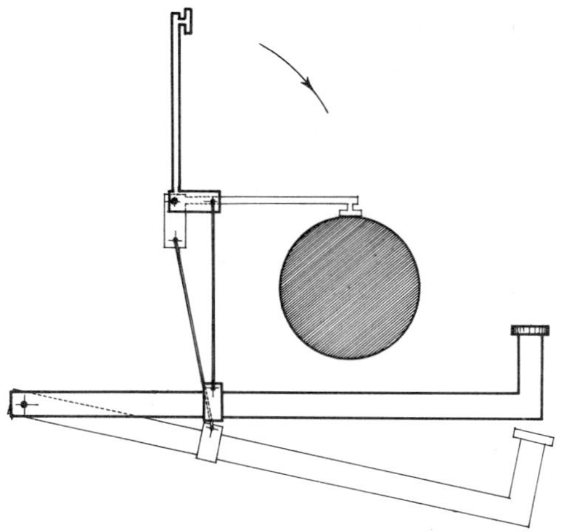

171. The down-stroke action. Type-bars were variously located behind the platen, as in the illustration . . .

the printed word to be visible. Here the type-bars stand vertically and are brought downwards on to the printing point. There are obvious advantages to this configuration, although many early manufacturers exaggerated the virtues with wild claims of visibility of typing: on some machines (Bar-Locks, early Salters, etc.) the operator could read the written word with ease only if he placed the machine on the floor and typed with his toes. To complicate the problem still further, some of these designs permitted the reading of little more than the letter just typed,

172. . . . or in front of the platen, as on the Rofa.   (AC)

the impressions being almost immediately covered by guide-plates and the like, to say nothing of the rest of the page which might easily have disappeared irremediably.

Of the down-stroke designs, the first to appear were those on which the type-bars were positioned in front of the carriage, between the platen and the keyboard. Later, they were placed to the rear behind the carriage, as on the Fitch. From the

193

point of view of visibility, this was a far better arrangement. On the grasshopper machines, the type-bars were either in front (Maskelyne) or back and front (Williams), but they rested horizontally rather than vertically. Despite this, they belong with the down-stroke instruments because the bars lift in the course of the 'hop' and strike downwards before reversing the procedure and returning to their places of rest.

An alternative configuration is the lateral down-stroke, like the Oliver, where the type is carried by inverted U-shaped bars banked on both sides of the printing point. Then there are the odd circular down-stroke machines on which the type-bars form a full circle, as on the Daw and Tait, Crary, etc., and finally, those made for typing in bound books which were straddled by such machines as the Fisher.

The problem of just what to do with the paper was a recurrent nightmare which manufacturers of many down-stroke designs merely passed on to their clients.

173. The down-stroke Fitch, with type-bars behind the platen. The blank page can be seen rolled up in the front basket. (WC)

Those with type-bars in the rear, as well as the grasshoppers, are the worst offenders from this point of view. After all, if a machine has the bars at the back, where

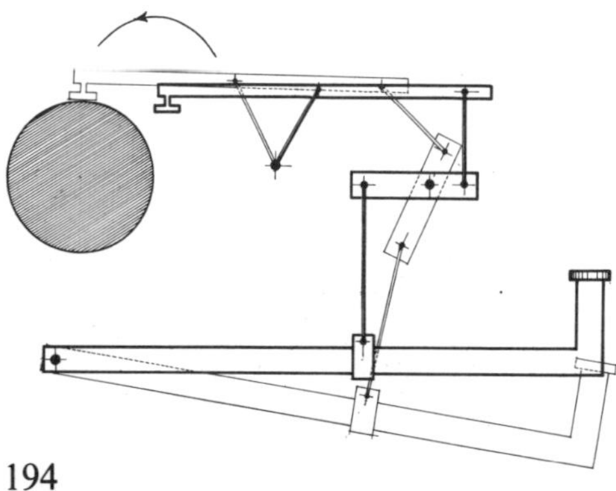

174. The grasshopper action.

is the sheet of paper to exit if not forwards? And if it comes forward, and continues to come, will it not inevitably foul the keyboard? And what about the grasshopper where type-bars are both in front of and behind the platen? Here the paper can travel neither forwards nor backwards. The stock solution was to curl the paper up in open cylindrical frames, although this was hardly conducive to visibility. The blank page was rolled and inserted into one frame and it rolled itself up into the other as typing progressed.

Interestingly enough, the idea was not conceived by down-stroke inventors but was first used commercially on the Hammond, with the difference, however, that only the unused portion of the page was left rolled up in the basket. The rest was exposed and fully within sight. The Hammond's hammer made this design necessary: more about this later.

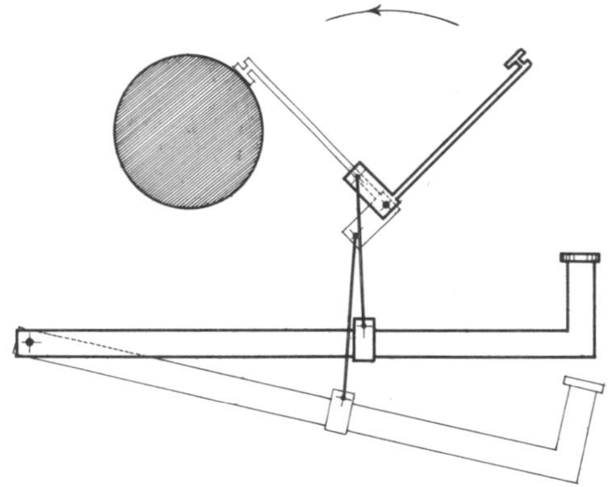

175. The oblique front-stroke movement.

The final group of type-bar instruments is the front-stroke, a design to which considerable attention has already been given. This is the Horton style of machine. The advantages are obvious, insofar as the writing is immediately visible and remains visible for the rest of the page. In fact, so obvious are the advantages, and so accustomed have we become to this design, that it seems hard to understand

176. The first oblique front-stroke machine, the Horton. It was rare enough in its own time and is now all but extinct—Don Sutherland photographed what is probably the only extant specimen.

195

that companies should have continued to manufacture other types after the introduction of the front-stroke principle. But this is retrospective wisdom, for as late as 1909 'whether the front-stroke will ever become a universal favourite is hard to prophesy'.[27]

With front-stroke machines, type-bars may be either horizontally or obliquely (45°) positioned in front of the platen. The only grasshopper in this group is the Visible Yost. Here the type-bars rested against the pads beneath the platen and were brought up to the printing-point by the familiar hopping as against swinging action.

2 INDEX CLASS

These were characterized by their relative simplicity, few moving parts, and modest price. Today we find it hard to believe that people took them seriously, but they did. The instruments in this class all shared the essential feature of the 'index', meaning that the type was moulded or fixed to a single plate which was moved till the desired letter was brought to the printing-point. The shape of this plate determines the group classification. Hence, the Velograph is a circular

177. The circular index Simplex. This 'Souvenir' model was issued in 1903 on the occasion of the centenary of the Louisiana Purchase and is adorned with pictures of Jefferson and Napoleon. (AC)

index machine, the World employs a sector of a circle, the Hall a square index, the Morris a rectangular, the Lambert a spherical and the American Visible a linear index. Most, but not all, of these machines have rubber indices for the flexibility that this material offers: obviously if a flat plate is brought over the printing-point but only one small section of that plate (the letter being printed) is desired, then the plate must be flexible enough for just that portion to be lowered below the level of the others. However, there are certain machines in this class, such as the Odell, which have metal type and which overcome the difficulty by using a linear index traversing a cylindrical platen of small section at right angles:

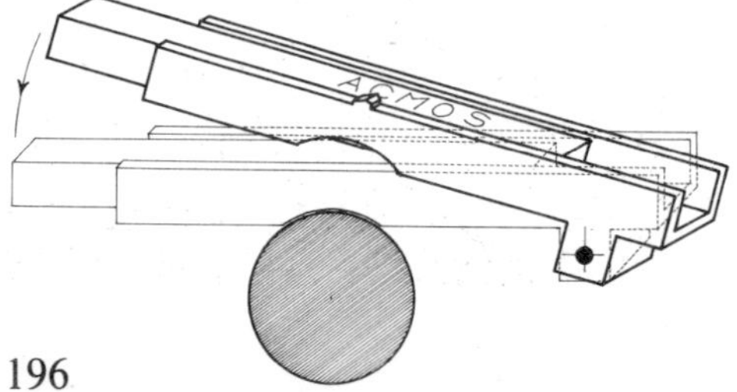

178. The linear index principle.

196

since the index is tangental to the platen, there is only one point of contact at any time.

3 TYPE-WHEEL CLASS

With keyboard machines using a type-wheel, problems centred around the best means of turning the wheel to the desired character; with manually operated

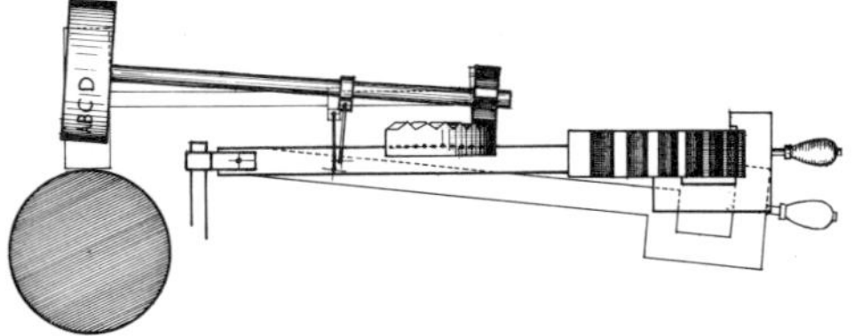

179. The type-wheel, with indicator for letter selection.

instruments, the indicator was simply geared to it. The type-wheel itself was usually small—for optimum performance, the smaller the better, not only because of inertia but also because of the shorter circumferential distance it was required to travel. Turning the wheel had to be done as efficiently and directly as possible, otherwise the operator's speed might exceed that of the machine. This problem was analogous to that faced by inventors of type-bar instruments, the speed of operation of which was limited by clashing of bars at the printing-point. In the case of type-wheels, proof of the efficient solution to the problem is the justifiable success enjoyed by the Blickensderfer and the Hammond, both of which belong

181. The prototype Cantelo had a three-row keyboard and type-sleeve . . .　　　　(AC)

180. The Edelmann. Note the toothed disk below the type-wheel, used to ensure accurate alignment.　　　　(AC)

182. . . . while the diminutive Junior combined the keyboard with a shy type-wheel hidden under the name plate.　　　　(AC)

to this class. The first is in the wheel group and the latter is a sector machine; an example of the type-sleeve design is the Crandall. On these keyboard instruments, the type-wheel is made to return to a point halfway along its course after a key has been depressed, thereby keeping travel to a minimum. In the case of those instruments using a letter index in place of a keyboard (Mignon, Edelmann, etc.), this operation is dispensed with, since the wheel is geared directly to the indicator, and the operator's speed cannot outstrip that of the machine.

A number of different systems was employed with type-wheels, thanks to the inherent versatility of the design. In the Hammond, Chicago (and others) a hammer struck the paper against the type; depression of a key first selected the corresponding character and then tripped the hammer. There were also the early typewriters and printing telegraphs in which the type-wheel rotated continuously, impelled electrically or by clockwork. Alternatively, the hammer might be dispensed with and the carriage made to rock forward against the stationary type-wheel. On most

183. The Keystone used a swinging sector with a hammer striking from the rear.   (LC)

designs, however, the wheel was first spun mechanically till the selected character was above the printing point and then brought down on to the paper, as on the Blick.

4 PLUNGER CLASS

This classification has traditionally been ignored as a separate and complete entity. With the exception of the Hansen Writing Ball and others of this immediate design (Foucauld, Kohl, Schade *et al.*), the instruments which operate on a plunger principle have usually been included among the index machines, when they have been classified at all. Others such as the Lambert, included[40] in the plunger class, are clearly there under false pretences.

Briefly, what distinguishes these machines is that the characters are at the ends of straight lengths of steel, each one sliding in its own hole and radiating out from the printing-point, or else carried to it by some suitable means. When struck, plungers are hit against the paper and are returned to their initial position either by a spring (Hughes) or, rarely, by gravity, sometimes assisted by a further blow in the opposite direction, as on the Edison Mimeograph. Several machines of this kind were discussed in Chapter Three: these primitive examples sometimes suffered from deformation of plungers and type-face as a result of blows which the materials could not withstand. It is immediately apparent, however, that the plunger action is intrinsically different from that of the index machines and must be classified separately.

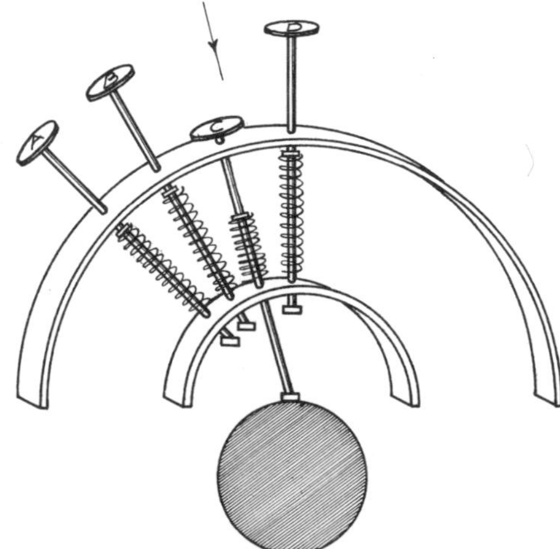

184. The radial plunger principle.

The Writing Ball-type instrument constitutes the first group of plunger machines. This is the radial plunger principle, and it is applied to two sub-groups: spherical and circumferential. In the former, the characters are at the lower end of rods which stick out porcupine-fashion all over a semi-sphere, the keys being attached to the outer extremities. A push on these rods brings them down to the paper and a spring returns them to their places. It is a simple, direct and effective design which economizes greatly on moving parts.

In the circumferential sub-group, the plungers stick outwards like extensions to the spokes of a wheel—radially, as above, but on the same plane. The wheel is spun to select the character and the plunger is struck from the inside outwards. Devincenzi conceived such an arrangement, among others, and the Cookson uses it.

These machines are not to be confused with those in the next group, in which plungers are disposed in a concentric ring around the periphery of the wheel, but in such a manner that each plunger is parallel to its axis, not radiating outwards from its centre like one of a number of spokes. Fig. 185 illustrates the difference. Here, the plungers slide vertically in holes around the perimeter of the wheel which is first spun till the desired letter is opposite the printing-point, whereupon

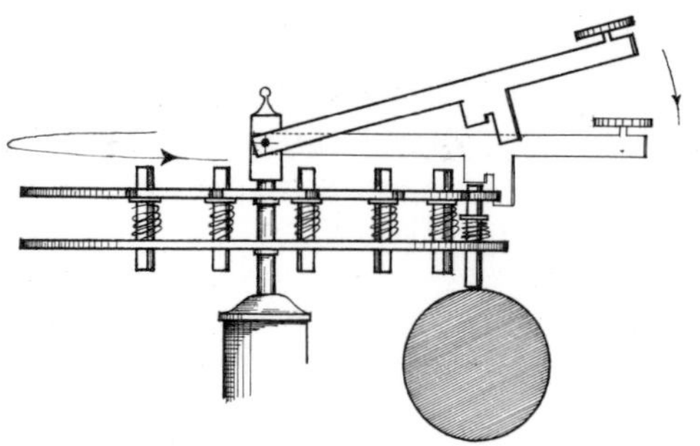

185. The peripheral plunger action.

186. The linear plunger Merritt, with index and indicator. The plungers struck up from beneath the platen, which had to be raised to reveal the typing. (LC)

some form of hammer brings type and paper together. Numerous precedents of this design date back to early days and the first of these instruments to have been actually manufactured was the Hughes in the 1850s. With one exception the design was limited to plungers striking downwards against the paper; the exception is the Edison Mimeograph in which the plungers are struck upwards.

A small third group has the plungers in a straight line. The Merritt is an example.

5 THRUST CLASS

Thrust-action machines constitute a select group which merits classification on its own by virtue of a special quality permitting 'silent' typing: the characters are not *struck* against the paper, but are *pressed* against it. The Rapid was the first machine employing this principle, but the best-known was the Wellington, in its various forms. A similar machine was the Kanzler, which differed from the others

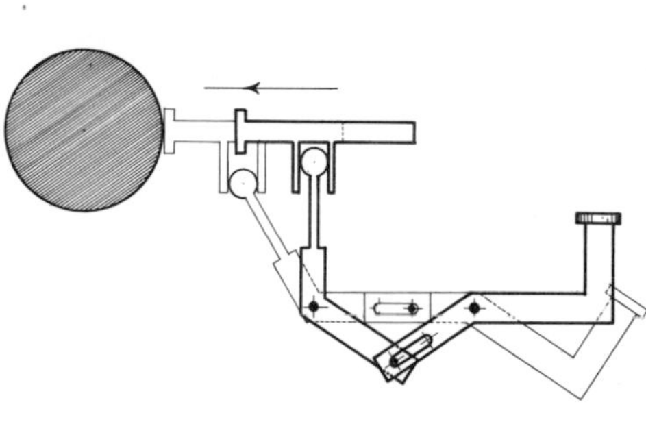

187. The thrust action.

188. The thrust-action Kanzler, a massive Germanic machine with eight characters on each type-bar, one of which can be seen approaching the platen. (AC)

200

in having eight characters to each type-bar: depression of a key not only thrust the bar forward but also located it vertically at the level corresponding to the selected letter. A simple lever action was used on the early examples of thrust instruments; later 'noiseless' machines incorporated an overthrow weight: upon depression of a key, the fully-extended linkage brought the type-face to within 0·15 inch of the paper, whereupon the momentum accumulated by the weight thrust the type into pressure contact with the paper.

The full classification is as follows:

1 Type-Bar Class
    (a) Up-stroke group
        (i) swing action
        (ii) grasshopper
    (b) Down-stroke group
        (i) from the front (Anterior)
        (ii) from the rear (Posterior)
        (iii) grasshopper
        (iv) from the sides (Lateral)
        (v) circular
        (vi) book typewriters
    (c) Front-stroke group
        (i) horizontal
        (ii) oblique
        (iii) grasshopper

2 Index Class
    (a) Circular group
    (b) Sector group
    (c) Square group
    (d) Rectangular group
    (e) Linear group
    (f) Spherical group

3 Type-Wheel Class
    (a) Wheel group
    (b) Sleeve group
    (c) Sector group

4 Plunger Class
    (a) Radial group
        (i) spherical
        (ii) circumferential
    (b) Peripheral group
    (c) Linear group

5 Thrust Class

Chapter 10

Some Further Technical Considerations

Other aspects of the technical development of writing machines have been mentioned in previous chapters and some of these are important enough to warrant separate consideration.

Inking long proved to be a serious obstacle to efficient typewriting and was as taxing on early pioneers as were the mechanics of their inventions. The very versatility of methods proposed, tried and abandoned is eloquent testimony: the range is all the way from machines which merely embossed by making holes in the paper, as did Foucauld's and Mitterhofer's and later machines for the blind, to those elaborate schemes whereby characters were printed through greased paper and then dusted over, as in Gonod's case. But most early efforts revolved around the use of different types of carbon paper, patented by an Englishman called Ralph Wedgwood in 1806. Many formulae were evolved for getting the lampblack or graphite to adhere to the paper, with varying degrees of success. Despite this, carbon paper proved the most popular solution to the inventor's problem; alternatively, liquid ink was used with different types of applicators.

In later years, as the machines themselves became more sophisticated, the first attempts at home production of typewriter ribbons are recorded. The invention is invariably attributed to the Italian Giuseppe Ravizza in 1855 but Alexander Bain's patents from the early 1840s establish priority (see page 72). Many materials were used in those days—fats and dyes and powders and inks and so on—and the results were not always satisfactory, but it became evident to men of vision that the ribbon was the best method yet devised. There is a graphic account[43] of what had to be done in 1870, before typewriter supplies were manufactured: '. . . it became necessary to visit the nearest dry goods establishment and select a bolt of silk or satin ribbon . . . and, having purchased it, we would buy a pint bottle of black ink and pour it into a wash bowl, and after unrolling the bolt of ribbon we would immerse it in the ink and allow it to remain until it was thoroughly saturated, and then towards evening before going home we would take it out of the ink and string it back and forth over the chairs and other furniture, and leave it to dry overnight. It was anything but a pleasant job . . . but in those days of rough bare floors, box wood stoves, sawdust cuspidors, Windsor chairs and smoke-blackened walls such operations could be carried on . . .'

Many improvements were later introduced to increase the longevity of the ribbon and to make the impression more effective. A number of machines used extra-wide ribbons, up to almost two inches in width, and the ribbon was made to zigzag so that the whole width could be used and not just a narrow strip in the middle. Another big step forward was the introduction of automatic ribbon reverse; prior to this, the operation had to be performed manually, either by pressing a knob on one side of the machine or else by loosening a clamp screw over the full spool and tightening it on the empty one. These operations were performed when the operator saw the ribbon had reached the end of its run; a careless typist, or one too deeply involved in what she was doing, had to pay the consequences for negligence. And on non-visible machines, many is the case of whole paragraphs typed without a ribbon when this ran off its empty spool or else jammed at the very end and had a hole drilled into it by the type. The invention of the automatic ribbon reverse came later and is a matter of some dispute; Remington claims to have been the first to introduce it in 1896, but independent sources[27, 42 etc.] concede priority to the Bar-Lock, invented by Charles Spiro.

189. The tiny Taurus, the shape and size of a pocket watch. It used two inking rollers, one on either side of the printing point.   (CSM)

But the ribbon as a means of impression was not universally accepted until late in the development of the machine, and it had two serious rivals from the earliest days: the inking roller and the inking pad. The first of these was used on such best-sellers as the Blickensderfer, the latter on Yost and others.

The principle of the roller was that, as the type moved, it passed over a small absorbent cylinder saturated with ink (sometimes two cylinders, as on the Junior,

190. The Sun No. 2 used the rare combination of type-bars and an inking roller and then—what is even rarer—added a larger roller as a reservoir against which the smaller one rubbed to keep itself inked.   (AC)

Taurus, etc.) so that it was moist by the time it touched the paper. The system was most commonly used with type-wheel machines, the theory of a large wheel turning a smaller one by friction being a concept which obviously lent itself well for the purpose. It was also used with type-bars, however, and an ingenious system was offered on early models of the Sun: the type-bars brushed the inking roller on their trajectory, at the same time thrusting the roller against a larger one which acted as a reservoir, keeping the smaller one moist. This in part overcame one of

191. Yosts never abandoned their preference for the inking pad, despite radical model changes. In the illustration, a type-bar can be seen on its hop from the pad to the platen.   (AC)

the principal defects of the roller method, namely its limited capacity; it required re-inking with annoying frequency when compared with ribbon machines. Irregularity of impression was another defect of the system.

Similar in concept was the inking pad. Here the type was left at rest against a moist pad from which it was lifted upon depression of a key. The system was also used with index machines such as the Velograph, with the pads attached to spring metal pressing them against the characters. On type-bar machines, the pad retained ink longer than the roller, the surface being larger, but by and large it suffered from the same defects. However, especially in the early days, rollers and pads offered an important advantage in that there was no ribbon to obscure the printing; other machines were obliged to resolve this difficulty first by the use of a manual ribbon-lifter and later by an automatic method very like that on conventional modern machines.

Both rollers and pads had their supporters and continued in use until as late as the 1930s, but they never truly challenged the general popularity and widespread use of the ribbon. Much the same may be said of the many keyboards which were offered to the public with varying degrees of success, none of which, however, managed to supplant the inefficient and unsatisfactory four-row standard

QWERTY arrangement as we know it today. In different chapters, mention has been made of keyboards and their development and it has been noted that the conventional keyboard and letter-order with which we are lumbered today is an unfortunate heritage left by Sholes and associates who developed it empirically as a means of overcoming type-bar clash in their early machines. Initially, their letter order was alphabetical and unfortunate remnants of this are retained (F, G, H, J, K, L, etc.). However, in order to prevent the sluggish type-bars from jamming at the printing-point when the speed of the operator outstripped that of the machine, the immediate solution was to place the most frequently used type-bars as far away from each other as possible in the type-basket (*not* on the keyboard). This failed to resolve the problem, but it helped, and the resulting letter order was retained almost *in toto* by Remington when they began manufacturing the machines.

Not everyone agrees that type-bar clash was responsible for the present keyboard. Some attribute it to the manufacturer's desire to make the salesman's job easier for him: since he had to demonstrate the machine to sceptical clients,

192. The Sholes and Glidden printed only capital letters and was never marketed in any other style. However, an undecorated experimental machine with upper and lower case used a rod on the carriage for change of case in days before the shift key had evolved.   (SIW)

placing the letters of the word TYPEWRITER on the same line made them easier for him to find, with corresponding increases in his typing speed . . . and volume of sales.

The Sholes and Glidden as well as the Improved Model One were upper-case instruments printing in capital letters only, as were several other contemporary machines such as the first Caligraph. The need for both cases, however, was suspected early in the game. An experimental Sholes and Glidden, equipped with upper as well as lower case, had a primitive shift mechanism operating by means of a bar on the carriage which was drawn towards or pushed away from the operator for the change of case. The movement first released a shift lock, allowing the carriage to traverse to the alternative position, whereupon it was re-locked. This is one of several prototypes to have survived; later, the Improved Type Writer

SIW

Model Two offered both cases commercially, although at the time the factory was convinced it would sell ten upper-case machines for each one with both. This, of course, was not so, and every manufacturer who at first offered only capitals was soon obliged to offer both cases in order to stay in business.

There were two basic approaches to the problem of combining upper and lower case in a single machine. The first, and the one which was eventually standardized as the conventional method, was the shift principle, invented by Byron A. Brooks in 1875. Much as on modern machines, each key served for both cases, which were selected by a separate shift key. The obvious advantages of this system were so apparent that a double shift was soon offered; combined with a more compact three-row keyboard it had two shift positions: the 'rest' position was lower case; the first position selected upper case; and an additional position selected figures and characters. On some machines, these positions were banked in the order just mentioned; in others, the 'rest' position was between the two, so that both upper case and figures were no more than one position away from lower case. Numerous mechanical alternatives were devised to make this possible. On some machines, the shift key moved the platen to its corresponding position; on others, the type-bars or -wheels were shifted. These alternatives are retained to this day in different makes.

The second approach to providing both cases on early machines was to offer what were called 'full' and 'double' keyboards. Critics of the shift system claimed that it suffered from unsatisfactory alignment, flimsy construction, unnecessary complications, additional operator fatigue, etc. Their solution was to offer a separate key for every character—large and often clumsy machines with 70 to 100 keys were the result. The first of these was the Caligraph: it had a 'full' keyboard with the lower case letters in the middle surrounded by upper case, figures and characters in what appears to be random order. Its disadvantages were soon apparent, for the operator was obliged to learn not only the lower-case letter

193. The double keyboard Hartford dispensed with the shift and assigned a separate key to each upper and lower case character. It was a blind machine, and its platen was spring-loaded to facilitate raising. (AC)

order, but also the individual position of each and every upper-case letter, and so on. A more efficient and obvious alternative was soon offered by Yost, Bar-

Lock, Hartford and others which had a 'double' keyboard consisting of identical rows of keys for both cases.

Letter order has already been mentioned earlier. The standard QWERTYUIOP order imposed itself on the others gradually and they succumbed to it one at a time. There were many machines however, which persisted with their own alternative, offering the 'Universal' as an option. Its chief rival was the 'Ideal' DHIATENSOR configuration: over 70 per cent of all English words[27] are composed of these ten letters, which were positioned on the same row, thereby making this an eminently more efficient system than the other. Such great machines as the Hammond and Blick used it but it failed all the same, as did other alternatives with less valid qualifications (some of which were listed in Chapter One), to say nothing of staggered, circular and porcupine keyboards which presented problems all their own. A full discussion of letter order is long and complicated, however; different arrangements were evolved in different countries and for different markets and one is at times uncertain as to the origins of contradictory letter orders on two machines of the same make. They were obviously destined for different purposes or countries, but it is often difficult to be more specific.

Three- and four-row keyboards dominated shift machines from the first, the former compensating for shortage of keys by providing double shift, as has already been explained. There were many designs, however, which used two-row keyboards, the rows, more often than not, being curved rather than straight. There were also curved three- and four-row keyboards. It is anatomically unnatural for the hands to be straight in front and parallel to each other, as anyone who learnt the piano as a child will attest. The more natural position is for them to angle inwards, and, utilizing the elbow as the axis, to operate obliquely rather than in a straight line. The curved keyboard was therefore a sensible design, although it would be thoroughly incorrect to believe that this consideration was foremost in the minds of manufacturers of such machines. They were certainly more likely to have been impressed by the relative mechanical simplicity of an instrument on which all keys and key-levers radiated outwards from a central area; the fact that this was also easier on the hands was coincidental. In any case, a considerable number of such machines was offered: the Hammond was the first, with a two-row semi-circular keyboard; Salter, Imperial and others used a three-row semi-circular one. On the other hand, the Crandall's keyboard was a gentle arc with two rows, and the three-row Williams was similarly curved, as was the four-row Kanzler. This design is a comfortable and pleasing one to use and it is unfortunate that typewriters have become so stereotyped and unimaginative that no true choice is any longer offered to potential clients.

The problem of standardization is a serious one indeed. A letter order common to all makes is clearly advantageous, for instance, and few people mourn the passing of haphazard keyboards which were the product of a manufacturer's whim. But why must the standardization of an unsatisfactory system continue to be perpetuated? In an era in which we are bidding farewell to feet and inches, pounds and

194. The beautifully built Typo, French version of the British Imperial Model B, used a comfortable three-row keyboard—but with a letter order that was almost as bad as QWERTY!
(AC)

ounces, shillings and pence, and a host of minor nightmares such as acres, rods, roods, perches, bushels, pennyweights, drams, furlongs, chains, guineas, crowns, stones, hundredweights, pints, quarts, etc. (to say nothing of long and short tons, gallons and so on), surely the time has come to offer serious consideration to any one of a number of highly advanced keyboard designs which have been offered over the years but never taken seriously enough. These were scientifically evolved for most major western languages, basing themselves on truly valid considerations: on letter frequency ('e' being the most frequently used letter in the English language, followed by 't', 'a', and so on), on letter combinations ('th', 'he', etc. the most frequent digraphs) and on human physiology. The result has been several keyboards in which the most commonly used letters are centrally positioned, operated by the strongest fingers (index and middle), in such a way that each hand has roughly 50 per cent of the work load, favouring the right wherever unavoidable since the majority of people are right-handed, and the letters in the most frequent combinations are operated by alternative hands. 'E' and 'r', for instance, should not be operated by the same hand, much less 'e' and 'd' by the same finger, while assigning 'a' to the little finger of the left hand is close to criminal. An added sophistication is a design in which each side of the keyboard is slightly curved, in much the way that the finger-tips of that hand describe a gentle arc. But it has all been to no avail; standardization on a sensible keyboard would mean alteration of existing machines, re-education of typists, and so on; and merely being sensible has never been sufficient incentive for the world to make much effort. Maybe, one day, when we all drive on the same side of the road, when we all use the same metric measures, when we deal in the same currency and speak the same language . . .

Maybe, when that happens, someone will have a brainwave and invent a typewriter with keys like those on a piano. Controversy surrounding the relative merits of such an arrangement can be traced back to the earliest days. The preference of inventors for this system was said[27] to be a 'fatal fascination. By the adhesion to this defect, the production of the perfect machine was ultimately

208

delayed.' However, the fatal fascination was pandemic, infecting virtually every major pioneer of the machine, including such men as Ravizza and Sholes himself. Wheatstone, for instance, used it on his first model, dropped it on his second, but returned to it on his third. Compared to modern machines, early ones were undeniably slow and clumsy, inefficient, poorly constructed, ill-conceived—but not because the keyboard modelled itself on the musical instrument. Consider the advantages of such a device, with a piano-style keyboard, an electric one, perhaps, with a sensible letter order and a foot-operated switch for spacing, carriage return and change of case. Suddenly we will realize that men are born with ten fingers and not eight, and two feet as well, so why not use them all? The boss will love it, because his secretary will type more quickly and efficiently. The trade unions will support it, because their members will suffer less fatigue. For if we merely use all ten fingers, then we can cover 'ETAONIRSHD', the ten most frequent letters,[24] without moving our hands at all, and with those letters we can write about three-quarters of the words in our language. A little switch at our feet, as on an electric sewing machine, is for spacing (the space key is the most frequently used of all), for change of case to keep our piano keyboard short, and for carriage return. After all, if a pianist can comfortably handle over 1,500 to 2,000 strokes a minute (the equivalent of 300–400 w.p.m.) on a much less compact keyboard than the one described, and without trying to break world speed records, either . . .

Finally, there are a number of minor aspects of typewriter development which will not be considered here, either because they have been dealt with in other chapters or else because they do not warrant separate mention. Folding typewriters are an example: for the sake of increased portability, several manufacturers brought out such models (Standard, Corona, Fox, Hammond, etc.). They are

195. A number of portable typewriters was designed to fold up—or down—making them easier to carry. On the Fox Portable, the carriage dropped back behind the machine to give it a flatter profile. (AC)

described individually in Chapter Twelve. A further feature of importance, namely the interchangeability of type, is discussed in Chapter Eleven under Bilingual Machines.

There is, however, one matter which has worried manufacturers almost from the start, and which has only recently reappeared in a modern machine. The debt which early inventors owed to the pianist has already been discussed at length; the other obvious influence upon the machine's development was the printer. The typewriter and the printing press are related instruments, the only real difference between them being that one prints a letter at a time while the other reproduces a complete page of material that has previously been composed. But there is another difference of which we are usually unaware, so accustomed have we become to the sight of a typewritten page which is really very ugly when compared to a printed one. A typewriter gives no consideration to the obvious fact that an 'l' requires a vastly different amount of space from 'm'. Every letter, figure and character is treated with complete equality, as behoves us, perhaps, in these modern times. But this equality is anathema to the printer to whom els and ems and ens are the very foundations upon which his entire empire rests. 'M' and 'w' resent overcrowding; they require more territory than most. On the other hand, 'f', 'i', 'j', 'l', and 't' are quite out of place in the wide open spaces of 'm' and 'w'. The printer, therefore, gives each letter exactly the right amount of room, and a number of early typewriters attempted to do the same, with varying degrees of success. The feature is called differential or variable or proportional spacing. It was quite an issue at one time, and passions ran high on the subject. Jenkins[59] dedicated an apparently disproportionate amount of time to it in his general survey, a fair indication of the degree of controversy surrounding it in the 1890s. Not only is it wrong to dismiss the difference between 'l', 'm', and 'n' simply by an arbitrary progressive increase in space, he said, but also the precise ratio of one letter to another is of major importance. He believed Maskelyne's

196. The Maskelyne boasted a grasshopper movement and full differential spacing with special provision for typing diphthongs. A later model came out with the most magnificent of all grasshopper movements, the type-bars performing a perfect somersault from pad to platen.

(CSM)

2:3:4 ratio 'does not give nearly such good working as Waverley's 1:2:3'. (He might be forgiven for this prejudice, if it be borne in mind that he was one of the inventors of the Waverley; he cannot be forgiven, however, for chosing to ignore that the marvellous Maskelyne used the '1' in its ratio for composing diphthongs!) He further associated the whole issue with 'terminal' spacing, which is the printing technique of so justifying the ends of lines that the right-hand margin is straight. On a typewritten page this, together with differential spacing, would undeniably present a strikingly clean aspect.

In their 'Informal History' handout, IBM are informal enough to state that *they* invented differential spacing; this of course is as false as their claim to the invention of the type-wheel, or Remington's claim that they invented the typewriter itself. Differential spacing was first applied by Progin in 1833; he allotted wider spaces for upper- than for lower-case letters. Bidet made non-mechanical provision for it in 1837. Thurber, eight years later, incorporated it in his abortive handwriting machine, and Jones did the same in his fine device of 1852. Harger used it in 1858, Mitterhofer in 1864, Hall in 1867, and Fontaine in 1869 applied it to all alphabets represented on his machine. As for the production instruments, Automatic,

197. The Automatic, a tiny up-stroke machine with type-bars an inch and a half long. This rare and desirable object incorporated differential spacing. (GC)

Crandall, Columbia, Daw and Tait, and Maskelyne all offered this feature in the 1880s and Waverley a few years later.

What *is* true, on the other hand, is that differential spacing was abandoned from around the turn of the century until IBM reintroduced it some forty years later. It was eventually deemed unnecessary as far as the final appearance of the work was concerned; this is quite untrue, of course. A more likely explanation is that it involved mechanical complications which manufacturers found difficult to handle and for which clients were not prepared to pay extra. Furthermore, correction of a typographical error on a differential spacing machine can be a messy affair: if the typist, intending to write 'mm', writes 'nn' instead, or even worse, ',,', then she will simply not have enough space in which to insert the correction. But, for all that, differential spacing has always kept its few supporters. They were considered cranks until the principle was readopted commercially whereupon they suddenly became men of vision.

Hopefully, the same will one day apply to the 'crack-pots' like Dvorak who have tried to reform keyboard design.

Chapter 11

Special Purpose Writing Machines

1 MUSICAL TYPEWRITERS

The construction of a successful typewriter for printing musical scores was perhaps the greatest challenge a writing-machine inventor could face. The piano, as we have seen, exerted a strong influence upon the development of the typewriter, yet successfully combining the functions of the two instruments proved an elusive goal. The earliest efforts were described in Chapter Two, dating back to Unger in 1745; his long tract solved all its mechanical problems (those at least which it acknowledged) by theorizing them away. And problems abounded—far more serious ones than those that beset the inventors of other forms of writing machines.

Basically, when we think of music, we think of a few notes from A to G, or Do to Ti, with a few semi-tones in between. But a machine to those specifications would be capable of typing music no more complicated than 'Twinkle, Twinkle, Little Star'. In fact, a musical typewriter must be able to handle not only the notes of the scale, and the semi-tones, but also a vast number of octaves, notes of different lengths from breves to hemi-demi-semiquavers, including dots and the like; different clefs, keys, sharps and flats and naturals and rests of different lengths; chords of any number of notes; slurs, staccato, pizzicato, trills, tremolo and so on, and only the most immediate and obvious have been mentioned.

So great were the obstacles, in fact, that relatively few efforts at constructing such a machine were made, and it was already well into this century before the first successes were scored. Initially, the instrument was conceived as an accessory to the piano, a fact which few will find surprising. Progin's suggestion (1833) that his Plume Ktypographique be used for typing music appears to be the first instance of an autonomous musical typewriter, but with a total of merely sixty-six or sixty-seven characters, he might have been able to progress from 'Twinkle, Twinkle' to 'Three Blind Mice', but not much further. Berry (1836), Dujardin (1838), and other contemporaries may be similarly dismissed.

The first serious, commercially distributed machine was the Tachigrafo Musicale introduced by Angelo Tessaro in 1887. It had the European market to itself till about the turn of the century; this, however, meant little, for the instrument was hardly likely to become a best-seller. But it won plenty of prizes, which means quite a lot when there is little else to hang on to. It was sold in Italy by Ricordi, a firm which is still in business.

A series of abortive efforts followed. As late as 1894 Cantelo took out a British patent on a device to be attached to a piano keyboard. In 1903, I. F. and L. C. Badeau, two brothers from New York, were granted the first of several patents, and the following year proved even better for these vintage instruments: the Burnham was exhibited at the St Louis exhibition, and died there, and a Hammond-inspired machine called Wirt's Musical Typewriter printed 280 characters in all, and was reputed to have got a foothold, but it still failed to make the mark. And then there was the Kromarograph, yet another retrogressive instrument attached to a keyboard and printing in dots and dashes, this being exactly the machine which Unger and others had proposed almost two centuries earlier. And as if it had still not been done to death by then, yet another attempt was made to reincarnate it in the 1929 device of Roger Visbecq, with virtually the sole exception that electricity was now used to power it.

But long before this, the problem had to all intents and purposes been solved by the 1910 Noco-Blick. This was essentially a Blickensderfer typewriter with type-wheel and keys, fitted into a somewhat larger body, with a sliding carriage controlled by an independent unit on the left. The right hand depressed the keys, selecting type of note or sign, etc., while the left hand moved a pointer over a scale on the separate unit to the position that the corresponding note should occupy. This unit was attached to the carriage, which rode towards or away from the operator, thereby locating the printing-point on precisely the musical line or space required. Treble and bass could be written in this way over seven and a half octaves in all. The idea was most ingenious and worked well and the demise of the machine was brought about only by the termination of Blickensderfer production in 1917. Meanwhile, an improved model of the Noco-Blick had appeared; the operator could now draw the five music lines on the paper (instead of having to buy the paper already lined), and, by means of pedals, could brake the carriage to permit the printing of chords etc. And as if the faithful old Blickensderfer had still not given enough of itself, Noco-Blicks could be used as ordinary typewriters merely by changing the type-wheel and locking the carriage at a given level.

2 ELECTRIC TYPEWRITERS

It is almost impossible to leave the Blickensderfer alone, once one has one's teeth in it. For its 1902 electric model 'represents almost the last word in power driven machines of the type-wheel class'[27] or any other class, for that matter, when those words were written in 1909.

On the Electric Blick a small motor running on household current powered the type-wheel shaft, which was kept stationary by means of a clutch while the motor was running. Depression of a key released the clutch and the 'golf ball' spun to its desired position and then forward to make contact with the paper, returning immediately to its place of rest.

Sound familiar?

The procedure was the same, under power, as it was on the manual Blicks, with the added advantage that depression of the keys was no longer mechanically responsible for the impression. 'The lightest and most fairy-like touch will produce

198. 'The machine does it!' Blickensderfer announced the first commercially successful electric typewriter in 1901 and marketed it the following year. The illustration is from the instruction manual; as far as is known, none of the machines has survived. (LC)

just the same quality of impression as would a sledge-hammer,' eulogized an independent contemporary historian,[27] '. . . a slight and almost imperceptible "click" is heard and behold, the letter is printed . . . the writing is as even as though from a printing press . . . the operator indicates what he wants the machine to do, *and the machine does it!* (sic) . . . Then there are two keys on the right of the keyboard, marked "R" and "L" respectively. Touch the one marked "L", and the carriage returns automatically to the commencement of a new line, shifting the paper as it does so. Touch the "R" key, and the carriage travels slowly to the right, enabling any margin to be made. Work the two fingers alternatively on these keys, and the carriage sways to and fro like a thing possessed, but always under the most perfect control . . . If the machine under notice contained no further novelty than these two keys, the makers would be fully justified in placing it on the market, on its own merits, in competition with all others . . .'

Not everyone agreed. Charles Oden in his notoriously biased and inaccurate book[31] published in 1917, made the following memorable *faux pas*: 'The electric typewriter would eliminate to a great extent the human element, and for that reason it is not likely to become popular.' (At the time, Underwood, for whom Oden was writing, had not succeeded in perfecting an electric model.)

For all that, the electric Blick worked, and worked well; while it was certainly not the first electric typewriter, it was at least the first one to prove successful, printing telegraphs excluded, of course, for the application of electricity to this branch of the machine dates back to Morse, Bain, Wheatstone, Brett, House and others whose contributions were examined in Chapter Three.

Devincenzi in 1854 patented the first of the non-telegraphic 'electrics' but in the following half-century, until the appearance of the Blick, little was done to promote the use of 'electric fluid' in ordinary machines. There was of course, Edison's historic patent of 1872 which was described in Chapter Six, and Hansen's Writing Ball appeared with an electric carriage-release from 1867, a minimal advantage which so increased the cost of the machine that it was retained for only a few years before a fully mechanical carriage was again offered.

And others followed. In surgery, every so often, one hears of a successful operation from which the patient died; the next 'electric' was successful, but production folded after forty units had been built at fantastic cost. Dr Thaddeus Cahill took out patents from 1893 on, and the Cahill Writing Machine Company was founded in 1901. Their product was to all intents and purposes a Remington No. 2, both in appearance and in mechanical conception. The type-bars, however, were electromagnetically operated, as were the space bar and other moving parts. Many considered this non-visible 'electric' the perfect machine at last—but alas, many more didn't.

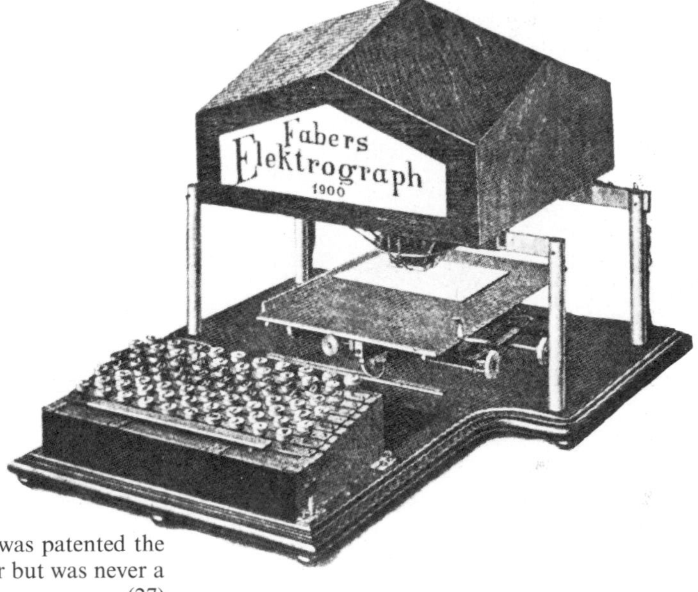

199. The Faber Elektrograph was patented the same year as the Blickensderfer but was never a practical proposition.   (27)

A half dozen or so additional attempts were made to break into this field between Cahill's and Blickensderfer's efforts. Some were AC-, others DC-powered; light touch and slight key depression (down to one millimetre) were their common features. Of these, the only one worth separate mention was one that looked like a miniature computerized parking-lot. The Faber Elektrograph, as it was called, was patented in Germany in 1900. Concealed under an architectural roof were plungers of Writing Ball type, beneath which paper on a flat platform rode front and back and right to left. A keyboard with a full complement of eighty keys controlled the contraption, and electricity supplied the power. The force and hence the depth of the impression was ingeniously controlled by the current: the more carbon copies you needed, the more volts you passed through the motor—an idea which, if taken a step further, might be successfully employed for embossing slabs of marble.

Chronologically, the Blick came next; it was not till 1921 (that is, nineteen years later), that the Mercedes Elektra appeared and satisfied the Germans[28] that they had, after all, produced the first *erfolgreiche* electric, once again disposing of the Blick as they did when they pronounced the practical musical typewriter to be a purely German invention, too.

3 PNEUMATIC TYPEWRITERS

Electricity provided the most practical source of power available to small machines. The other two sources were steam and compressed air, of which the former (to the best of the author's knowledge) was never used at all, and the latter only sparingly. This parallels the similar application of power to other appliances (phonographs, gramophones, etc.) at around the turn of the century.

The first attempt at using air appears to have been an ingenious device invented in 1891 by a Londoner called Wier, whose pneumatic thrust-action machine is

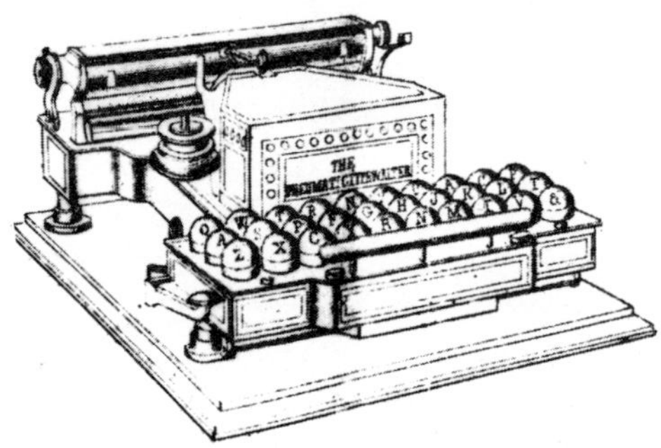

200. The thrust-action Pneumatic harnessed jets of air produced by the compression of hollow rubber balls on the keyboard. But it failed, we are told, because it raised blisters on the fingers of the operator.   (35)

reputed to have been capable of fine work. The keyboard consisted of three rows of hollow rubber balls attached by means of tubing to small type-pistons. Compression of the rubber ball forced sufficient air through the tubing to press type on to paper, but despite its smooth and silent operation the machine failed. Variations in atmospheric conditions were considered[27] at least partly responsible, though how this can have affected it is hard to imagine. A more practical approach[28] attributed the failure largely to the machine's unfortunate habit of raising blisters on the fingers of the operator.

This accusation could not be made against Soblik's machine, on which the fingers merely covered little holes in the keyboard. Invented in 1898 by a German engineer of that name, it utilized a foot-operated treadle pump for the supply of compressed air needed to keep its type-wheel constantly rotating. Later patents substituted motor-driven pumps for the foot operation. A company was founded in 1912 to exploit the patents, but few machines were made.

Several other abortive efforts at producing compressed air machines are recorded, but they are hardly worth detailed treatment. J. P. Moser invented one in 1900 which was to be called Germania: it sported a piano keyboard and pedal shift, each key compressing a rubber ball, on Wier's principles. But it was an American, the Revd H. J. Otto, who came up with the last word in pneumatic systems in 1907: why not install a compressed air pump in the cellar and connect all the typewriters in the building to it by means of concealed tubing?

4 MACHINES FOR THE BLIND

The extent to which the blind influenced the development of the typewriter has already been documented in previous chapters, and machines built specifically for them made impressive contributions. Turri's in 1808 was the first of which concrete evidence remains; Foucauld's and Hughes's achieved limited manufacture in the 1850s, and there were many others on both sides of the Atlantic, of which these are but the most prominent.

201. The dual purpose Diplographe was one of the many attempts at designing a machine that would simultaneously print and emboss, thereby permitting a blind typist to 'read' what he was writing.   (49)

As Foucauld well realized, inventors of typewriters for the blind faced the dilemma of having to decide whether their machines would reproduce print for the sighted or embossing for the blind. Foucauld himself attempted to resolve this problem by designing his dual-purpose instrument (see page 70), and others after him kept trying, but without appreciable success. One such device is the Diplographe, described[49] as 'unique of its kind' because of its dual-purpose feature but of course it was not unique at all and, in fact, appeared rather late on the scene (1877). It consisted of two type-wheels mounted on the same horizontal shaft, the one embossing braille cells on a flat sheet of paper on the left while the other on the right printed on another sheet, using roller inking.

Inventions for embossing braille cells began to proliferate as the nineteenth century drew to a close, and the first of what one might call the 'conventional' typewriters for the blind was a machine invented in 1891 by Frank H. Hall (no relation to Thomas Hall) of the Illinois Institute for the Blind. His Braille-Writer had six piano-type keys and a centrally-placed spacer and embossed a maximum

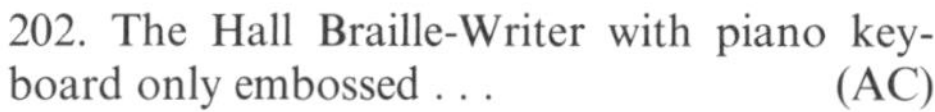

202. The Hall Braille-Writer with piano keyboard only embossed . . .   (AC)

203. . . . as did the later but not dissimilar Picht.
(AC)

braille cell of six dots: ⠿, being so designed that the three left-hand keys embossed
the left column, and the right keys the right column. Any number of keys could
be depressed simultaneously; the simplicity and conciseness of this mechanical
system are immediately apparent. Hall's was not the first machine to use the
concept of simultaneous depression of different keys for cell embossing, but it
was certainly the best, and its combination of six keys, with platen, was an important
step forward. Picht in Germany came out with a similar instrument, as did Stainsby
and Wayne in England, Perkins in the United States, Cereseto in Italy, Constançon
in Switzerland, etc. Various modifications and improvements were introduced,
including a series of shorthand machines using a similar principle and embossing
on rolls of paper. At this point the design had reached finality.

In addition to the braille writers listed above, a whole series of converted type-

204. The Gerda was designed for one-handed
operation and printed but did not emboss. The
'window' below the curved index permitted the
operator to feel the braille cell corresponding to
the character selected.   (AC)

writers was also introduced to allow the blind to type for the sighted. These were standard machines with braille keyboards: World, Kosmopolit, Edelmann, Kneist, Hammond, Mignon, Adler, to mention but a few. Picht also brought out a type-wheel instrument with a braille index and pointer.

The sudden increase in the numbers of blind following the First World War created a large market for these machines. Some, like the Gerda, were mass-produced and achieved wide distribution. Others again tackled the old problem of providing copy simultaneously for the blind and the sighted. But they were more sophisticated than the Diplographe: some were electrically powered, others were four-row front-stroke machines. Gadgets were manufactured which made it easier for the blind to operate them: attachments for insertion and alignment of paper, blocks to prevent overtyping at the end of lines, and so on. Nothing intrinsically new was offered, however.

5 SHORTHAND MACHINES

Nothing intrinsically new—this appraisal applies with repetitive frequency to virtually all branches of writing-machine history. One arrives at a point (usually alarmingly close to the beginning) at which everything seems to have been done; from then on, it is essentially the same old fare time after time, merely warmed over and served in plates of different sizes and colours.

Stenographic machines are no exception. They began to appear early in typewriter history, second only to machines for the blind. Here, however, the inspiration was provided solely by the inventor's preoccupation with speed.

Time and again, as we saw in Chapter Three, performance was measured in terms of the velocity of the machine as compared to that of handwriting, the latter being credited with the most widely divergent and purely arbitrary figures. Ordinary typewriters were never fast enough, nor could they hope to be, for even when they matched and surpassed the speed of longhand, what of the speed of speech? This was the ultimate standard by which the writing machine was judged, and the ordinary typewriter fell far short of the mark. It suffered from the intrinsic burden of having to supply copy that was not only universally legible, but aesthetically pleasing and orthographically perfect.

In its very essence, the stenographic machine suffered no such handicaps. By and large, no one really needed to understand it except the operator, for the copy was then transcribed on to a standard machine. Orthography could be ignored, vowels dropped, sounds printed with phonetic economy, abbreviations made according to the will of the operator, characters reduced to a minimum by eliminating duplication of letters and sounds like 'S' and 'Z', 'C' and 'K', and so on.

But for all this, it is still doubtful whether the shorthand machine would have achieved its objective had it not been for an essential principle which set it apart from the standard typewriter. This element was first introduced in 1827 by the Frenchman Gonod who, like almost all his contemporaries, drew his inspiration from the pianist. It was not merely the musician's facility in the manipulation of keys which impressed him, however: it was his ability to play more than one note at a time in the form of a chord. If one could devise a machine where whole syllables or even words were printed in one simultaneous depression of numerous keys, then operating speed would be dramatically increased. Combine the 'chord' principle with complete orthographic freedom and you have the stenographic typewriter.

Gonod's was the first, a primitive affair with its characteristic greased inking. Galli, Drais, Dujardin, Michela, Livermore, Bryois and others in the course of the following half-century or so developed the machine in one way or another. Some of them invented complete shorthand alphabets to go with their machines, thereby contributing yet another step to the development of stenography. An essential mechanical element was added with the introduction of the endless roll of paper tape to replace the conventional page, this system lending itself far better to 'chord' typing.

Stenographic machines took longer than typewriters to achieve success and widespread acceptance, despite the fact that all the essential elements had already been developed at such early dates. Meanwhile, some peculiar machines put in an appearance. There was Züppinger's and later Gentilli's Glossograph to reproduce the sounds of speech automatically by utilizing the different movements of the tongue, lips and mouth: part of the machine was therefore held in the mouth and the sounds recorded by lines drawn on paper tape by six pencils held in type levers; there was Schoch's more conventional patent, which used pens instead of type, and so on. But there were also some good ones. There was Maggi's Logomatografo patented in 1872 which grew into the Clavigrafo of a decade later: it was a nice machine which took its name from its piano keyboard (like Michela's, a separate one for each hand) and which wrote 'without the help of any of the great modern forces like steam and electricity', according to the blurb put out by the Società Clavigrafica (Maggi & Cia.) in 1881. Maggi, it will be recalled, borrowed a Cembalo Scrivano from Ravizza that same year and returned it in 1884, presumably having used it for study and testing (see page 128). Then there were the inventions of the prolific Italian Dario Mazzei, beginning with the Stenotiposillabica patented in 1879; his third machine (1883) is said to have achieved the highly creditable rate of 200 to 220 unabbreviated words per minute. There was the Anderson Shorthand Typewriter of 1889 which promised that 'a corps of typewriter copyists can be kept busy transcribing while a single Anderson operator is taking', although twenty years later it was laconically reported that 'the prophecy has not yet commenced to mature',[27] and so on, through numerous other efforts.

As late as 1894, however, Jenkins[59] dismissed stenographic machines which, he said, 'have never gone much beyond the experimental stages of their existence', and Rochefort-Luçay[35] echoed similar sentiments two years later. But the design had already reached finality and in the years immediately after they expressed

205. The Anderson Shorthand Typewriter failed to honour its many promises, perhaps, but at least it was a start in the right direction.
(MPM)

206. The record-breaking Grandjean Sténotype.
(AC)

these critical sentiments, several machines of worthy design and construction established themselves: the Hardy Stenotyper (1897), the Bivort Sténophile (1906), the Grandjean Sténotype (1910), Ireland's Stenotype (1911), National (1916), and a number of others. Many manufacturers of typewriters for the blind (Stainsby-Wayne, Picht, etc.) soon introduced stenographic models, for it quickly became obvious that here was a skill and a profession that the blind could master even better than the sighted.

A word about speed. In the early days of stenographic instruments, specially on the Continent, unsubstantiated claims of terminal velocities were relatively commonplace. Gonod modestly claimed that his machine could merely equal dictation—Galli's could exceed it sixty times, while Drais's was supposedly capable of a thousand characters a minute. More recently, the claims were only minimally less impressive. Michela's was 225 w.p.m., Mazzei's up to 220, Bivort's a colossal 320! On the other hand, by way of comparison, a French record of 180 was set in 1912 on a Grandjean, which was then excluded from competition because nothing could keep pace with it. The record increased to 220 by 1920, 250 by 1939, and 290 by 1947, record speeds remarkably similar to those of shorthand writing.

6 PRINTING TELEGRAPH

In introducing the subject of this book, an attempt was made to place the whole question of writing machines and their development in its proper historical perspective. The claims of early typewriter fanatics were rejected and it was made clear that our machine was not the most important development of the last century, nor anywhere near it. In fact, it comes quite modestly low on the list of important inventions. It was, however, intimately associated with another innovation, the telegraph, which comes preciously close to grabbing top honours, and the typewriter probably owes a greater debt to telegraphy than to any other single influence in its history.

The telegraph was extensively theorized from as early as the second half of the eighteenth century, and many fantastic schemes were proposed for transmitting messages between two distant points. Research was intensified in the early nineteenth century, and progressed hand in hand with developments in the many applications of electricity. In fact, man's fascination with the 'electric fluid' characterizes this period, much as his earlier fascination with mechanics led to the vogue for complicated automatons and so on. The wildest example is perhaps the discovery of electrolysis and its application to telegraphy by Sömmering in 1809: as the inventor would have it, the message was transmitted in the form of

207. Siemens and Halske printing telegraph transmitter, as it appeared around the turn of the century. From an old trade handbook.

electrical impulses which were to be used to liberate hydrogen at the receiver. As many wires were used as there were characters to be transmitted: sending an impulse over the 'A'-line would cause bubbles to rise above that letter at the receiver.

Telegraphy, in its practical form, however, followed close on the discovery of electromagnetism, and the contributions of Morse, Bain, Cooke, Wheatstone and others have been described in Chapter Three. These were only some of the more famous offspring sired by Ampère in 1820 when he successfully transmitted signals electromagnetically over twenty-six wires, one for each character—and without using bubbles, either.

A history of telegraphy is not, however, within the scope of this book, except where these early systems inspired printing telegraphs. For it was almost immediately evident that a permanent record of messages was essential. Initially, the operator received transitory signals, either visible or audible, which he had to transcribe. Both were fraught with problems: the visual method, consisting of a scale of characters on which a moving needle picked out the message, was slow not only of transmission but also of reception, for the operator had to watch both the scale and the piece of paper on which he was writing. The audible method was hardly better, for while speed of operation was increased, highly skilled operators were required. In both methods, the presence of the operator at the receiver was essential at all times; if any part of the message were missed, it was gone for ever, and there was no means of checking and correcting mistakes.

A telegraphic system which would leave a permanent record of the message was clearly the solution to all these problems. Ideally, and in the form it assumed when it reached finality, the receiving apparatus should print out the message in Roman characters simultaneously with its transmission from a machine which would require no greater skill from the operator than the use of a typewriter keyboard. The development of such a printing telegraph was no mean feat, however, and a detailed study of every invention is beyond the scope of this book. To give some idea of the magnitude of such a task, Zetzsche[44] claimed that he selected only about fifty of the most important inventions prior to 1877 and left out the rest for lack of space!

The earliest permanent record system was electrolytic, the coded message being recorded by discolouration of a strip of litmus at the receiver. This was a decade or so before Morse and his associates produced only marginally more satisfactory records by embossing a code of long and short lines on a paper tape moved at constant speed by clockwork. And then, within a matter of years, virtually every major country in the world had its telegraph system, using to a greater or lesser extent the contributions of Bain, Cooke, Wheatstone, Brett, House and others already mentioned. Almost all of these were made obsolete by the Hughes machine of 1855 which underwent transformations in many countries—in France by Froment, Germany by Siemens and Halske etc.—and achieved almost universal acceptance.

Essentially, this printing telegraphic principle had now virtually reached finality. Numerous other developments were made, by Higgins, Steljes, Baudot, Hoffman, Essick, Barclay, Burry, and so on, but these were merely refinements of existing devices in various combinations. In essence, the operator typed the message at one end on the transmitter, while one or more receivers picked it up and typed it out at the other end. A somewhat late yet good example of this kind of instrument was the Yetman Transmitting Typewriter, introduced in 1903. Although it enjoyed only a brief existence, this type-bar machine was reported to have left a precise and perfect record of the message such as might be obtained from any typewriter, while simultaneously transmitting it in Morse.

This kind of machine, however, suffered from several disadvantages, not the least of which was the fact that utilization of the telegraphic line was limited to the speed of the operator—clearly, the line would take messages at a faster rate than a man was capable of typing them out. Several ways of increasing line capacity were devised, firstly by the simultaneous transmission of several messages on the same line, and then by the use of the automatic telegraph system, 'automatic' meaning that messages were transmitted by mechanical rather than by manual means.

Essentially, the 'automatic' principle involved perforating the message on a tape which was subsequently fed through the transmitter. The speed of transmission no longer depended on the speed of the operator, who could perforate at his leisure, but solely on the technical capabilities of the machine. Several perforators could therefore be employed to feed a single transmitter; on the other hand, this system involved the additional labour of feeding the tape into the machine, and was at first limited in its use to those lines where traffic was heavy. Otherwise, the additional staff requirement made it uneconomical.

Automatic systems began with primitive instruments such as those introduced by the ubiquitous Wheatstone in 1867. A small perforator with three keys for dashes, dots and spaces punched the message out in Morse on the paper tape; at the receiving end, the signals were recorded as inked dots and dashes on similar tape. The method was basic, yet transmission speed was increased from roughly twenty-five words per minute (on manual systems) to seventy w.p.m.

More sophisticated instruments soon followed; the obvious step, taken by Allan, Siemens, Creed, Kotyra, Kleinschmidt, Gell and others, was to utilize a typewriter keyboard perforator in which the Morse combinations of dots and dashes were automatically punched out as the corresponding key was depressed. And there were further refinements as well: Creed first introduced a re-perforator reproducing the original perforated tape at the receiving end (thus making re-transmission easy and automatic), and then a printer into which this re-perforated tape was fed and which automatically typed the message in Roman letters. As a result of his and other developments, by Baudot, Murray, Krumm, Siemens and Halske and others, transmission speeds reached 600 words per minute by the turn of the century.

However, manual systems continued to have their applications despite the advent of automation, and, in fact, are used to this day, while towards the end of the nineteenth century, a machine was introduced which appeared destined to revolutionize manual telegraphy. Till then, manual systems suffered either from the slowness inherent in the early 'step by step' principle, or else from difficulty in achieving and maintaining accuracy in timing transmitters and receivers in the synchronous systems. This was the great failing of the Hughes-based instruments.

In solving these problems, 'start-stop' telegraphy was conceived, and from it was born the teleprinter and teletype as we know them today. Very simply, the start-stop system is one in which both transmitter and receiver are in phase in their positions of rest, and return to this position after each signal. Both instruments thus operate normally during transmission, except that a special signal to 'start' precedes the transmission of each character signal, whereupon a 'stop' signal orders the printer back into its rest position.

The invention of start-stop telegraphy has been incorrectly attributed[19] to Krumm in 1907, and the American Morkrum machine which developed out of

208. 'The most wonderful of all': Kamm's Zerograph, with its contact arm to the extreme left in the 'zero' position. Note the segmental comb. (AC)

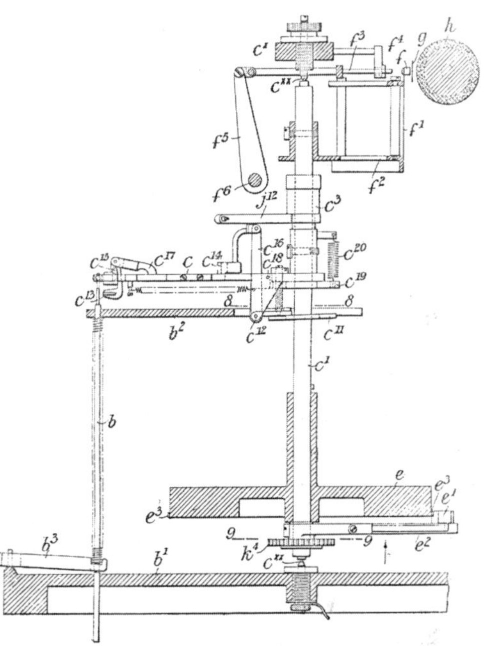

209. Details of the Zerograph printing action. (BP)

his device was introduced around 1920. About the same time Creed came out with a similar instrument in England.

However, the machine which was to initiate this telegraphic revolution was neither of the above. Mares[27] writes: 'Wonderful as the previously mentioned adaptations of the typewriter (to telegraphy) may be, probably the most wonderful of all is the machine to which we now make reference, namely, the Zerograph of Mr Leo Kamm, of London.' This was written in 1909, the year in which Karrass[25] described the Zerograph in detail—it had apparently been set up in the Berlin Postal Museum in 1898 and was still there at the time of his writing. Further back, the *Scientific American*[67] gave a detailed account of the instrument in 1903.

Kamm took out numerous patents (1895, 1898, 1901, 1905) and worked on the machine as well as on its promotion over a twenty-year period; considering the excellence of its design and the enthusiasm of its reception it is remarkable that nothing concrete materialized. It was never manufactured; Kamm's appears to be the honour, however, of having built the first start-stop printing telegraph. The instrument was of the swinging sector group, employing a comb not unlike that of Wheatstone. An electromagnetic 'hammer' was also employed, and carriage return and line spacing were achieved automatically at the end of a line, a clock spring keeping the carriage under tension. A semi-circular two-row keyboard printed upper case only.

The start-stop mechanism, from which the machine derived its name, was in some respects not unlike that of a Hammond: on depression of a key, the type-sector was released and swung, or rather was pushed, around by a spring until its arm made contact with a protruding pin corresponding to that character, whereupon the hammer struck the type against the paper and the sector was pulled back to its rest or 'zero' position electromagnetically, in readiness for the depression of the next key. The 'zero' position was at the extreme left of the type-sector instead of in the centre, as it was in the Hammond. An identical machine at the receiving end performed the same operation: its type-sector was released and stopped by the signal transmitted either by telegraph or by wireless. One of Kamm's instruments is illustrated in figure 208: this unit is perhaps the only example of his work to have survived.

'In view of the possibility of development of this machine, there would seem to be no reason why a man sitting at his Zerograph in London, may not, in the future, be able to hold written converse with his correspondents in the furthermost parts of the globe, without the intervention of any actual physical connection.'[27]

7 CIPHER MACHINES

Man has been fascinated with cryptology for thousands of years and the art has played an important part in military and diplomatic history for almost as long as this has been recorded. The advent of the practical and successful typewriter during the last century should have created a veritable revolution in cryptography; after all, the desire to print a ciphertext automatically must have been one of the dreams of cryptologists for centuries. A typewriter with a standard keyboard but random letter order on type-bars or -wheel should do it: merely typing out the message would automatically encode it as it is being printed.

It sounds too good to be true—and it is! Unfortunately, cryptanalysis had reached advanced and sophisticated levels by the time the typewriter appeared. A ciphertext such as the one produced by the above-mentioned machine could have been read off by even an amateur cryptologist in only slightly more time than it would have taken him to read off the plaintext. Such simple substituion ciphers are worthless, and the reason is easily understood. If 'E' on the keyboard of our cipher typewriter prints 'g', or '+' or '6' or whatever character we have chosen (and it makes no difference what this character is), then every time we depress the 'E' key, we shall print the same symbol, and so on with every other letter of the alphabet. A cryptanalyst has merely to add up the frequency of the different characters in the ciphertext, substitute the letters according to that frequency, and the message can virtually be read off. If the 'E' on our keyboard prints ' + ', then there will be more ' + 's' in the message than any other symbol, for it is known that 'E' is the most frequent letter in the English language. Even without going through the motions of a letter count, an experienced cryptanalyst can spot this by merely glancing at the page; he can immediately assume that the combination $-°+$ which might appear on several occasions is the plaintext 'the' for instance, the more so since ' $-°$ ' and ' $°+$ ' represent the two most frequent digraphs in English. And so on.

The typewriter and cryptology made a bad marriage, then. But even so, several attempts were made: Dujardin's Tachygraphe (1838) printed a selection of dots over which a special calibrated rule had to be applied in order for the text to be readable—the use of such a rule in cryptology was a historical curiosity by that time. Baillet de Sondalo and Coré (1841) suggested modifications to their Compositeur in order to print a monalphabetic substitution cipher and, a few years later, Pape came out with a machine which printed only one character, again decipherable by applying a scale to the line. Labrunie de Nerval virtually reinvented Bramah's type-wheel machine, the principles of which were used in the Jefferson wheel cipher, and our old friend Charles Wheatstone, an enthusiastic cryptologist, invented a respectable cipher-wheel device without, however, adapting it to his typewriters.

MP

210. Kohl's Cryptographe was a barely disguised Hansen Writing Ball telegraph model. (AMP)

Monalphabetic substitution machines were the only ones offered for the rest of the last century. Numbers of them were made: Hansen offered his Writing Ball as a cipher machine in 1872; a printing cipher device is reported[24] to have been made by Vinay and Gaussin in 1874, but details have apparently not survived;

another using a type-wheel by de Viaris appeared in the mid-1880s. At the same time, Alexis Kohl of Denmark virtually re-invented the telegraphic model of Hansen's Writing Ball as a cipher machine. It failed to get off the ground and was eventually donated to the Museum. According to correspondence in the archives, the inventor made improved models but they never appeared in public. (Morelli[29] is badly confused over this instrument, dating it 1843 and calling it the predecessor of Hansen's machine.)

AMP

By this time, the Velograph was offering a cipher model, as was Merritt, and, somewhat later, the Diskret, Hammond and others followed. None was any good as a cipher machine. And the ultimate in this line of useless instruments was suggested in 1911, although why it took so long to appear is hard to understand: sets of removable caps were to be placed over the keys of a standard typewriter, thereby changing the letter order at will and instantly converting the typewriter into a monalphabetic substitution machine.

However, it was necessary to depart from the typewriter principle entirely in order to arrive at a satisfactory printing cipher device, which eventually materialized in the 1920s. The seeds had been sown much earlier: characteristically enough, Charles Wheatstone is credited with inventing the basic instrument in 1867, although it was first conceived and built fifty years earlier by an American called Decius Wadsworth. The idea is both simple and ingenious: essentially, it consists of plaintext and ciphertext alphabets of unequal lengths geared together. If the plaintext has twenty-seven elements (the twenty-six letters plus a blank for spaces) and the ciphertext has twenty-six only, and these are stamped on to two inter-meshing wheels with the corresponding number of teeth on each, then every time the larger wheel completes one revolution, the smaller one will complete one revolution plus a tooth, or letter. Encipherment is therefore progressive: plaintext A would be ciphertext A on the first revolution, B on the second, and so on, unless the ciphertext alphabet were jumbled, as indeed it was on Wheat-

211. 'The most ingenious mechanical creation in all cryptography' was the Hagelin. On this Cryptos model, the wheel on the left selects the plaintext character; lowering the handle on the right prints the letter on the left-hand tape and its ciphertext equivalent on the tape to the right. The six rotors, right front, indicate the cipher sequence. (AC)

228

MP stone's Cryptograph, which used fixed alphabets but geared pointers to achieve the same ends.

Such is the essence of the sophisticated instruments which began to appear in this century. There were the Enigma, Kryha, Damm, and that 'most ingenious mechanical creation in all cryptography',[24] the Hagelin Cryptos. That source had taken the trouble to calculate that the crude early models of these machines, using five rotors, would not repeat themselves in sequence until after more than ten million successive encipherments. 'The special merit of the rotor system springs from its outpouring of cipher alphabets in such hemorrhaging profusion as to provide a different alphabet for each letter in the plaintext longer by far than the complete works of Shakespeare, War and Peace, the Illiad, the Odyssey, Don Quixote, the Canterbury Tales and Paradise Lost all put together.'

But lest that sound excessively impressive, the 'little jewel of a cipher machine'—the Hagelin—took the system even further: gearing was almost infinitely variable, and six keywheels with 26, 25, 23, 21, 19 and 17 letters respectively meant that the sequence would not repeat itself now until over 100 million successive letters had been enciphered. This figure is ten times greater than that previously mentioned. The Hagelin was the most famous and popular cipher machine of them all; used extensively during the Second World War, it remains today standard equipment with sixty governments, to say nothing of private and commercial clients. Wadsworth's idea has been refined beyond recognition: in making it possible for so many countries (and not always mutually friendly ones) to use the same machine, the manufacturer apparently[24] calculates the number of variable elements as more than twenty-four quintillion quintillion quintillion quintillion, whatever that may mean.

What typewriter could hope to compete against such specifications? The security which these machines offered was absolute, and the codes were perfect—except that cryptanalysts quickly proceeded to crack even those. Nowdays, computers sometimes do the job. And they break *them* as well. The codes, that is.

8 SYLLABLE MACHINES

Another branch of typewriter development which never achieved success or acceptance was the syllable machine. To have fulfilled its purpose even modestly, it would have required such a proliferation of keys that it would have resulted in a most unpractical contrivance. After all, the Brackelsberg with 132 keys couldn't make it, even though 'the number can be almost indefinitely extended'.[27]

Two approaches to the problem occupied the efforts of inventors and manufacturers. One was the 'chord' principle already discussed in the section on stenographic machines: the simultaneous depression of two or more keys to produce the syllable or word. The other approach was to take the most frequent letter combinations and short words and provide each with its own key. Neither system was any good, for it failed in its basic purpose of increasing typing speed.

Efforts go back to 1841, when Baillet de Sondalo and Coré proposed rebuilding

212. The syllable typewriter was a lost cause from the very beginning, but the Dennis Duplex put up a valiant albcit suicidal struggle. The fallacy was that typing two letters simultaneously would double the speed.　　　　(MPM)

their Universal Compositor as a syllable machine capable of printing all 500 syllables in the French language. Peeler (1866), Ravizza (1882), Mergier (1892) and Toepper (1895) are among those whose inventions never left the drawing-board and patent offices.

The first one actually to reach the market was the Dennis Duplex, invented by A. S. Dennis and patented in 1895. This machine was equipped with two key-boards, one for each hand, and two keys could be struck simultaneously. It was of the up-stroke class: the type-bars were suspended in two semi-circular baskets the centres of which were one space apart. The machine never caught on, but a curious device on it is worthy of special mention: the carriage could be shifted a fortieth of an inch so that when a line or word was retyped, it looked as if it had been shaded.

The Duplex has the dubious honour of being the most successful of the various syllable-machine failures. The 132-key Brackelsberg which appeared in 1897 offered *four* keyboards, making it possible to type up to four-letter words with one chord. More could be added at will. The Universal of 1904 used the other approach: it provided a normal three-row keyboard, then a row of digraph, followed by three rows of trigraph keys. Depression of a syllable key automatically moved the carriage through the corresponding number of spaces. Several modifications of the machine were offered without success, and the Bennington of 1903, a down-stroke machine using the same basic approach, was no better received although its inventor was more tenacious, for he is heard of as late as 1922 organizing the Xcel Typewriter Co. to promote the same basic design. Nothing came of it, nor of any of the other ten or twelve efforts, the most remarkable of which was the Schiesari which permitted the simultaneous impression of up to six characters from a double shift keyboard of only twenty keys. Down-stroke from the rear, like North's or Brooks, each type-bar had provision for an accumulative eccentric lateral displacement of $2\frac{1}{2}$ mm per character up to a maximum of six, conditional, of course, on the letters in the word appearing in alphabetical order. The paper was advanced automatically for the same number of spaces as of keys depressed.

9 BILINGUAL MACHINES

If machines printing two or more letters at a time were doomed to failure, what hope was there for machines typing two or more languages? The concept, however, was entirely different: these machines developed directly from the versatility inherent in certain designs, particularly of the type-wheel and -sector group.

Most established manufacturers, even those with limited sales, offered machines for a variety of languages. This entailed the use of alternative keys and letter orders for those languages using Latin characters, as well as completely different arrangements for Greek, Russian, Hebrew, Arabic, and other oriental languages. Ideographic tongues presented special problems which were virtually insoluble on a practical level: typewriters for Chinese and Japanese with their thousands of different basic ideograms were, in fact, developed, and their general appearance might be described as resembling that of a square or circular index machine with three or four thousand characters on it. Their use was justified by the fact that these languages are very slow and difficult to write by hand, each ideogram consisting of anything up to twenty-five strokes delicately executed by brushwork, so that the typewriter is hardly likely to have proved slower to operate.

In the case of other languages, however, the problems are not so severe. A mere change of type-wheel or -sleeve, for instance, offered a new language on a Helios or a Chicago—the change took a few seconds. Other designs are less fortunate in this regard, the machines being virtually condemned to remain in the language with which they left the factory. With the exception of a couple of makes specializing in easy disassembly (Demountable, for instance), type-bar machines were often impossible for the layman to change over; where it was deemed possible, it was a long, messy job. Most index instruments also required screwdrivers and/or wrenches. Type-bar machine manufacturers sought alternative solutions to the problem: Ideal, for instance, came out with their Polyglott model in 1902: in its simpler form, it offered two combinations of upper-case characters (for example, German-Greek), keys with letters in both languages on them, and the usual shift mechanism for selecting the language. A more sophisticated model with forty-two keys and double shift offered 126 possible characters, giving upper and lower case in the two languages. The double-keyboard Jewett offered kits of interchangeable type and keys, also a bilingual model with 122 keys so arranged that each half of the keyboard printed a different alphabet. Even clumsier is Imperial's solution of basically bolting two conventional typewriters together side-by-side, with a single carriage traversing both machines, one of which has standard type while the other is fitted out to the client's specifications.

Such methods appear desperate, however, when compared with machines like the Hammond, the Blickensderfer, and the Mignon which offered literally hundreds of alternatives. Mignon had matched pairs of letter index and type-sleeve, both of which slipped on and off the machine in a second. Hammond and Blick, having keyboards, had to provide an index as a guide to the new sector and type-wheel.

213. A bilingual Hammond Multiplex, with Greek and Roman keyboard. The Greek character is in red on each key below its black Roman 'equivalent'. Each language is on a separate shuttle and is simply selected by turning the spindle through 180°.   (AC)

To hold this index, Hammond even supplied the machine with two little clips as standard equipment. Caps were also available: these were slipped over the keys to make them conform to whatever type-wheel was in use, and—as a bonus— after the first model, all Hammonds had provision for two type-sectors of the user's choice, and selection of one or the other was performed merely by turning the spindle through 180°. The operator could choose one sector of block type and the other of script, for instance, allowing him to italicize with the mere flick for a wrist; alternatively he might choose one of large and one of small print. Or he might want two separate languages: if he ordered this, the keyboard came with characters in both languages, one printed in red and the other in black. Or there might be a sector of scientific and technical symbols.

Bilingual machines died out, we are told,[27] when typewriters became cheap enough for a man with special requirements to buy a separate machine to fill each one of them. We can therefore do little more than mourn the passing of the days in which one so-called 'expensive' machine would do the job for which we

214. The Blickensderfer Oriental is somewhat different: designed for typing languages which read from right to left as well as conventional Western tongues, its carriage is fitted with reverse escapement, with a selector lever at the rear. (AC)

are now obliged to buy five or ten so-called 'cheap' ones. This applies particularly to machines such as the special bilingual models with reverse escapements which allowed those languages (Arabic, Hebrew, Turkish, Persian, etc.) written from right to left to be used in conjunction with conventional western languages where the typing is from left to right. A small lever reversed the direction of the carriage, depending upon which of the languages was selected. By comparison, our modern typewriters are indeed dull, graceless and inadequate instruments. Of course, they require less concentration on the finer points—a simple mistake with the !siht ekil enil a htiw eno sevael 'latneirO' na no revel tnemepacse

10 NOISELESS MACHINES

As the typewriter developed into a commercially acceptable and desirable accessory, it became increasingly evident that one of its principal drawbacks was the amount of racket it produced. An office of moderate size, properly equipped with writing machines, might well have provided a satisfactory soundtrack to a First World War movie! And this was in an age, unlike our own, in which silence was still largely ubiquitous; in those days man's, and particularly the boss's, threshold of endurance to noise was quickly reached.

There were several reasons for the clatter emitted by early machines. Design was largely responsible: carriages spaced in jerks; type-bars with a long throw to overcome initial inertia and gain momentum as they reached the printing point slammed into platens—and sometimes hammers did the job instead; and the long travel of shift actions, grinding of ratchets on carriage return, and slamming of metal against metal as the carriage reached its stop—all these were noise-creating elements. But apart from design, the early machines were built to very comfortable manufacturing tolerances which, combined with a little wear in the hands of ever more proficient operators, set up sympathetic vibrations in everything within the immediate vicinity, except in humans whose vibrations, as we can imagine, were anything but sympathetic.

Many methods of reducing or eliminating the noise were tried: smoother carriage escapements, rubber type-face—even enclosing the entire machine in a sound-proof glass-panelled case, leaving only the keyboard exposed. But by and large, it was more careful design with shorter type-bars requiring less percussive action, together with tighter manufacturing tolerances, which largely resolved the problem. Eventually, front-stroke machines, if not silent, at least emitted a tolerable noise level.

Despite progress with conventional methods, a separate 'noiseless' design was

AMP
MPM

215. The Noiseless thrust-action machine developed by the inventor of the Wellington. (QC)

evolved, and its popularity continued for several decades. This was based on the thrust-action described on page 200, in which the type was pushed rather than swung against the platen. The Rapid was the first machine to use this system, which was more widely diffused in the Ford, the Granville, and the numerous makes which evolved from the Wellington. Kidder, who designed that machine, used essentially the same movement in a later instrument, appropriately called the Noiseless. Manufactured by the Noiseless Typewriter Co., the machine was perfected in 1912 after which it enjoyed increasing success; Remington took over the operation in 1924 and the design continued to be used in a variety of makes both in the United States and in Europe until well into the 'thirties.

11 BOOK TYPEWRITERS

A small category of machines was built for people with special book-keeping problems not solved by the mere yard-long carriage offered by Blick. Several companies manufactured what were called Book Typewriters for use with bound volumes rather than separate sheets, although they were also suitable for the latter. The two best known of these instruments were the Fisher and the Elliot Hatch. Others included the Dukes, Johnson, Marriot, etc.

234

Bascially, the concept of them all was very similar: the machine was mounted on a frame, provision being made for it to move from back to front and left to right. This frame was then placed over the book and typing was performed by down-stroke on to the horizontal page. The entire machine—keyboard and type-bars—moved for letter and line spacing by rolling along the frame in these two planes.

The ancestry of this principle can be traced back to Progin in 1833. Sherman's 1877 patent was the first of the next generation's entries, but it was not until 1896, when the Fisher appeared, that the book typewriter became a commercial reality. Early models had a full keyboard, which was later replaced by a four-row. A year later, the Elliot Hatch was introduced; in 1903 the two merged to form the Elliot Fisher Co., thereby virtually collaring what small market there was for this type of machine.

216. The Elliot-Fisher, complete with its special desk. The pedal lowers the table-top for paper insertion. One of the keys is shown on its down stroke. On book typewriters, the whole unit rides on rails from left to right over the stationary paper—with the typist in hot pursuit!   (AC)

Chapter 12

Complete Catalogue of Unconventional Typewriters

(This list of historic machines includes all typewriters manufactured up to the 1930s other than front-stroke, type-bar machines with four-row keyboards. By that date, standardization on the conventional design had become ubiquitous; some otherwise famous names, therefore, fail to qualify and are excluded. All inventions never marketed are also listed up to the year 1885, by which time such a repetitive proliferation of abortive designs was patented that only those of intrinsic merit or special interest have been selected for inclusion.)

Abeille: A machine offering visible typing, introduced in France in 1904 and described[28] as having a small removable keyboard of thirty keys printing ninety characters by double shift, using a ribbon.

Addey's Typograph: A precocious type-wheel machine by this name is described[27] and dated 1889, but was apparently never manufactured. It was small: 14″ by 10″ by 6″; a three-row keyboard of thirty-six keys controlled a sphere 2″ in diameter on which there were 118 characters in six rows, converging at the poles. The platen pressed paper against character. Inking by roller. It was equipped with a back-space key which was not generally offered until some years later.

Aders: German patent granted in 1907 to Franz Aders for a pneumatic typewriter which was never manufactured.

217. Klein-Adler   (AC)

Adler: A German bicycle etc. manufacturer called Adlerwerke vorm Heinrich Kleyer A.G. introduced this machine in 1899, the first model of which was No. 7. It was virtually identical with the *Wellington*, having been made under licence from the inventor of that machine, Wellington P. Kidder. Many subsequent models of the thrust-action Adler were produced, including the four-row keyboard Model 15 in 1909, the portable *Klein-Adler* in 1913, etc. The *Klein-Adler* was also marketed under the names *Adlerette*, *Adlerita*, *Adler Piccola* and *Aigle*. A Model Seven in TMT is labelled *Crown*. (Fig. 217)

Adlerette: See *Adler.*

Adlerita: See *Adler.*

AEG: See *Mignon.*

Agar: See *Juventa.*

Agar-Baby: See *Juventa.*

Aigle: See *Adler.*

Ajax: See *Imperial (a).*

Aktiv: A cheap type-wheel machine produced in Germany in 1913 by Gustav Tietze A.G. Printing was accomplished by rotating the horizontal wheel to select

218. Aktiv   (DTU)

the desired upper- or lower-case character, which was brought to the paper by depression of a handle. The same company manufactured the *Famos*. (Fig. 218)

Albertson: A four-row, front-stroke folding machine patented in 1913 by Ernest Albertson of New Jersey, whose Albertson Folding Typewriter Company appears to have folded before manufacture began.

Albus: One of the many three-row folding portables built after *Standard Folding* patents, this one was manufactured in Vienna by the firm of Carl Engler from 1909. It was made largely of aluminium and weighed a mere 6 lb. Also produced as the *Emka, Engler, Proteus* and *Atlas*. In 1912, the company was bought out by Clemens Müller A.G. and moved to Dresden where the machine was produced as the *Perkeo*.

Alissoff: Type-wheel machine patented in 1872 in France and in 1874 in England by a Russian of the same name. It was displayed in the Philadelphia Exhibition of 1876 but never manufactured. The type-wheel was mounted co-axially with a letter index and selector handle. Upper- and lower-case characters (and other alphabets as well, if required) were placed in grooves around the thickness of the wheel so that each set or alphabet formed a separate complete circle, its position relative to the printing-point being selected by turning a worm. Inking by roller. The paper was fitted around a platen and lowered by a handle or pedal on to the type-wheel for printing. (Fig. 147)

Allen (a): A front-stroke machine with three-row keyboard and double shift invented by Richard Uhlig and manufactured from 1918 by the Allen Typewriter Company of Allentown, Pa., from which the company took its name. It was mechanically good but commercially unsuccessful for which its three-row keyboard has been blamed.

Allen (b): US patents were granted to R. T. P. Allen of Kentucky in 1875 for a plunger machine with circular keyboard and in 1876 for an up-stroke design with five-row keyboard. These are the Allen machines erroneously listed in other sources and mentioned on page 121. (Figs. 219 and 220)

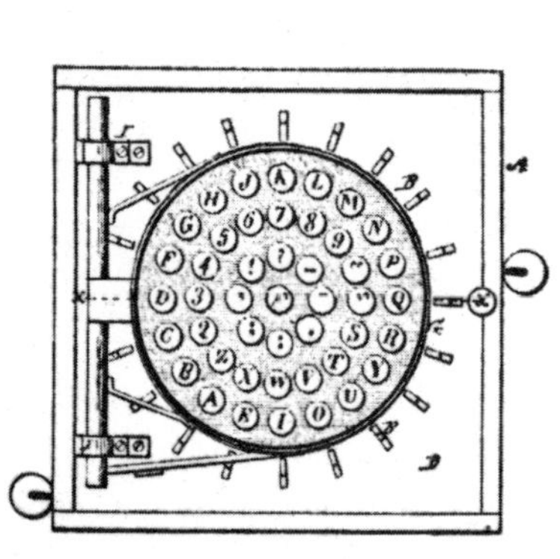

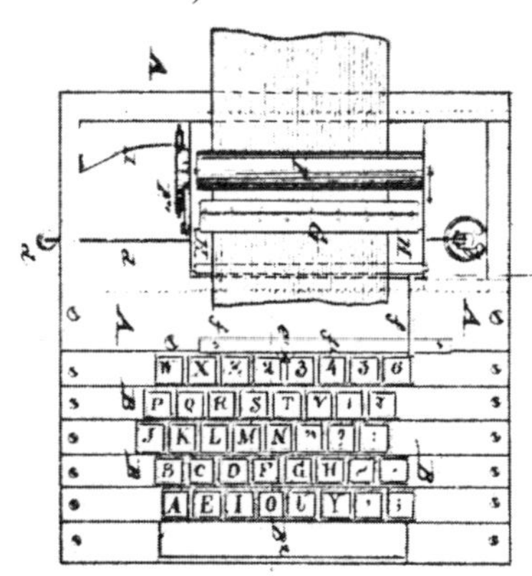

219. Allen (b), 1875   (USP)   220. Allen (b), 1876   (USP)

Alphabetaire Typographique: French patent granted in 1875 to a man called Cabanel for a type-wheel machine with letter index.

Ambler: J. A. Ambler of Mass. was granted a US patent in 1884 for a machine consisting of three concentric circles of type-plungers above a platen; ribbon inking.

Ambrosetti: An improved model of Mazzei's *Stenotiposillabica* built by Valerio Ambrosetti in 1883 and reportedly capable of 200 to 220 w.p.m., but nothing further came of it. It followed Bussadori's similarly unsuccessful attempt with his *Tachigrafica.*

America: See *World.*

American (a)*:* Second machine produced by the American Typewriter Co. to whom the inventor, L. P. Valiquet of N.Y., assigned his 1893 patent. The company, already manufacturing the *American Visible,* introduced the present machine in the same year as the patent was granted. It was of swinging-sector design, with a large arc of rubber type mounted on a vertical shaft beneath a corresponding letter index. A pointer was swung to the desired character and a key on the left of the machine was depressed to bring the rubber letter in contact with the paper. Inking by roller. Locking holes provided alignment. Also marketed as the *Globe* in England and the *Champignon* on the Continent. (Fig. 15)

American (b)*:* An up-stroke machine with a three-row universal keyboard and double shift introduced in 1899 by the American Typewriter Co., which had already manufactured the *American Visible* and the *American (a)* swinging-sector machine. This type-bar model was patented by N. R. Marshman and Lee Burridge of N.Y. in 1896 and assigned to Halbert Edwin Payne of that city, Payne being sometimes listed as inventor. The most distinctive feature of the machine was its economy of parts (only 325). This was achieved by attaching the keys directly to the ends of the type-bars which rested in a cradle beneath the platen and radiated outwards to the keyboard, being pivoted on the way. This

221. Armstrong (American (b))   (AC)

design would perhaps have been more acceptable had the keyboard been curved, but as this was not the case, depression of a key at the far right or left caused the lever to veer outwards as it descended, an unpleasant sensation to which one might or might not have become accustomed. The carriage had to be raised to reveal the typing. Inking was by means of a ribbon. It was one of the smallest up-stroke machines, and appears to have enjoyed considerable success, for it was marketed as the *Armstrong* in England, the *Herald* in France and *Europa* in Germany. It also appeared under the names *Eagle, Elgin, Favorit, Fleet, Mercantile, Pullman* and *Surety*. Manufacture ceased in 1915. (Fig. 221)

American (c): See *Daugherty*.

American Double Wheel: A remarkable patent granted to the inventive team of N. R. Marshman and L. Burridge in 1898 and assigned to the American Type-

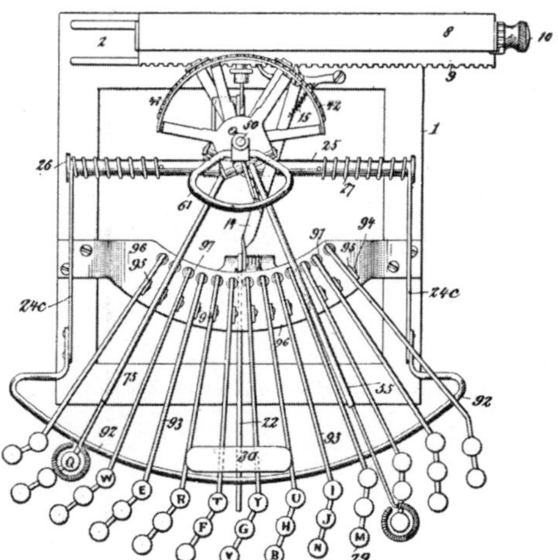

222. American Double Wheel   (USP)

writer Co., this design looks like a cross between all their other machines! Two swinging sectors are mounted one above the other on a vertical axis next to the platen. Each sector corresponds to half the 'keyboard' and has its own selector arm and ring. The three-row 'keyboard' serves merely to locate the rings: in mounting one or other of them over the desired character, the type-sector is swung till the corresponding letter is opposite the printing-point, whereupon depression of ring and key brings the sector into contact with the paper. Pad inking. Unfortunately for later collectors, this unique design was never manufactured, but it has all the feel of an American Typewriter Co. product. The name does not appear on the patent, but was supplied by a secondary source[6] which presumably took it from contemporary publicity material. (Fig. 222)

240

223. American Flyer   (LC)

American Flyer: A toy typewriter by this name is in LC and is dated circa 1930. It was perhaps made by the original manufacturer of a line of toy electric trains by the same name. The machine had a sliding indicator geared to a type-wheel, with a pointer over the wheel showing the selected letter. Printing through ribbon, by depression of the sliding indicator knob. Made by American Flyer Mfg. Co., Chicago. (Fig. 223)

American Pocket: The prototype of a small portable by this name is reported[22] to have been produced in 1926 after a design by A. M. Leggett. It was never manufactured. It appears to be similar to a small portable produced in 1923 by a corporation called Rochester Industries, which is reported[42] to have bought up in the previous year another similar machine known as the *Leggatt.* The numerous similarities between all these component elements seem too great to be caused by chance alone. See also *Rochester.*

American Standard: See *Jewett.*

American Visible: First product of the American Typewriter Company, reported[22, 28 etc.] to have been introduced in 1891, although the machine in AC bears a transfer stating that the company was 'founded 1893'. The American Visible had a rubber linear type-index which ran in a groove parallel to the platen. Attached to it was a sliding thimble over a plate on which the characters appeared in slightly oblique columns, so that drawing the thimble down a column displaced the type-index in a lateral direction. Printing was achieved by depression of the entire plate, with a separate space lever at the front of the machine. Inking was by pads on both sides of the printing point; the pads were fixed to a flexible strip which also served as a mask preventing letters other than the one required from being printed. (Fig. 13)

Amka: See *Bing.*

Anderson (a): A shorthand machine patented in the United States in 1889 by G. K. Anderson of Tennessee. Some modifications to the original specifications were filed in subsequent patent applications. It was a small and compact machine which wrote phonetically using the chord principle, the printing being on a paper tape in Roman characters. It had fourteen keys, and from 1898 onwards was equipped with a shift key. It was fairly well received, in days before the stenographic machine was fully accepted, and was later taken over by the Dictatype Shorthand Machine Co. and marketed under the name *Dictatype.* (Fig. 205)

Anderson (b): J. Tate Anderson of Philadelphia was granted a US patent in 1877 for yet another constantly rotating type-wheel controlled by a pin barrel.

Anderson (c): An interesting up-stroke type-plunger design was patented by H. and F. Anderson of N.Y. in 1878. The plungers were carried on a swinging sector attached to an indicator arm, which selected the characters from a curved scale. The sector swung beneath the platen, the plungers therefore striking upwards.

Archo: Another thrust-action double-shift machine, similar to the Wellington, was produced in Germany by the Archo works of Winterling & Pfahl, later Inh. Carl Winterling. The first model with three-row keyboard was introduced in 1920 and a standard four-row model in 1935. The manufacturer offered this as a musical typewriter called *Archo-Norma* in 1937—with forty-four keys and double shift, the Model Ten printed a total of 132 musical and Roman characters while Model Eight, with single shift, wrote only music. The most noteworthy mechanical feature of the machine was that sectors meshed with the ends of the type-bars to propel them forward, these sectors being connected to the key levers by means of a linkage which turned them as the key was depressed.

Archo-Norma: See *Archo.*

Ardita: See *Juventa.*

Arlington: Richard Uhlig invented this machine, naming it after the town in which he lived at the time. A 'visible' with three-row keyboard and double shift, it was manufactured in 1914 principally to compete in export markets against higher-priced standard machines.

Armstrong: See *American (b).*

Arnold: A primitive swinging sector machine with curved letter index was patented in 1876 by Benjamin Arnold of Rhode Island, but the device was never manufactured. The patent model has survived (MPM). (Fig. 224)

224. Arnold   (MPM)

Atlas (a): This small portable, with oblique front-stroke action, three-row key-board and double shift, was yet another of the machines invented by the prolific Richard Uhlig. An improvement on his previous Arlington, it was manufactured in small quantities in 1915 by the Atlas Typewriter Co. of N.Y.

Atlas (b): See *Albus.*

Atlas (c): See *Standard Folding.*

Autist: See *McCall.*

Autocrat: See *Harris.*

Automatic (a): An outstanding non-visible machine of very early manufacture, the Automatic was the smallest up-stroke ever made. It was invented by Major E. M. Hamilton of New York and is reported[22] to have been produced from 1881 to 1883. However, this information is almost certainly incorrect. The first patent covering the design, assigned to the Hamilton Type-Writer Co., was filed in 1884 and granted in 1887, and the machine was still being retailed in 1891.[62] Made almost entirely of brass, it had forty-eight keys in three rows operating type-bars a mere $1\frac{1}{2}$ inches long. These rested on an ink pad under the platen; upon depression of a key, the corresponding type-bar lifted off the pad, moved towards the centre of the type circle and then up to the printing point. The carriage had to be raised to reveal the printing, and the platen was polygonal in shape, having flats corresponding to the lines being printed. It incorporated differential spacing and printed upper case only. Dimensions: approx. 11″ by 8″ by 4″. Also marketed under the name *Hamilton.* (Fig. 197)

Automatic (b): The Automatic Word Writing Machine was designed in 1929 to type up to 1,000 words a minute! It consisted of a standard typewriter keyboard around which were 164 word-keys which printed whole words at one stroke. Several word-keys could also be depressed simultaneously. It was never produced.

Auto-Typist: See *McCall.*

Aviso: A front-stroke three-row double shift machine produced in Germany by Otto Schefter in 1923.

Axial: Two US patents granted in 1887 to J. H. Waite of Mass. were assigned to the Axial Typewriter Co. of Maine. One covered a circular index design and the other a machine with horizontal type-bars in a complete circle, pressed down on to the paper by a selector knob. Neither was manufactured.

Baby Fox: See *Fox.*

Baby Practical: See *Simplex (a).*

Baby Rem: See *Blickensderfer* and *Remington.*

Baby Simplex: See *Simplex (a).*

Badeau: US patents for a musical typewriter were granted from 1903 to Isaac Badeau of New York, and later jointly to his brother Louis, but the machine was never manufactured.

Bailey: A shorthand machine with thirty-four keys which, depressed separately or in chords, perforated paper tape according to a special code. Patented in US in 1896.

Baka 1: See *Moya.*

Baldridge: A US patent for a four-row type-bar machine with electric assistance was filed by G. W. Baldridge of Missouri in 1883 and granted in 1886.

Balkan: See *Senta.*

Baltimore: See *Munson.*

Bamberger: See *Helios.*

Bar Let: See *Mitex.*

Bar-Lock: A fine machine of down-stroke design patented by Charles Spiro in 1889 and (according to an original advertisement) introduced the same year. The first models, with moulded and embellished type-bar surround, were simply labelled *Bar-Lock*; later models, circa 1900, were called *Columbia Bar-Lock*, with this name in relief on an otherwise smooth pressing. These machines, up to Model Fourteen, had double keyboards—there is little to differentiate between them, and dating is difficult since virtually no two sources agree. The problem is aggravated by the fact that many models were available simultaneously, different carriage lengths often being the only distinguishing feature. Thus, Models Three, Seven, Nine and Eleven were wide-carriage models, and so on. Model Twelve, 1907, was the first to offer a major design departure in the form of a standard four-row keyboard with shift. A similar Model Fourteen followed in 1910. These two machines were called *Columbia*, the 'Bar-Lock' being dropped. Up to and including Model Fourteen, all machines were down-stroke, with two semi-circles of type-bars standing vertically in front of the platen.

Model Fifteen appeared in England as an oblique front-stroke machine in 1911–13 and Model Sixteen, of conventional front-stroke design, in 1921. Previous models marketed in England were called *Royal Bar-Lock*, but from Model Fifteen the name was once more simply *Bar-Lock*. Manufacturer: Bar-Lock Typewriter Co.

225. Bar-Lock   (AC)

Charles Spiro, the New York watchmaker and inventor of the machine, had previously designed the *Columbia* type-wheel device, produced by the Columbia Typewriter Manufacturing Co., which was also responsible for US production of the Bar-Lock.

The machine got its name from a small semi-circle of pins between two of which the type-bars were 'locked' at the printing point. This system prevented them from jamming, and assisted with alignment. It was claimed to be the first visible type-bar instrument and, while this is not strictly true, the design did offer visibility to all except short operators or those who 'sat far back',[42] for the semicircles of vertical type-bars could impede full visibility. It was certainly better than the up-stroke design, however. It was also claimed to be the first machine with automatic ribbon reverse.

The Columbia Typewriter Co. is said[6] to have announced the development of a cheap version of the Bar-Lock to be called *Sterling*, but it failed to appear. (Figs. 164 and 225)

Barlow: Alfred Barlow was granted a British patent in 1874 for a primitive linear index machine.

Barrett: Glenn J. Barrett, one of the inventors of the *Fox* typewriter, is reported[27, 30 etc.] to have produced a prototype machine under his own name, but details are lacking.

Bath: A British patent granted in 1869 to John Bath of London for a type-bar machine with a keyboard of forty-eight keys and a flat paper-table.

Baudouin and Bombart: These two Frenchmen patented a shorthand machine in 1876. It had a keyboard of twenty keys which controlled a swinging sector of flexible metal blades bearing the type. Printing was by means of a ribbon.

Bavaria: Front-stroke German portable with three-row keyboard and double shift manufactured in 1921 by Bavaria-Schreibmaschinenfabrik Gebr. Siegel & Co., later Ria Büromaschinenfabrik Gebr. Siegel & Co., GmbH.

Beckenridge: J. W. Beckenridge of Indiana was granted a US patent in 1880 for an attachment to existing machines or alternatively for a self-sufficient instrument in which four keys for each hand could be depressed in different chord combinations to print the entire alphabet.

Beethoven: A musical typewriter invented in France by Louis Aillaud in 1928. First four models were based on the double-keyboard Smith Premier No. 10, to which an additional piano keyboard was attached. The fifth model used a conventional shift mechanism. It was never manufactured.[28]

Beke: See *Bing*.

Beko: See *Bing*.

Bellows: Benjamin Bellows of Nebraska was granted an 1878 US patent for a three-row machine with a hammer striking the type against the platen.

Bennett: See *Junior (a)*.

246

Bennington: A down-stroke syllable machine with a five-row keyboard and single shift introduced in 1903 by the Bennington Typewriter Co. of Dayton, Ohio, the patent granted the following year. Similar to the Bar-Lock in appearance, the

226. Bennington (27)

machine had separate keys for the most frequent syllables and short words. The undertaking failed and the company was liquidated in 1905. The inventor, W. H. Bennington of Kansas, later tried the same idea with the *Xcel.* (Fig. 226)

Berger: A machine for the blind with six keys embossing a sheet of paper held around an aluminium cylinder was devised by Professor Octave Berger, himself blind, in 1919.

Berni: See *Bing.*

Berolina: See *Meteor.*

227. Berwin Superior (QC)

Berwin Superior: A toy type-wheel machine manufactured in the 1940s. The realistic-looking keyboard is actually painted on the tinplate. (Fig. 227)

Besermenje: A musical typewriter invented by Michael Besermenje in 1925 after seven years of development never reached production, although a company is reported[28] to have been formed in Agram (Zagreb) to exploit it.

Beyerlen: Angelo Beyerlen, who later became a Remington representative in Germany, invented a writing machine for the blind in 1882. It embossed braille cells by the use of six keys.

Bijou: See *Erika*.

Bing: A four-row oblique front-stroke machine manufactured in Germany by

228. Bing   (AC)

Bingwerke A.G. in the 1920s. It was a compact portable weighing less than seven lb. complete with case, but it was built exclusively of pressings without castings and was consequently of fragile construction. The first model had roller inking while the second used a ribbon. It was also marketed under the names *Amka*, *Beke* (or *Beko*), and *Berni*. (Fig. 228)

Blick: See *Blickensderfer*.

Blick Bar: See *Blickensderfer*.

Blickensderfer: One of the great machines of typewriter history. First patented in 1889 by George C. Blickensderfer and produced by the Blickensderfer Manufacturing Company, the small type-wheel portable with three-row keyboard and double shift, marketed in 1893 as the Model Five, was an almost instant success and set the pattern for the company's future models for the following twenty-five years. It featured literally hundreds of instantly interchangeable type-wheels for all purposes and languages. Inking by roller was used throughout. Upon depression of a key, the type-wheel was rotated to the corresponding character and brought down on to the platen, displacing the roller on the way. Upon re-

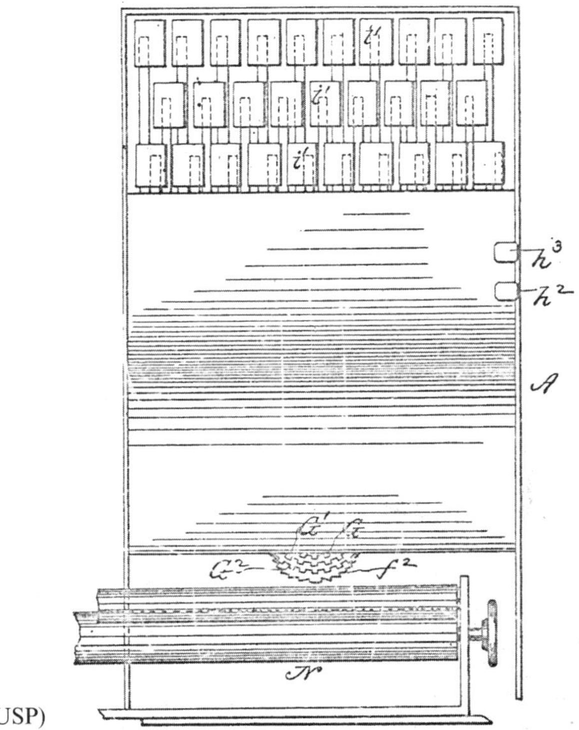

229. Blickensderfer, No. 3   (USP)

lease of the key, the wheel returned to its neutral position. Later models differed in detail only—Model Six: 1896, Model Seven: 1897, Model Eight: 1907, Model Nine: 1917. The Ideal keyboard was offered on all models with the Universal optional from 1910. Lightweight aluminium models were marketed under the names *Featherweight Blick* and *Featherweight Blickensderfer*, also the musical typewriter *Noco-Blick* (1910), *Blickensderfer Electric* (1902) and *Blickensderfer Oriental* (the last three machines are discussed in Chapter Eleven). The Blickensderfer was marketed in France under the name *Dactyle* and elsewhere as *Blick*, *Weltblick*, *World-Blick* and *Service Blick*.

The make is so intimately associated with the above type-wheel design that other models are often ignored. However, the Model Five was *not* the first machine launched by the company. Previous models were not very successful and few have survived. The Model Three patented in 1891 had the flat compact appearance of the *American (b)*. Later, type-bar models were introduced: the *Blick Bar*, a four-row front-stroke design, was introduced in 1916 and appeared set to establish itself in the market when George Blickensderfer died the following year and his death threw the organization into confusion. The type-wheel models were discontinued and replaced in 1919 by an unsuccessful three-row front-stroke portable called the *Blick Ninety*, later (1922) the *Roberts Ninety*. Production ceased altogether in 1924. Harry Smith had briefly marketed the Blick Bar as the *Harry A. Smith*—he specialized in buying up the remains of dying typewriter concerns.

The Blick type-wheel design was resurrected in 1928 by Remington, presumably after Blick patents had expired, and was unsuccessfully marketed for a short time as the *Rem-Blick* and *Baby Rem*. (Fig. 28)

An index machine using certain Blick components was manufactured by the company in 1902 under the name *Niagara*. (Figs. 168, 198, 214, 229)

Blind: A machine with this title is merely named[42] without details. Webster's 1898 Catalogue also mentions it.

Blitz (a): A type-wheel machine of this name is reported[6] to have been invented by Otto Ferdinand Mayer, but no other details are offered. The inventor was one of the men responsible for the design of the *Kneist* which is, however, an index and not a type-wheel instrument.

Blitz (b): See *Toepper*.

Bocquet: A Frenchman blinded in the First World War attempted to combine a standard typewriter with one for the blind by wiring them together electrically, but the experiment, dated 1923, was not a success.[28]

Bonita Ball Bearing: See *Sholes Visible*.

Book Electric: A telegraphic typewriter invented by S. V. Clevenger was to have been built and marketed by the Book Electric Typewriter Co. of Delaware in 1906, but the project apparently failed to materialize. The inventor also designed a different machine which he named after himself. See *Clevenger*.

Bosch: A machine on which the blind could correspond with the sighted was designed in 1893 by Robert Bosch, the well-known later manufacturer of electrical and other equipment. His instrument was similar to the *Hall Type-Writer*.

Boston (a): See *World*.

Boston (b): The invention of this machine appears to be incorrectly attributed[22] to John Becker whose *World* was also marketed under the name *Boston (a)*. Becker may have been concerned with the manufacture and marketing of the *Boston (b)* but the design would appear to have been patented by D. E. Kempster of Boston in 1886, with improvements filed the following year and granted in 1889 to Blount, Kempster, Currier and Dore. The machine was produced in 1888

250

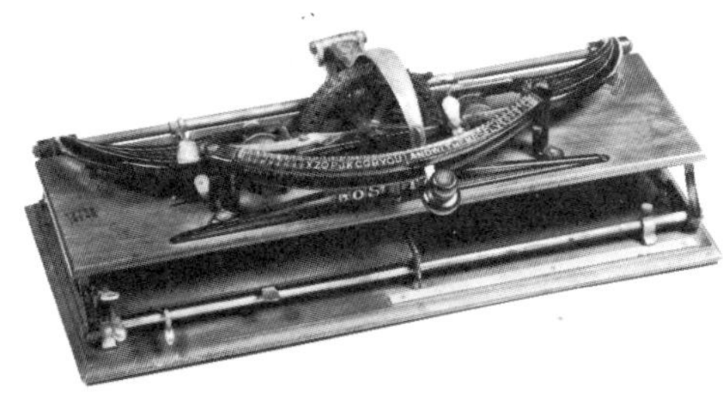

230. Boston (b)   (MPM)

by the Boston Typewriter Co. and used a vertical type-wheel made of metal, rotated by means of a selector and brought down on to the paper below. (Fig. 230)

Brackelsberg: A syllable typewriter designed in 1897 by a German of the same name. It was composed of units which could be assembled side by side, to virtually any number. Each unit comprised one alphabet and consisted of a column of several rows of keys with its own type-sector. Thus, if four such units were fitted side by side, the four type-sectors could be operated by the chord principle to

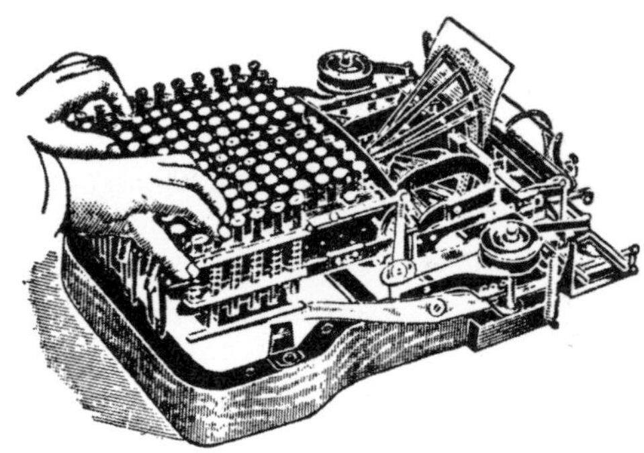

231. Brackelsberg   (27)

print syllables or words up to four letters in length at a single stroke. The machine was apparently never manufactured. Its inventor had previously designed the *Westphalia.* (Fig. 231)

Brade: Julius Brade of Germany patented in 1902 an adaptation of a typewriter to a shorthand system of his design. It printed syllables above and below a line, according to their position in the word.[16]

Bradford: An electric syllable typewriter to be manufactured by the Electric Power Typewriter Co. of Bradford, Canada, was announced in 1906. Apart from rendering everything then on the market obsolete, the 300-keyed machine was to be guaranteed for twenty years, which probably constitutes something of a record.

251

232. Brady and Warner   (MPM)

Brady & Warner: These two men were granted a patent in 1878 for a machine with a horizontal selector and a circular letter index, the patent model of which is in MPM. (Fig. 232)

British Empire: See *Wellington.*

British Oliver: See *Oliver.*

Broadway Standard: See *Daugherty.*

Brooks: This down-stroke machine patented in 1885 by one of the pioneers of the typewriter, Byron A. Brooks, was the first commercially successful design on which the type-bars were positioned behind the platen. The inventor is credited with having designed the system of placing upper- and lower-case type on the same bar while working on the development of the Sholes and Glidden. The Brooks had a three-row keyboard with double shift and, because of the positioning of the type-bars behind the platen, offered visible typing. It was thus one of the Union Typewriter Co.'s two 'visible' machines (the Monarch was the other) and production continued until the other members of the Trust (Remington, Smith Premier and Yost) 'went visible' in 1908, whereupon the Brooks, together with the Caligraph and Densmore, was put to sleep, having outlived its usefulness. See *Eclipse (a)*. (Fig. 166)

Brown: An interesting grasshopper design with pad inking was patented by C. T. Brown of Chicago in 1879. It was something like the front half of the later Williams machine, except that the keys were located *above* the type-bars.

Brown-Wood: A machine by this name is listed[6] as having been produced in the United States in the early 1890s. It was supposed to have looked like the

American (a) or the Parisienne, except that the carriage was in the middle rather than to one side. No trace could be located in patent records.

Buckner: Listed[42] without details.

Burbra: A pneumatic machine with this unlikely name was said[28] to have been invented in 1914 by Juan Gualberto Holguin of Mexico.

Burg: A printing cipher machine was invented in 1906 by a German called Hubert Burg. Of the type-wheel group, it simultaneously printed a copy of the ciphertext and plaintext.

Burlingame: A telegraphic typewriter which transmitted the message at the same time as it was being printed was the subject of a British patent granted in 1908 to the Burlingame Typewriter Co. of San Francisco. It was reported[42] due for production by the United States Wireless Printing Telegraph Co. the same year, but nothing further was heard of it.

Burnett: See *Triumph Visible.*

Burnham: A musical typewriter displayed in 1904 at the St Louis Exhibition. In appearance, it is reported[28] to have looked like a standard machine.

Burns: A double keyboard up-stroke machine patented by Frank Burns in 1889. The Burns Typewriter Co. was formed in 1890 to exploit the machine but few were made. One is in MPM.

Burridge: A remarkable rocking radial type-plunger machine in which each plunger had three keys and each type-head nine characters. Unmistakably a Lee Burridge invention, patented in 1897. (Fig. 167)

Burt: A circular index machine consisting of a disk sliced into segmental teeth with the type at the ends, controlled by an indicator knob, was patented by S. Burt of Chicago in 1885.

Büttner: A swinging sector machine with rubber type and letter index. The sector was mounted obliquely, as was the platen. The character was selected by an indicator which was then depressed, whereupon the platen with the paper swung upwards and struck the type. Invented by Dr Büttner of Darmstadt. No date.[6]

Cadmus: A US patent granted in 1872 to Eugene Cadmus of Washington. The inventor was a telegraphist and his piano keyboard type-wheel machine merely used a variation of the pin barrel principle. A sewing machine treadle beneath the unit kept the barrel revolving. (Fig. 233)

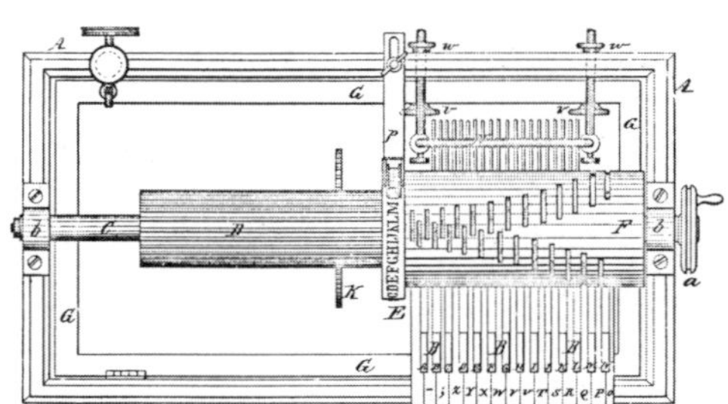

233. Cadmus   (USP)

Cahill: An electric machine, patented from 1893 by Dr Thaddeus Cahill of Washington. It was strongly reminiscent of the Remington 2 except that all moving parts, including carriage and type-bars, were operated by means of electromagnets. It was apparently manufactured from about 1901 by the Cahill Writing Machine Co. but only forty units were built at a total cost of $157,000!

Calahan: An 1876 US patent was granted to E. A. Calahan of Brooklyn for a design consisting of an indicator at the end of a type-wheel shaft. A lever brought paper strip into contact with type. Roller inking.

Caligraph: George Washington Newton Yost was the prime mover of this up-stroke machine. After his separation from Remington interests, he organized the Caligraph Patent Co. in 1880 to handle the new design he had worked out with Franz Wagner. The first patent on a Caligraph detail was filed by Yost in 1880 and granted three years later. Others followed; however he still had to pay royalties to Densmore for some of the features. Other improvement patents, in Walter Barron's name, date from 1884.

The first Caligraph was introduced in 1881 and printed only capitals, but was soon replaced by a second model (1882) with both upper and lower case—this was the first machine with a 'full' (not 'double') keyboard: white lower-case keys were positioned in the centre with black upper-case keys to the sides. It was a bad design because it meant having to learn the special position of each upper case letter, as against 'double' keyboards which reproduced identical rows of keys for both cases.

The machine extended towards the operator, with an inclined covered area between him and the keyboard, ostensibly to rest the hands on, but actually to conceal the ends of the key levers which were pivoted at the operator's end of the machine, with the keys positioned further along them, thus reducing depth of key depression. The first models had polygonal platens with flats for printing surfaces.

In 1885, Yost broke away from the American Writing Machine Co., which had

254

taken over manufacture, and proceeded to develop another machine called the *Yost*. Meanwhile, production of the Caligraph continued; in 1898 the first major design change was made with the Model Five, which retained the up-stroke movement but substituted a double keyboard for the previous full one. The name of the machine became *New Century Caligraph*, later simply *New Century*. It was one of the machines controlled by the typewriter trust called Union Typewriter Co. which phased it out in 1906 or so, when Remington, Smith Premier and Yost, the major products of the trust, were about to go visible. (Fig. 151)

Cantelo: J. L. Cantelo of England filed his first patent for a typewriter in 1887 and over the years followed it with a further nine, all of which were abandoned, until 1903 when a patent was finally granted to him. He had meanwhile been successful with a patent application for a musical typewriter to be attached to the keyboard of a piano. This was granted in 1894, and a final patent for a braille embossing machine appears in his name in 1905.

The typewriter, subject of the successful 1903 patent, was a small portable aluminium type-sleeve machine with three-row keyboard and double shift. The type-sleeve was located obliquely above the printing point; depression of a key brought it down on to the platen, simultaneously locating the corresponding letter by sliding it back and forth along its axis and twisting it to the required position. The characters on the vulcanite sleeve were in nine rows. Inking by roller. The machine illustrated is the patent model. (Fig. 181)

Caramasa: This machine was invented in 1885 and described[36] as having type-bars in two pieces with a cylindrical coupling so fitted as to allow the bars to turn on their axes. When a key was struck, a special guide led the type-bar through 180° on its way to the printing point. Pad inking.

Carissima: An anachronism in typewriters, this small German machine of the

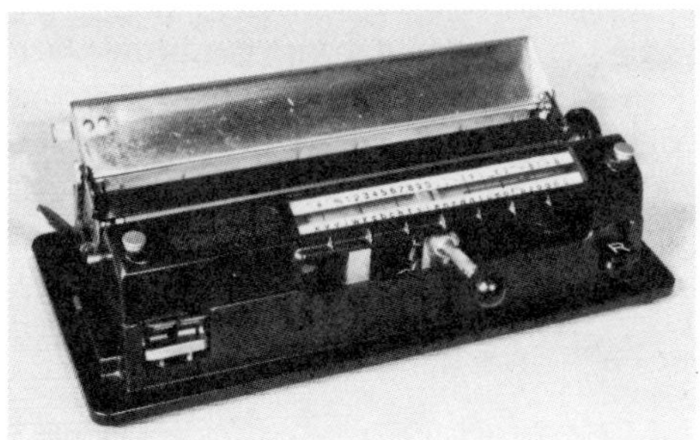

234. Carissima   (TMV)

same approximate dimensions as the Junior (a) was introduced in 1934 by Th. Knaur-Höel & Denk. It weighed only 4 lb thanks to its bakelite housing. The machine had a letter index with sliding indicator, two shift keys and back-spacer. Inside the machine was a small metal type-wheel which turned to the desired character as the indicator was moved and raised for capitals or figures by the shift keys. Printing by ribbon. (Fig. 234)

Carlem: See *Sun*.

Carmen: Small German front-stroke portable with three-row keyboard and double shift, of conventional design. The manufacturer went through three changes of name before settling for Carmen Werk A.G. Introduced in 1920.

Carmona: Several patents were issued from 1897 to Manuel Carmona of Mexico City for a typewriter employing five keys and two levers to select characters from among ninety-three around the edge of a type-wheel.

Carpenter: A stenographic machine patented in the United States in 1888 by W. M. Carpenter of Missouri. It was a complex design with three sets of keys, printing its own shorthand alphabet.

Carsalade: A Belgian inventor by the name of Paul Carsalade du Pont is recorded[28] as having built a total of ten different syllable machines between 1905 and 1912, none of which was successful. This is hardly surprising, judging from a description of his fifth model, which had a keyboard twenty inches high to accommodate the complex linkage.

Cartograph Tessari: Recorded[42] as exhibited in Venice in 1907, but not described. There was a Tachygraph (*Tachigrafo*) by a Tessaro which was still around by 1907, and the casual way in which foreign 'funny sounding' names were handled in those days—in any language—leads one to suspect that it may well be the same machine.

Cash: A down-stroke machine which typed on a flat paper-table instead of a platen. It was patented in 1887 by Arthur Wise Cash of Hartford, and marketed first by him, then in 1896[6] by the Typograph Co., the name of the machine being

235. Cash   (MPM)

changed to *Typograph*. It had a four-row keyboard controlling a semi-circle of vertical type-bars in front of the printing point. A ribbon was used for impression. (Fig. 235)

Celtic: A French manufacturer of conventional four-row machines, S. A. Celtic, introduced a folding portable model in 1923. Weighing less than six lb, the machine had a three-row keyboard with double shift and, when unfolded, was of the down-stroke class, the type-bars forming an arc in front of the printing point. In its folded state, it was only an inch and a half high.

Centaur: A portable machine by this name is reported[28] to have been announced in London in 1925. Details lacking.

Century (a): A conventional front-stroke machine with three-row keyboard and double shift was marketed by the American Writing Machine Company, which had previously produced the Caligraph and New Century. It was virtually identical with the *Remington Junior*, introduced in 1914 and reportedly designed by Fred Sholes. Until the advent of the Century, the American Writing Machine Co., which formed part of the trust called Union Typewriter Co., appears to have remained dormant from 1906, when the New Century was withdrawn from production.

Century (b): The third machine invented by Thomas Hall: the first employed type-bars, and the second, named after him, a square type-index. On the Century, the inventor substituted a type-sleeve for the flat plate but apart from this, the machine was similar to the *Hall Type-Writer*. The sleeve had ten rows of ten characters each, thereby offering a hundred characters which presumably gave

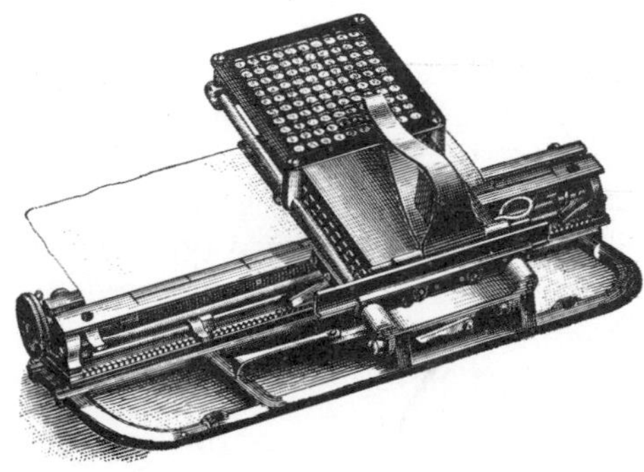

236. Century (b)   (LC)

the machine its name. An index and selector like the *Hall*'s were used, and printing was effected by a similar depression of the selector. Inking by roller. (Fig. 236)

Century (c): Two patents were granted in 1897 to Lee Burridge and assigned to the Century Machine Co. of New York for a design featuring a curved three-

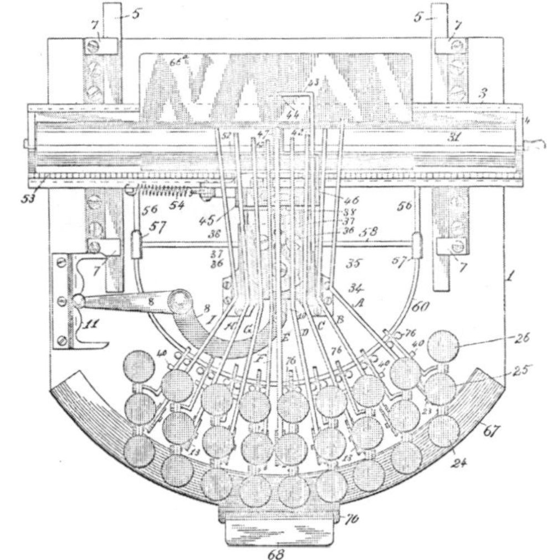

237. Century (c)   (USP)

row keyboard of twenty-seven keys on only nine key levers, so that there were three rocking keys per lever. The nine type-bars had three characters on each of three faces, the bar rotating to present the desired face to the paper. Inking by roller. The machine is sometimes referred to as *Hess and Stoughton,* after the two men who patented some subsequent improvements on Burridge's characteristic design. (Fig. 237)

Cereseto: A machine for the blind invented in 1894 by Vittorio Cereseto of Genova and exhibited in Venice in 1907. The fingers operated keys arranged according to frequency; each thumb had its own key projecting from the keyboard towards the operator. This sensible design was inspired by the anatomy of the hand. One thumb key was for spacing, the other for carriage return.

Chambonnaud: Shorthand machine designed by a Frenchman of this name, with sixteen radially disposed type-bars operating on the thrust principle: the instrument, in fact, looked like a small Wellington. It utilized the 'chord' principle in conjunction with four additional identification keys, and printed in Roman characters. Several models appeared but the machine was not manufactured.

Champignon: See *American (a).*

Champion: See *Peoples.*

Chancellor: See *Kanzler.*

258

Chicago: See *Munson.*

Childress: A cipher machine reportedly[42] patented by H. P. Childress of Memphis in June 1922 is probably an erroneous reference to a patent issued to that man in July 1902.

Christiansen: B. F. Christiansen was granted a British patent in 1886 for an electric machine which was never produced.

Cienant: Paul Cienant of Buffalo is reported[28] to have invented a musical writing machine in 1902. It operated by clockwork and printed the score automatically as it was played.

Cipher: A *Hammond* typewriter formed the basis for this cipher machine introduced in 1919 by the International Cipherwriting Co. of Chicago. Invented in 1916 by Frederick Sedgwick, it encoded by altering the position of the type sector relative to the keyboard.

Cito: See *Stoewer.*

Clavigrafo: Isidoro Maggi was granted an Italiàn patent for a shorthand machine in 1872. Called *Logomatografo*, it hàd a separate piano keyboard for each hand and printed its shorthand code on paper tape. It was not a success and a further eight years of developmental work went into perfecting it. Renamed the *Clavigrafo* and patented in Italy in 1880, it retained the piano keyboard, but printed in Roman characters. A company was formed to promote the machine but the project collapsed upon the death of the inventor in 1884.

Clerk: A man by this name was granted a US patent for an electrical typewriter in 1890.

Cleveland: See *Hartford.*

Clevenger: An inventor of this name designed a machine to be marketed for $10 but the project failed. The same man invented the telegraphic typewriter called *Book Electric* and was also granted a patent in 1911 for a book machine using a novel arrangement of plungers operating a spherical index similar to the Lambert.

Coffman: G. W. Coffman of Garden City, Kansas, was granted a US patent in 1902 for a small linear index machine, with upper- and lower-case characters cast on a rubber belt. It had an indicator for selecting the letters from a two-row index on the upper surface of the machine. Designed to be carried in the pocket, it measured approximately 10″ by 2″ by 2″. (Fig. 9)

Cole: A simple circular index machine with a knob for selecting and printing, patented by J. W. Cole of Kansas in 1884.

Columbia (a): A small type-wheel machine incorporating differential spacing. It was patented by Charles Spiro, a N.Y. watchmaker, in 1885 and marketed by the Columbia Type Writer Co. of N.Y. The vertical type-wheel with characters around its periphery was rotated by means of a knob on its right; this wheel was geared at 90° to a pointer which indicated on a circular scale the character immediately above the printing point. Depression of the handle not only brought the type-wheel into contact with the platen, but also advanced the carriage prior to impression by means of a linkage whose travel was regulated to coincide with the width of the letter to be printed, thus providing the differential spacing. Inking by pad. The first model, printing only upper case, was followed almost immediately by a second which had both upper and lower case around the periphery of the type-wheel, although a machine in SML has separate wheels for each case mounted parallel to each othcr. (Fig. 163)

Columbia (b): See *Bar-Lock*.

Columbia Bar-Lock: See *Bar-Lock*.

Commerciale: See *Velograph*.

Commercial Visible: A type-wheel machine with a three-row keyboard and double shift, it was one of the many inventions of Richard Uhlig. Covered by a patent filed in 1897 and granted two years later, it was first introduced in 1898 by the Visible Typewriter Co. of N.Y., the name later changing to Commercial Visible Typewriter Co. It was of limited success. Also marketed as *Fountain.* (Fig. 162)

Competitor: A machine by this name is listed[42] without details.

Complete: Index machine listed[6] without further details.

Conde: Invented by Samuel L. Condé of Illinois and patented in the United States in 1894, this machine with four-row keyboard had a semi-circle of vertical type-bars in front of the platen, but was not a full down-stroke machine, like the Bar-Lock, by virtue of the fact that the platen was placed on a level high enough to prevent the type-bars having to descend to a horizontal position. This, of course, made the typing more readily visible. Very few were made. (Fig. 238)

260

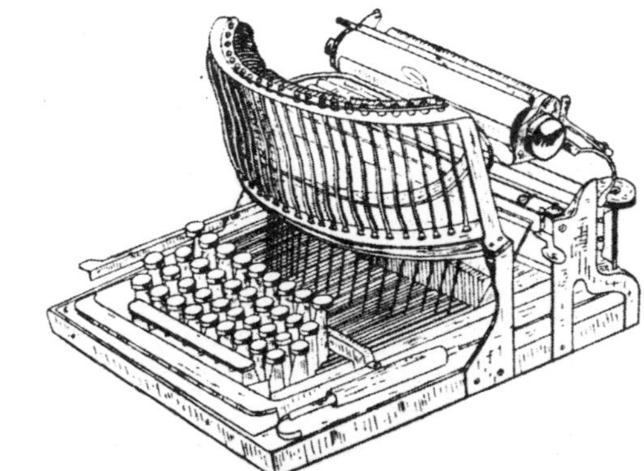

238. Conde   (14)

Conover: See *Munson.*

Conrad: Hiram Conrad of Pennsylvania was granted a US patent in 1876 for a machine in which the characters were placed on the ends of flexible tongues radiating outwards from a horizontal disk pivoted in the centre.

Constançon: A machine for the blind, invented by Maurice Constançon of Switzerland in 1910. It embossed braille cells on both sides of a page by using the space between lines on one side for embossing on the other. A special device on the machine made it possible for a blind operator to perform this operation. It was manufactured in two models.

Cookson: F. N. Cookson of Wolverhampton, England, was the inventor of this beautiful machine, classified as anonymous by SML. His British patent of 1885 differs from the existing model in minor details, the latter having improved

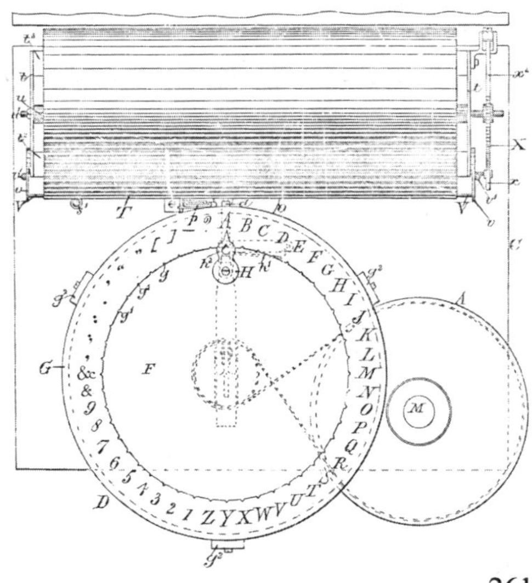

239. Cookson   (BP)

letter-selection mechanism. It is a plunger machine with the type radiating out from the periphery of the type-wheel. A hammer presses plungers against platen. (Figs. 120 and 239)

Corona: See *Standard Folding.*

Corona Piccola: See *Standard Folding.*

Corradi: A machine for the blind, invented in 1904 by Piraino di Corradi of Palermo, having six keys in two rows.

Correspondent: See *Rofa.*

Cosmopolit: See *Kosmopolit.*

Cosmopolitan (a): See *Kosmopolit.*

Cosmopolitan (b): See *Crandall.*

Courier: See *Oliver.*

Couttolenc: A French patent granted in 1902 to Eugène Couttolenc for a machine in which the type-wheel was positioned *beneath* the platen—a rare example of a non-visible type-wheel design.

Craigmiles: A shorthand machine, for which a US patent was granted to Edwin Lee Craigmiles of Tennessee in 1893. It had keys in pairs for depression either singly or together by the same finger. Positioned according to the anatomy of the hands were four pairs each hand for the fingers, one for each thumb, and one for each palm.

Cram: A combined typewriter and adding machine listed[42] as having been introduced unsuccessfully in 1907.

Crandall: This was the first type-sleeve machine and was patented in 1879 (Great Britain) and 1881 (US) by Lucien Stephen Crandall; it was one of a number of instruments designed by this pioneer typewriter inventor. Manufacturer was the Crandall Machine Co. of Groton, N.Y. The type-sleeve had six circles of characters

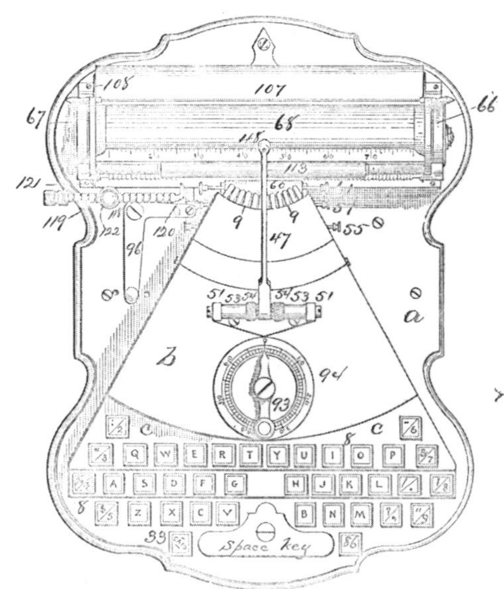

240. Improved Crandall   (USP)

around it and was mounted obliquely horizontal on the first model and vertical on the two subsequent models. Depression of a key rotated it and moved it along its axis to select the corresponding letter, bringing it down to the platen and locking it by engaging a pin in a hole. Printing by ribbon. The first model appeared in 1881 and had a three-row keyboard with double shift. It was followed by a second model with a curved two-row keyboard while a third, labelled *Universal Crandall*, returned to the straight three-row design and was introduced in 1893. Model Four is reported[22] to have been introduced in 1906 by the only source which offers a definite date, but an 1898 ribbon catalogue (Webster Co.) already lists a fourth model. Other noteworthy features of the Crandall included provision for differential spacing.

A different machine altogether was patented in 1895 and called *Improved Crandall*. It was of three-row keyboard design with thirty-six keys and single shift, employing twelve horizontal type-bars pointing radially towards the platen. On the underside of each bar were six characters; depression of a key caused one of the bars to slide forward till the corresponding character was above the printing point, whereupon a hammer striking from above produced the impression. Roller inking, a later model using ribbon. It is reported[27] to have been offered in England under the name *Cosmopolitan*. (Figs. 158, 159 and 240)

Crary: A down-stroke book typewriter patented in 1892 by J. M. Crary of New Jersey, with three rows of keys describing complete concentric circles around the enclosed type-basket. It printed upper and lower case on the flat sheet of paper

241. Crary   (MPM)

or page beneath, the whole unit moving laterally as typing progressed. Only a small number of these machines was made. (Fig. 241)

Crown (a): The machine which took its name from its shape was patented by an American called A. G. Donnelly of N.Y. in 1887 and introduced by the Crown

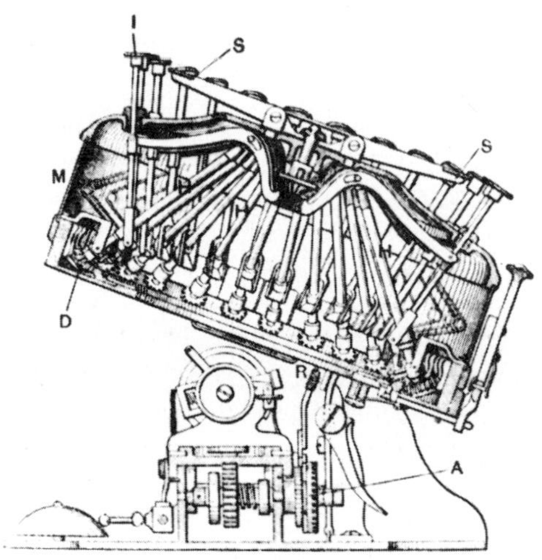

242. Crown (a)   (59)

Typewriter Manufacturing Co. of Albany, N.Y. This down-stroke design specified a drum mounted obliquely over the platen; a single circular row of keys protruded from the upper surface of the drum and, upon depression, lowered the corresponding type-bars which were pivoted at the outer base of the drum and formed a frustum. Upper and lower case were offered by means of a unique double shift arrangement: the lower end of each type-bar was located by means of a small wheel geared to a toothed ring which circled the lower periphery of the drum, so that lateral displacement of the ring turned each of the type-bars to the corresponding shift position. The type, therefore, was on three faces of the end of the type-bar. Ribbon for inking. A truly desirable collector's piece of which no

264

example is known to have survived. Also known as *Donnelly*. (Fig. 242)

Crown (b): The second machine produced by Byron A. Brooks, this one used a type-wheel geared at right angles to a pointer which selected the letters from a

243. Crown (b)   (GC)

straight scale. Upper and lower case were located by double shift and the type-wheel was brought down to the platen by depression of a lever, inking itself from a roller. The inventor's type-wheel patents date back to 1883, but the present machine was not marketed until some years later by the National Meter Co., N.Y. (Fig. 243)

Crown (c): See *National (a)*.

Crown (d): See *Adler*.

Culema: Conventional German front-stroke machine with three-row keyboard and double shift, named after its inventor, Curt Lehmann, and introduced in 1919 by Schreibmaschinenfabrik Gebr. Lehmann K. G. Manufacturer's name changed four times during the seven years in which the machine was made. Model Two appeared in 1920, and Three in 1922, but changes were minor. Also marketed as *Hansa* and *Orientalis*.

Dactygram: A diminutive French type-wheel machine with five keys and double shift, printing ninety characters by the 'chord' system. Invented in 1920 by Georges Moulin. The fingers of one hand operated the keys without changing position: depression of the index finger only printed A, index and middle printed D, middle only E, etc. Dimensions: approximately 8″ by 3″ by 3″.

Dactyle: See *Blickensderfer*.

Daisy Point Writer: US machine for the blind invented in 1865 by Joel Smith, so called because the six radially positioned keys above the cell pricker resembled the petals of a flower. The machine was improved and a more compact model introduced in 1880—the large frame previously used was replaced by a winding knob and detent for line spacing.

Damon: A shorthand machine patented in the United States by Charles Damon in 1889. It had a key for each of the eight fingers, plus a central key for use of either thumb and a space key. Not manufactured.

Dandy: Listed[28] but without details.

Darling: A small type-wheel machine introduced in 1910 by Robert Ingersoll and Bros of N.Y. Several models appeared under this name, the smallest consisting of nothing more than a hand-held type-wheel and indicator attached to a

244. Darling   (AMP)

thin frame that held the paper. Another version was similar but could be clamped to a table. The most luxurious model appears to have been the one illustrated in Fig. 244. Also marketed in Europe under the name *Trebla*. (Fig. 18).

Dart: A type-wheel machine marketed by the Dart Marking Machine Co., designed for printing signs in large letters as on crates, etc. It consisted of a horizontal

245. Dart   (6)

266

index with a knob and pointer geared to a vertical type-wheel, the whole thing in a frame running on wheels. Turning the knob selected the letter; depressing it performed the printing and drove the machine to the right to the next space. Inking by roller. The type-wheel was approximately eighteen inches in diameter. It was reportedly introduced in the US in 1899, although a patent to L. Dart assigned to the Type-Writing Machine Co. of Conn. for this instrument is dated 1890. (Fig. 245)

Dattilo-Musicografo: Italian musical typewriter invented in 1921 by Andrea Ferretto and manufactured for several years.

Daugherty: Patented in 1891 by James Daugherty of Pennsylvania, this four-row machine was the first manufactured front-stroke 'visible'. Its shape was characteristically long, narrow and flat, except for the carriage, which went

246. Pittsburg Visible (Daugherty)   (AC)

straight up at the back of the machine, giving it an elongated L-appearance. It was characterized by easy disassembly, and the type-bars and keyboard lifted out as a unit.

Manufactured from 1890 first in the Crandall plant in Groton, N.Y. and then by the Daugherty Typewriter Company in Kittanning, Pa., it was renamed *Pittsburg* in 1898, with Model Ten appearing in 1902 and Model Eleven in 1908. Model Twelve, in 1911, was of conventional appearance but the feature of easy disassembly was retained. The company went into receivership in 1913, and closed down completely eight years later. Later models were marketed in different countries under the following names: *American, Broadway Standard, Decker-Beachler, Fort Pitt, Reliance, Reliance Premier, Reliance Visible, Schilling,* and *Wall Street Standard.* A company in Pittsburg also offered it as the *Shilling Brothers* for a brief period. (Fig. 246)

Davis (a): See *Wellington.*

Davis (b): Edmund Davis of N.Y. was granted a US patent in 1877 for a slow type-wheel device.

Daw and Tait: An interesting down-stroke design, invented by T. G. and H. Daw and patented in England in 1884. It had a circular keyboard, similar in appearance to the *Crown* and *Crary*, but with two circles of keys. The keys were connected with type-bars pivoted around the lower perimeter of the circular frame and kept in a raised position by means of coil springs. Upon depression of a key, the type-bar struck down on to a cylindrical platen, printing through carbon paper placed *beneath* the page, so that the impression was not visible until the page was removed from the platen. The machine incorporated differential spacing. (Fig. 14).

Decker-Beachler: See *Daugherty.*

De Dedulin: A shorthand machine patented in France in 1872. It was of radial plunger design, printing through carbon paper on a tape around a roller which advanced along an endless screw. The formula appears to have been: De Dedulin = Foucauld + Wheatstone.

De Esse: See *Stoewer.*

Defi: See *Eagle (a).*

Delta: One of the many machines designed by Richard Uhlig. This one appears never to have been manufactured.

Dement: A number of patents were granted to Merritt H. Dement of Chicago in 1883 et seq. for a design featuring type sliding in grooves cut longitudinally in a cylinder.

Deming: Two 1876 patents for up-stroke machines similar to the Sholes and Glidden were among several patents granted to Philander Deming of N.Y.

Demountable: A front-stroke upright machine with a standard four-row keyboard manufactured in 1921 by the Rex Typewriter Co. and first labelled *Rex Demountable,* so named because it offered easy disassembly and interchangeability of keyboard, type basket, ribbon mechanism and carriage. It was widely exported, and many machines survive to this day in markets like Spain. The manufacturing company changed its name to Demountable Typewriter Co. in 1923.

Dennis Duplex: The first syllable machine to reach the market, this non-visible up-stroke was patented by A. S. Dennis in 1895 but produced earlier by the Duplex Typewriter Co. The original patent refers to and improves upon another granted in 1884 to a Henry Orpen of Missouri for modifications permitting two keys of a Remington or Caligraph to be depressed simultaneously. Its 100 keys were split into two halves, each hand operating its corresponding half. The left side printed a complete upper-case alphabet followed by a lower-case, and the same lower case alphabet was repeated opposite on the right, with figures and characters above it. Each half of the keyboard controlled its own type-bars, which hung in two semi-circles the centres of which were one space apart. In this way, both the left and right hand could depress a key simultaneously. The inventor hoped by this means to double the typing speed which, of course, did not happen. The machine used a pad for inking; subsequent models, simply called *Duplex*, used a ribbon. In Germany some units were labelled *Germania-Duplex*.

The later *Jewett* was a development of the Duplex. (Fig. 212)

Densmore: Amos and Emmett Densmore, Walter Barron and Franz X. Wagner collaborated on the design and development of this up-stroke machine with standard four-row keyboard which was produced in 1891 by a company called Densmore and Densmore. The design was noteworthy chiefly in that the type-bar pivots were made to run in bearings as was the carriage, and a novel type-bar arrangement was said to improve touch. It was a well-constructed machine which enjoyed considerable success. Model Two: 1897, Model Four: 1902, and Model Five: 1907; shortly thereafter, it was phased out by the Union Typewriter Co. (Fig. 153)

Destaillats: A French patent for a type-wheel machine was granted to an Abbé of this name in 1903.

Diadema: See *Juventa*.

Dial: A toy circular index machine manufactured by Louis Marx Co. in the 1930s. Full title was *De-Luxe Dial* Typewriter. A circular dial was geared to

247. Dial De-Luxe   (LC)

the type-wheel with characters around its edge. Printing was by depression of a key on the front of the machine, inking was by roller. (Fig. 247)

Diamant: Three-row front-stroke machine with double shift invented in Germany by Jakob Heil and manufactured 1922–26 by Diamant Schreibmaschinen GmbH. Sold in England as *Diamond.*

Diamond: See *Diamant.*

Dictatype: See *Anderson (a).*

Dictograph: See *Stenophile.*

Dietrich: Listed[6] without details.

Diplographe: An 1877 French machine for the blind which printed simultaneously in Roman characters for the sighted and embossed the same text for the blind. (Fig. 201)

Direct: Listed[6] but without details.

Discreet: See *Volksschreibmaschine.*

Diskret: See *Volksschreibmaschine.*

Dissyllabe: A Swiss shorthand machine produced by A. Persiaux in 1920. It had twelve keys and printed a form of combination type first devised by Livermore.

Dodge: M. C. Dodge of Michigan was granted a US patent in 1885 for a circumferential type-plunger design.

Dodson: An impractical multiple type-wheel device was patented by D. W. Dodson of Pennsylvania in 1884.

Dold: A French shorthand machine, invented in 1920, which permitted up to ten syllables to be printed per line—the usual machines then in use, printing on paper tape, were limited to one syllable per line. It was claimed that this additional printing capacity made it easier to read. Not manufactured.

Dollar: Primitive vertical type-wheel machine almost identical to the *Herrington.* It was patented in 1890 *et seq.* by Robert Ingersoll of N.Y. and manufactured by his company, which had already made a name for itself with its Dollar watches. (Fig. 248)

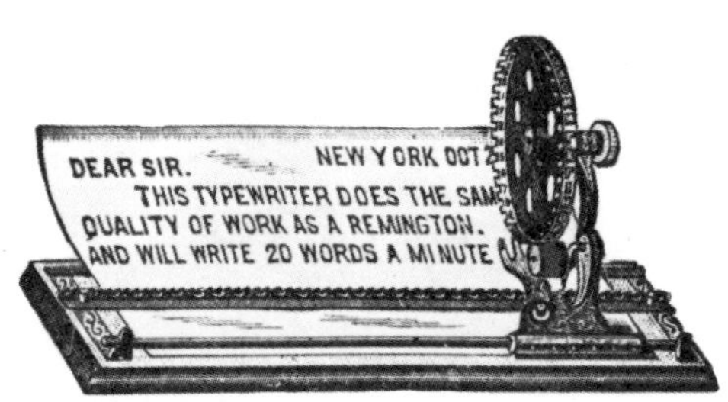

248. Dollar (LC)

Don: A machine by this name is listed[42] without details.

Donnelly: See *Crown (a)*.

Doropa: See *Meteor*.

Draper: See *Munson*.

Driesslein: An early patent for an electric up-stroke design granted to C. L. Driesslein of Chicago in 1879. Ribbon inking.

DS: See *Stoewer*.

Dukes: A dual-purpose machine developed in 1908 by Dukes Typewriter Co. of Little Rock. Of the down-stroke group, it had a standard four-row keyboard and was designed to be used either as a book typewriter or, by coupling it with the optional carriage, as a regular machine. Patents date back to one for a book

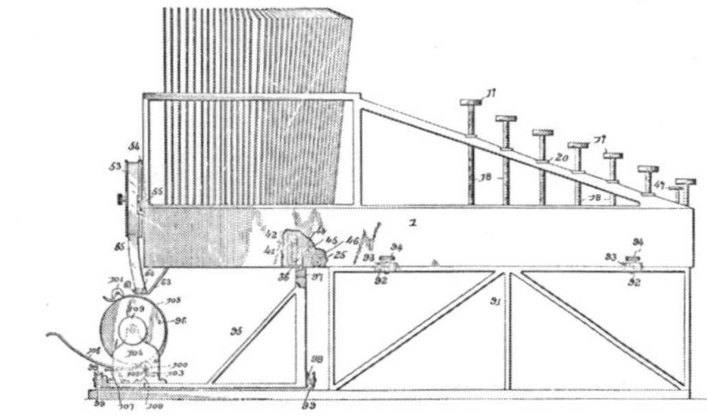

249. Dukes (USP)

machine granted to Harry S. Dukes of Missouri in 1897. Another patent, filed in 1900 and granted in 1906, is for an interesting 'Vertical Plane Type Writer', equally suitable for printing in books and for work held vertically in the machine. (Fig. 249)

Duographe: A machine for the blind invented in 1907 by a clergyman called

Stiltz, which simultaneously printed in Roman characters and embossed braille cells.

Duplex: See *Dennis Duplex.*

Duployé: A shorthand machine with fifteen keys in three rows invented in 1890 by Gustave Duployé.

Dupray: A French patent for a double-shift machine granted to a man of this name in 1903.

Duval: See *Express (a).*

DWF: Yet another thrust-action machine with three-row keyboard and double shift, patterned after the Wellington. This late arrival was introduced in 1923 by Deutschen Waffen- und Munitions-Fabriken of Berlin.

Eagle (a): Three-row double-shift machine with the type on a swinging sector, similar to the Keystone, it was patented from 1902 by Charles J. Paulson and

250. Defi (Eagle (a))   (MPM)

produced by the Eagle Typewriter Co. of New York in the same year. Printing through ribbon. Also marketed as *Defi.* See *Sterling (a).* (Fig. 250)

Eagle (b): See *American (b).*

Eckels: George Eckels of Chicago patented this complex type-wheel syllable machine in 1892 *et seq.* An example is in MPM where it is listed as *Eccles*, inventor and manufacturer unknown. The origins of this spelling are not given. Other sources[28] add even more confusion.

Eclipse (a): A machine in MPM which is virtually identical in every detail to the *Brooks*. They may be the same machines under different names.

Eclipse (b): See *Mignon.*

Edel: German thrust-action machine, invented in 1922 by Ernst Dettmann and Erich Laude, whose initials supplied the name. It was not manufactured.

Edelmann: German type-wheel machine introduced in 1897 by Wernicke, Edelmann & Co., later by several other companies with different names. An indicator at the front of the machine selected letters from a curved index, the type-wheel being geared directly to the indicator. Depression of the indicator key brought the type-wheel down onto the paper. Roller inking. Additional keys on the left were for spacing and for two shift positions. It was a solid and well-built machine produced virtually without change until the First World War. An Edelmann in AMP is labelled *Gladstone.* (Fig. 180)

Edison: An electric type-wheel machine patented in 1872. See page 147 and Fig. 134.

Edison Mimeograph: Invented by Thomas Edison and manufactured by A. B. Dick Co., Chicago, in about 1894, the patents dated the following year. The company was already producing a stencil duplicating device called Mimeograph, invented by Edison and widely used, and they decided to offer their own type-writer to go with it. However, competing typewriter manufacturers threatened to have the duplicating equipment boycotted unless the typewriter was withdrawn, and the company complied, so that very few were actually made.

The machine had a wheel mounted on a vertical shaft geared at the bottom to a disc with three indicators (labelled SMALL, CAPS and FIGS-CHARS) placed equidistantly around its periphery. A curved letter index was on the base of the machine in front of the operator and any one of the three pointers could be used to select the corresponding characters from this index. The indicator disc, geared to the vertical shaft, turned a wheel on its upper end; around the circumference of this wheel were type-plungers which struck upwards at the platen. This of course had to be raised to reveal the printing. The plunger was raised by being struck by a hammer swung upwards by the depression of a printing key on the left, and as the hammer fell (by gravity), a projection brought the plunger back to its original position. A space key was provided beside the printing key, and a ribbon was used for printing. Model One had seventy-eight plungers,

Model Two eighty-six and Model Three ninety. It was an awful design, given its late date and the calibre of the inventor, but from a collector's point of view, remains one of the most desirable machines ever made (Frontispiece).

Edland: Circular index machine patented in 1891 by Joe L. Edland and manufactured by the Liberty Manufacturing Co. of New York. The characters were cast into a metal housing, set obliquely in front of a platen; beneath, a disk with the type on flexible segments was attached to the indicator shaft. Teeth around

251. Edland   (QC)

the circumference of the plate locked the indicator in place at the desired character, and printing was done by the depression of the indicator key. A later model introduced in 1894 utilized a type-wheel mounted on a vertical shaft geared to the indicator. The letter index was now semi-circular; depression of the indicator brought the type-wheel against the platen. (Fig. 251)

Eggis: See *Velograph.*

Elektrograph: An electric machine of the type-plunger class invented by Dr Faber of Berlin in 1900. Paper was laid flat on a platform, provided with lateral and longitudinal movement, above which were type-plungers similar to those on the Writing Ball. These plungers were electrically controlled from a keyboard of eighty keys in front of the instrument. According to period descriptions, the machine was capable of making clear carbon copies, the number of which might be multiplied 'indefinitely'[27 etc.] by merely increasing the number of volts in the circuit! (Fig. 199)

Elgin: See *American (b).*

Elite: See *Stoewer.*

Elliot-Fisher: Product of the 1903 merger of the Elliot Hatch Book Typewriter Co. and the Fisher Book Typewriter Co. was this product, manufactured by the Elliot Fisher Co. See *Elliot-Hatch* and *Fisher.* (Fig. 216)

274

Elliot-Hatch: A down-stroke book typewriter patented in 1896 by George Crawford Elliot and assigned to Walter P. Hatch of N.Y. and introduced by the Elliot Hatch Book Typewriter Co. in 1897. Like its competitor the *Fisher*, it was a down-stroke machine in which the type-basket was located below the four-row keyboard. Once again, the whole instrument rode on rails laterally and longitudinally. It was offered in five models, according to the size of page to be covered, and was also suitable for typing on loose sheets, cardboard etc. In 1903, the company merged with the Fisher Book Typewriter Co. to form the Elliot Fisher Company.

Emerson: A front-stroke machine with a unique type-bar action, three-row key-board and double shift, this was one of the most successful designs of the prolific Richard Uhlig. Invented in 1907, it was marketed by the Emerson Typewriter Co. of Boston, later Chicago, with a factory in Momence, Ill. The Emerson had vertical type-bars standing in front and to the sides of the platen, with the characters

252. Emerson   (LC)

on the same horizontal plane as the printing-point. The type-bar pivots were therefore vertical, successively banked outwards and, on depression of a key, the corresponding bar swung around on its pivot to strike the platen. In 1910, the operation was bought up by Mr Roebuck of Sears Roebuck and vigorously promoted, the company changing its name first to the Roebuck Typewriter Co. and finally, in 1914, to the Woodstock Typewriter Co., which withdrew the Emerson in favour of the standard front-stroke Woodstock.

The machine was also placed on the market by Harry A. Smith, who made a living by preying on the remains of dying typewriter concerns. In this case, he most likely bought unsold Emersons from the Roebuck Typewriter Co., at the time when it was already marketing the Woodstock. The Emersons sold by Harry A. were labelled *Smith*. (Fig. 252)

Emka: For a time the *Albus* was marketed in Germany under this name which was formed from the initials of the agent in that country, Max Keller.

Empire: See *Wellington.*

Engler: See *Albus.*

English: An English down-stroke machine, patented in 1890 by Michael Hearn and Morgan Donne and marketed by the English Typewriter Co., Ltd. It holds the unique distinction of being a visible machine, which the inventors took pains to make non-visible by having the typed portion of the page roll itself up in a drum behind the platen! This was promoted as one of the advantages of the

253. English   (SML)

machine; anyone standing on the other side of it could not read what had been written, and the drum was elevated and so positioned as to obscure the line being printed from any but the operator.

There was a two-row curved keyboard with double shift, and the impression was by ribbon. There was no direct linkage between type-bars and key levers, which were kept in an upright position by an extension below the pivot acting as a counterweight. The key lever struck the type-bar, thereby swinging it down to the platen, whereupon the counterweight brought it back. (Fig. 253)

Ennis: Electric type-wheel machine invented by George Ennis of Troy, N.Y. and patented in 1901. It was to have been manufactured by the Electric Typewriter Co., but the project never left the ground. The type-wheel was mounted vertically above the platen and was brought down onto it electromagnetically. Inking by roller.

Erika: A front-stroke folding portable patterned after the *Standard Folding,*

276

with three-row keyboard and double shift, manufactured in the *Ideal* works of Seidel & Naumann A.G. Produced from 1910 until 1927, after which the Erika went four-row and dropped the folding feature. Also marketed as *Bijou* and *Gloria*.

Essex: Manufactured by the Essex Universal Typewriter Co. of N.Y., which apparently offered two models although information on the machines is confused and none is known to have survived. The first model is said to have had sixteen keys in two rows; later a three-row with double shift and twenty-seven keys was substituted. It would appear that the early model had rubber type on a swinging sector off-set to the right of the machine to permit visibility of printing. This model used a hammer to strike the type against the paper from inside the sector outwards. Later, the carriage was made to rock forward for impression. Inking on the first model by roller, later by ribbon.

Eureka (a): A simplified version of the *Hall Type-Writer* was introduced in Germany in 1898 by the sewing-machine manufacturer Karl Heinrich Kochendörfer, who later manufactured the German *Imperial* and also invented an electric machine. The major difference between his Eureka and the Hall was that the latter's index was square while the former's was eye-shaped. Otherwise, a pointer selected the character as on the American's machine, the type-index was likewise of rubber, the impression being made on paper held flat on the base of the machine.

Eureka (b): See *Simplex (a)*.

Europa: See *American (b)*.

Excelsior Script: A dual-purpose machine which printed either script or type without change in the machine was invented by an American called Halstrick and was to have been marketed by the Excelsior Script and Typewriting Machine Co. of San Francisco in 1899. Type case from the owner's handwriting was offered as a gimmick, although the chief purpose of the instrument was rather its ease with italicizing.

Express (a): An electric type-wheel machine invented in 1900 by a Frenchman called Jules Duval. It used a pointer to close electrical circuits within the machine by touching contacts at the characters on an index plate. It was highly commended at the time but was not produced. Sometimes called after the inventor, *Duval*.

Express (b): See *Liliput*.

Express (c): See *National (a)*.

Faktotum: German copy of the British Imperial Model B, introduced in 1912 by Fabig & Barschel. Later model (1913) replaced the curved keyboard by a straight three-row. Production ceased in 1916. Also marketed under the names *Leframa* and *Forte-Type.*

Famos (a): A cheap and primitive circular index machine manufactured by the German firm of Gustav Tietze A.G. from 1910 on. The index stood vertically over the printing-point; it was turned by means of a knob to select the desired character and brought down onto the paper by depression of a key. Roller inking. The paper was clipped to the platen, which then rotated for letter spacing so that the lines were typed *around* the platen, as on early inventions described in Chapters Three and Four. Marketed in France as the *Victoria.* The same company later produced the *Aktiv.*

Famos (b): See *Geniatus.*

Farmer: A swinging-sector machine comparable to the later *World* was patented in 1885 by L. C. Farmer of Minneapolis.

Favorit: See *American (b).*

Fay-Sho: See *Rem-Sho.*

Fay-Sholes: See *Rem-Sho.*

Featherweight Blick: See *Blickensderfer.*

Featherweight Blickensderfer: See *Blickensderfer.*

Fidat: See *Juventa.*

Finnland: See *Sampo.*

Fisher (a): A down-stroke book typewriter patented in 1896 by Robert J. Fisher of Tennessee and manufactured by the Fisher Book Typewriter Company of Cleveland, Ohio. The type-bars stood vertically in a semi-circle around the printing point which was located *under* the machine, to allow it to type on the pages of bound books. The keyboard was above the type-bars—the first model had a double keyboard, later models a standard four-row. The whole assembly rode on a frame placed over the page and moved laterally on rails for typing. An adding attachment was offered in 1902; the following year, the company merged with its competitor Elliot Hatch to form the Elliot-Fisher Company.

Fisher (b): An L. G. Fisher is reported[27] to have invented in 1904 a machine similar to the *Oliver* but using a standard four-row keyboard with single shift. It was apparently never manufactured.

Fitch: This down-stroke machine with type-bars located obliquely behind the platen was patented in 1886 by Eugene Fitch of Des Moines and produced in 1891 by the Brady Manufacturing Co. of Brooklyn. It was first produced with a two-row keyboard with double shift, later three rows. The type was made of vulcanized rubber and inked by brushing past a roller on its way to the paper, which was coiled in frames in front of and behind the platen. (Fig. 173)

Fiver: See *Oliver*.

Fleet: See *American (b)*.

Forbes: A stencilling machine patented by George Forbes in 1876, similar to Thurber's Chirographer. Not manufactured.

Ford: A thrust-action machine with a curved three-row keyboard and double shift, patented by E. A. Ford of N.Y. in 1892, and manufactured in 1895 by the Ford Typewriter Co. of N.Y. The movement differed somewhat from the earlier

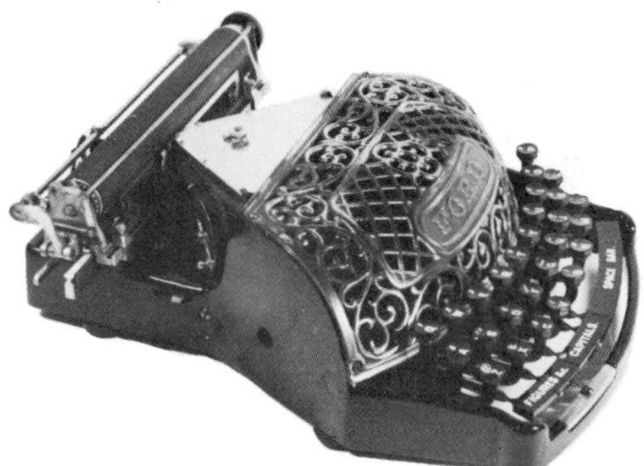

254. Ford   (LC)

Wellington—the type-bars also radiated outwards from the printing-point but instead of being impelled in a strictly horizontal direction, the Ford's described an arc on their trajectory to the platen. Printing by ribbon. The machine was made in two models, one of aluminium and the other of cast iron. It was assembled in France and called *Hurtu* and also in Germany and labelled *Knoch*, both of these names being taken from the firms which assembled and marketed it. (Fig. 254)

Forte-Type (a): See *Meteor*.

279

Forte-Type (b): See *Faktotum.*

Fortoni: A musical typewriter by a man of this name was first made known in 1922 after twelve years had been devoted to perfecting it. Four years later, a further improved model was announced: it now had a regular typewriter keyboard printing Roman characters, in front of which was a piano keyboard for the musical printing. The death of the inventor apparently prevented manufacture.

Fort Pitt: See *Daugherty.*

Fortuna: A German index machine patterned after the *Hall* and announced in 1908 by Hannoversche Apparatefabrik H.F. Tolle.

Fountain: See *Commercial Visible.*

Fox: A fine up-stroke machine patented in 1898 by W. R. Fox and Glenn J. Barrett and manufactured by the Fox Typewriter Co. of Grand Rapids, Michigan. No two sources agree on marketing date but all[22, 28, 42 etc.] list it as post-1900. However, an 1898 ribbon catalogue (Webster and Co.) describes the machine as being 'lately put on the market'. It was similar in design to other non-visible up-strokes, with the principal exception that the shift key moved only the platen and not the entire carriage. Various sub-models were offered until 1906, when the machine went visible with its standard Model Twenty-four. A small folding model called *Fox Portable* was introduced in 1917, also known as *Baby Fox.* It differed from the *Standard Folding* design in that the carriage folded back behind the machine, leaving it flat on top; nevertheless Fox was sued by the Corona Typewriter Co. over this folding feature which was subsequently dropped in favour of a conventional three-row double-shift front-stroke portable known as *Fox Sterling,* introduced in 1920. The manufacturer went into receivership the following year. (Fig. 195)

Fox Portable: See *Fox.*

Fox Sterling: See *Fox.*

Franconia: See *Standard Folding.*

Francus: An invention in which the type-bars were suspended above the printing point, as on the *Elektrograph,* patented in France in 1904 by Joseph Francus of Russia.

Franklin: One of the products of the inventive genius of Wellington Parker Kidder but less successful than his *Wellington.* The patent covering this down-stroke design was filed in 1889 (granted 1891) and assigned to the Tilton Manufacturing Co. of Boston. It was marketed by the Franklin Typewriter Co. of the same city until 1907 when it was bought up by the Victor Typewriter Co.

255. Franklin   (LC)

of N.Y. and replaced with a visible front-stroke machine of that name. Early models of the Franklin had a two-row curved keyboard and double shift, later models a similarly curved three-row with single shift and were labelled *New Franklin*. The patent covering this modification would appear to be one granted in 1897 to J. A. White of Boston. There was no direct linkage between key levers and type-bars, which were geared together in such a way that depression of a key rocked the lever on its fulcrum and, by means of intermeshing sectors on each, brought the type-bar down onto the platen. Impression by ribbon. The machine was decorated with a transfer of Benjamin Franklin, after whom it was named. Development of a square keyboard model was announced in 1906 but it apparently never reached the market.[27] (Fig. 255)

Frey: An electric machine for the blind with four keys and a space-bar, invented by Frederick Frey of Ohio in 1900. It was also suitable for telegraphic purposes.[6]

Frister & Rossmann: A German copy of the *Caligraph* introduced in 1892, under licence from the American machine. It was manufactured by the sewing-machine company Frister & Rossmann A.G., which later produced the *Moya* (for a short time) and the *Senta*.

Frolio: See *Gundka*.

Galesburg: See *Munson*.

Galiette: See *Perkeo*.

Galloway: An 1873 patent was granted to John Galloway of N.Y. for a linear index machine with scale and indicator, more than a decade before Burridge and Odell patented and manufactured almost identical machines (cf. *Sun* and *Odell*). The inventor then patented a shorthand machine of radial plunger design in 1882. All keys were operated by one hand and depressed in chords. The resulting code took the form of dots (up to eight) in a square format around a central point of reference: ∎:

∷ , etc.

Garbell: Portable thrust-action machine with three-row keyboard and double shift invented by Max Garbell of Chicago and manufactured by the Garbell

256. Garbell   (GC)

Typewriter Corp. of the same city, in 1919. Its design was similar to that used on the *Wellington*. Production ceased in 1923. (Fig. 256)

Gardner: A novel type-sleeve machine patented by John Gardner of Manchester in 1899 but reportedly manufactured by the Gardner British Typewriter Co. of that city in 1890[28] or 1893[40]. The machine had sixteen keys, two shift keys, and a space bar. Each letter key was divided down the middle, with a character in black

257. Gardner   (SML)

282

on one side and another in red on the other, with figures etc. in smaller print next to them. Simple depression of a key printed the 'black' letter; the 'red' required simultaneous depression of the key and the space bar. In addition, the correct shift position had to be selected from the separate shift keys, so that up to three keys had to be depressed simultaneously. The idea was to economize on size and price and mechanical complexity, although the product merely shifted these problems from the manufacturer to the operator. The type-sleeve stood vertically in front of the paper and impression was produced by a hammer striking the paper from behind. Inking by roller. There was a shield to prevent the paper touching the sleeve which had a 'playful'[27] habit of jamming, so that it received the typing rather than the paper. Production ceased in 1895. It was also marketed in France as the *Victorieuse* and in Germany as *Victoria*. (Fig. 257)

Garrison: US patent granted in 1885 to G. C. Garrison of Pennsylvania for a type-sleeve design with rectangular letter index and indicator.

Gefro: See *Gundka*.

Geka: See *Geniatus*.

Geniatus: Small German swinging sector machine manufactured in 1928. The type was on a vulcanized rubber belt divided into sections, like a tank track, with three characters one above the other per section. This belt was attached to the sector at both ends but passed over a pulley at the printing-point so that only the selected sector was presented for printing. Double shift. The characters were on a curved letter index on the front of the machine, and depression of the indicator key produced the impression through ribbon. Also marketed as *Famos*, *Geka* and *Gloria*. The manufacturer of this machine is apparently unknown, although the name *Geka* (*G.K.*) offers a clue—the *Gundka* came out with a model called *G & K* (in German: *G und K*) and an examination of the two machines side by side reveals that they may well have come from the same factory, even though the designs are somewhat different. Serial No. 7238 (QC) is dated 3/7/1928. (Fig. 26)

Gerda: A German machine manufactured in 1919 by Georg Emig. It was designed specially for the blind and one-armed victims of the First World War. A plate placed obliquely on the front of the machine was swung to select the letter, the pointer in this case being fixed while the index was movable. The type-wheel above the platen meshed with this plate. In addition, an aperture at the top of the plate revealed an embossed braille alphabet fixed to the housing beneath. By resting the hand on the plate and the index finger in the aperture, the blind could swing the plate to the desired character. Depression of the plate brought the type-wheel down onto the platen. Inking by roller. Two shift keys and space key were provided on the left and could be operated by the thumb—a one-armed operator who had lost his right hand could also have depressed them with his little finger, but not so easily, and there is no provision on the casting of the base for reversing the action for left-handed operators. The type-wheel was from the early model Blickensderfer. Also marketed as the *Melbri* (Fig. 204).

Germania (a): A pneumatic machine with a piano keyboard of fifteen white and fourteen black keys, this German invention by J. P. Moser looked exactly like a small harmonium, standing on its own legs and with two pedals for the two shift positions. It was to be manufactured in 1900 but the name could not be registered, having already been protected by another manufacturer, and nothing further was heard of this machine. The inventor next built a down-stroke machine of the same appearance as the Germania, calling it *Patria* but nothing came of this either, nor of a third inventive effort which was to have resulted in a similar book typewriter.

Germania (b): An electric machine by this name was exhibited in 1900. It was a prototype built by the German firm which assembled *Jewett* machines and was in fact an adaptation of one of these models. It was not manufactured, however.

Germania (c): See *Jewett*.

Germania-Duplex: See *Dennis Duplex*.

Germania-Jewett: See *Jewett*.

Gilman and Kempster: A type-wheel machine in which characters were selected along a linear scale by drawing an indicator along a rod. A cord attached to the indicator and wrapped around the type-wheel caused it to turn. (Cf. later *Index Visible*.) The patent was granted in 1884 to W. H. Gilman and D. E. Kempster of Boston; the latter invented the *Boston* some years afterwards.

Gisela: A German front-stroke portable with three-row keyboard and double shift, manufactured in 1921 by Gisela Schreibmaschinenwerk Günter & Co.

G & K: See *Gundka*.

Gladstone (a): See *Edelmann*.

Gladstone (b): See *Simplex (a)*.

Globe: See *American (a)*.

Gloria (a): See *Erika.*

Gloria (b): See *Geniatus.*

Glossografo: A machine designed to reproduce the sounds of speech directly as they were spoken was invented by Amadeus Gentilli of Italy in 1882. Part of the machine was placed in the mouth.

Glossograph: Invention for printing the spoken word directly by recording the movements of the tongue, lips and chin in a code of longer and shorter strokes, entailing part of the machine being placed in the mouth. Invented by a German called Züppinger. The Italian Gentilli tried to achieve the same goal some years later with another machine of the same name.

Gnom: See *Liliput.*

Golfarelli: A stenographic machine perfecting Lamonica's invention is reported[16] to have been made by Innocenzo Golfarelli in 1881 (see page 120).

Grant: An 1879 patent granted to H. M. Grant of New Jersey for a four-row design with convergent plungers printing 'combination type' similar to Livermore's but more complete. The inventor called it 'Monogrammic Alphabet'. The machine clamped to a table top.

Granville Automatic: Invented by Bernard Granville, who had previously designed the *Rapid,* this second thrust-action attempt by the same man proved hardly more commercially successful than the first. Patents date from 1891. Labelled 'Automatic' because all operations except insertion and removal of

258. Granville Automatic   (MPM)

paper were controlled from the keyboard, the Granville was manufactured in 1896 by the Granville Manufacturing Co., later by the Mossberg and Granville Manufacturing Co., both of Providence, R.I., and was mechanically similar to the *Rapid*. (Fig. 258)

Graphic: One of the numerous machines (Eureka, Fortuna, Kneist etc.) sired by the *Hall Type-Writer*. This one was introduced in 1895 by C. F. Kindermann & Co. of Berlin. It was too similar to the *Hall* to warrant separate description. A slightly improved second model was introduced in 1897, and also a model for the blind.

Griffen: A four-row keyboard, type-wheel machine patented in the United States by John Griffen of N.Y. in 1902. Not produced.

Grönberg: A German machine said[6] to be similar to the *Soennecken* and *Schapiro*.

Grundstein: See *Keystone*.

Grundy: The second invention of a standard front-stroke machine, this US patent

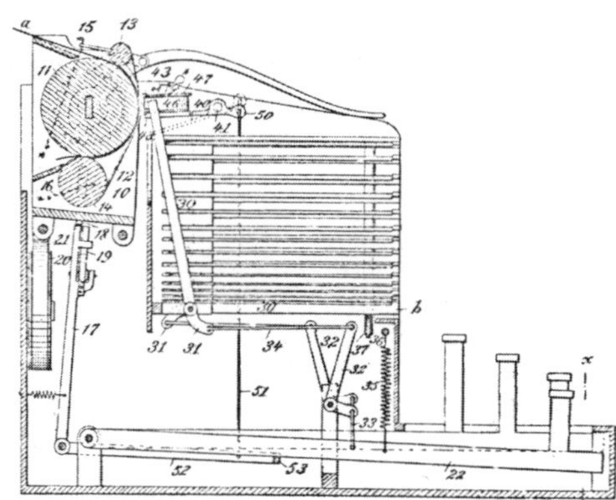

259. Grundy   (USP)

was filed by Arthur Grundy on 18 January 1887, thirteen days after the one by *Prouty and Hynes*. It was granted on 11 June 1889. The machine was not manufactured. (Fig. 259)

Gundka: Cheap German type-wheel machine invented in 1924 by Paul Muchajer and manufactured by Gundka-Werk, GmbH. It bore more than a passing resemblance to the solid and well-built *Edelmann*, with which it shared the curved letter index, indicator geared to the type-wheel shaft, double shift by keys on the left, and printing by depression of indicator key. The only actual design

260. Frolio (Gundka)   (AC)

departures (apart from quality) were its use of a ribbon and the location of its printing-point in front of instead of above the platen, as on the Edelmann. The Gundka was marketed in different countries under the names *Frolio*, *Gefro*, *G & K*, *M W*, *Perlita*, *Scripta*, and *Write Easy*, but not all these models appear to have been distributed simultaneously. Early *G & K* in AC (serial no. 7111), Fig. 27, is inferior to late *Frolio* (serial no. 86,259), Fig. 260. Note also the change in letter index and indicator: intermediate models used the early base pressings but covered the indicator slot with a wider index plate.

Gynee Cipho: A primitive machine with a vertical type-wheel similar to the *Dollar*, manufactured by the Gynee Cipho Typewriter Co. of London. Its price was one guinea, which gave the machine its name.

H: A small machine combining type-bars and an indicator in place of a keyboard is in MPM and there attributed to *Horton*, but without justification for this attribution. The only distinguishing mark on the machine is the large letter H. It is more likely, however, to have been an early experimental version of Hamilton's *Automatic*.

Haberl: A machine by this name is reported[42] to have been exhibited in Venice in 1907 but no further details are supplied.

Haegele-Ritter: A German patent from 1880 for a down-stroke machine designed on three levels: the upper floor was a completely circular keyboard, the mezzanine the type-bars, and the ground floor the platen. It was never produced.

Hagelin: Famous make of printing cipher machines, offering progressive encipherment by geared rotors. Type-wheel printing, roller inking. See page 228 and Fig. 211.

Halda: Up-stroke machine with four-row keyboard, introduced in Sweden by Halda Fickurfabriks A.B. in 1896. It was of the usual 'blind' design—the first models were made in small quantities only, but Model Four (1902) had a considerably increased production. The make went visible in 1914 with its Model Eight.

Hall Braille-Writer: The inventor of this machine was Frank H. Hall who was in no way related to Thomas Hall, inventor of the *Hall Type-Writer.* Together with G. A. Sieber and T. B. Harrison, he devised in 1891 a simple instrument on which three keys for each hand could be depressed singly or in chords to emboss braille cells. The keyboard was designed to resemble that of a piano, the black keys being the operative ones. A centrally placed space key was provided. It was an excellent inexpensive machine and enjoyed considerable success (Fig. 202).

Hall Type-Writer: The first of the index machines, this simple design was very popular and inspired numerous copies (*Eureka, Fortuna, Graphic, Kneist*) for over twenty years after it was patented in 1881 by Thomas Hall and produced the same year by the Hall Type-Writer Co. of Salem, Mass. It had a square rubber index connected to an indicator which selected the letter on a similarly square letter-table above. The index inked itself from a pad and everything but the character selected was masked off by a metal plate. Upon depression of the indicator, the top plate was lowered and a small block pressed the flexible rubber type against the platen. The paper remained stationary, while the whole assembly described above rode from left to right, a space at a time, by means of an escapement along a rod. A separate space key was supplied. In 1889, an improved model was introduced with metal type bonded to the rubber index plate, thereby allowing copies to be made. The carriage moved on this model, while the index remained stationary. The Hall continued to be sold well into the nineties, and perhaps even later. It dates from the days in which the machine was labelled Type Writer; the later version was sometimes labelled New Model (Fig. 161).

Hamilton: See *Automatic (a).*

Hammond: One of the great machines of typewriter history, this swinging-sector design was evolved by James Bartlett Hammond over an extended developmental period before the machine was finally patented in 1880 *et seq.* and marketed in 1881 by the Hammond Typewriter Co., N.Y. If the inventor dedicated the better part of his life to producing this machine, a student today could easily dedicate the better part of his own to establishing with precision when and in what order the many models appeared. There seems to be no authentic or conclusive record, and the profusion of models and sub-models which reached the market in the forty years during which the machine was produced—many of them not identified except by the name of the manufacturer—only adds more confusion. Furthermore, a number of different models were marketed simultaneously. This is one machine with which research has proved virtually fruitless.

The principal feature of the Hammond was its swinging sector. In all but the first model, this took the form of a wheel on to which two type shuttles could be fitted. One shuttle faced the platen and could be interchanged for the other merely by turning the wheel through 180°. There were literally hundreds of shuttles covering all languages, alphabets, types, scientific and other uses, and any two of them could be fitted. The first model was different in that it had two diametrically opposed segments, one for each half of the keyboard, and no alternatives could be fitted simultaneously.

Depression of a key first swung the sector to the corresponding character and then tripped a hammer located behind the paper; this struck it against the sector, with a rubber strip between the hammer and paper and a ribbon between the paper and type. The paper on all models was coiled in a basket and was fed by rubber feed rollers which did not, however, serve as platens.

Keyboards presented a problem with this make, for no other manufacturer catered to more individual tastes than Hammond. The first model was offered with only a curved two-row keyboard with ebony piano-type keys. From then on, subsequent models were offered either with a curved two-row keyboard, or else with a square three-row, the letter order at first restricted to the Ideal but later the Universal was offered as an alternative. An aluminium portable was offered, and the same machine was made even more compact soon afterwards by making its keyboard fold upwards. Additional models were introduced for special purposes—there were reverse escapements for Oriental languages, the machine was remodelled as a cipher machine, or one for the blind, etc. Approximate dates were as follows: Model One: 1880–1; Model Two: 1893; Model Twelve: 1905; Multiplex: 1915; Aluminium portable: 1915; Folding: 1923; Vari-typer: 1927. The Vari-typer suppressed all other Hammond production and became the company's sole model. It was originally manufactured by Hammond Typewriter Co.; from 1931 by Vari-typer Inc.; and later still by Ralph C. Coxehead Corp., N.Y. (Figs. 160 and 213)

Hammonia: This novel German linear index machine was invented by Andrew Hansen of London and covered by British patent in 1882. H. A. Guhl of Hamburg protected it the following year with a German patent and produced it in his sewing-machine factory, Guhl and Harbeck, which later brought out a machine called *Kosmopolit.* It was the first European typewriter and is usually described as resembling a bread slicer. The type was cast in one row on the underside of a brass blade on the top of which was the handle. The blade ran inside a slot in a housing set at 90° to the printing direction, a toothed rack providing lateral spacing upon each depression of the bread knife, which also did the printing. The characters were selected by sliding the index to the desired point, with teeth

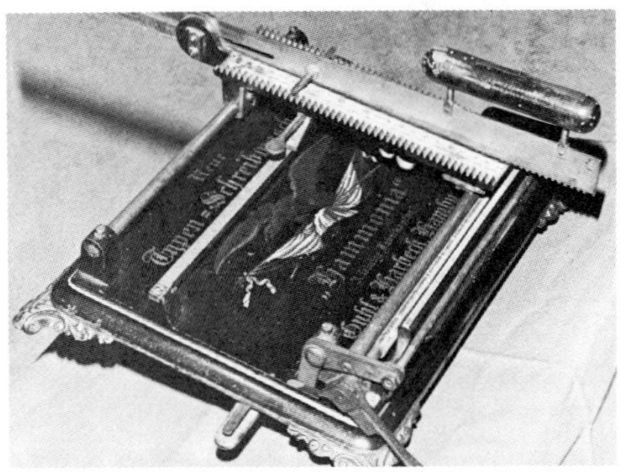

261. Hammonia   (TMT)

provided for accurate alignment. The whole affair rested on a base of sewing-machine design, a silk or paper ribbon providing the impression. (Fig. 261)

Hansa (a): See *Culema.*

Hansa (b): See *Kanzler.*

Hanson: This remarkable machine was patented by Walter H. Hanson of Milwaukee in 1899. It is best described as an oblique front-stroke design with a vertical platen; this revolved for typing and was raised a step at a time for line spacing. It was apparently[27] the product of considerable development but was never manufactured. The only known example is in MPM—there is part of another unidentified vertical platen machine in GC. A Revd Lee is said[22] to have worked on Hanson's designs after his death and the present machine is sometimes referred to as *Hanson-Lee,* probably owing to the fact that Hanson's patent was partly assigned to Ole Lee. See also *Nickerson.* (Fig. 262)

Hanson-Lee: See *Hanson.*

Harding: British patents for two machines were granted to G. P. Harding in 1870, one of them for a device with a wheel or segment bearing type-plungers which were depressed for printing, and the other for a conventional fixed type-wheel design in which the paper was struck against the type. Roller inking or carbon.

Harris: A front-stroke upright machine of conventional appearance and three-row keyboard with double shift was introduced in 1911 by the Harris Typewriter Co., and named after the inventor. It was also called *Harris Visible.* The manufacturer became the Rex Typewriter Co. from 1914 onwards and marketed its

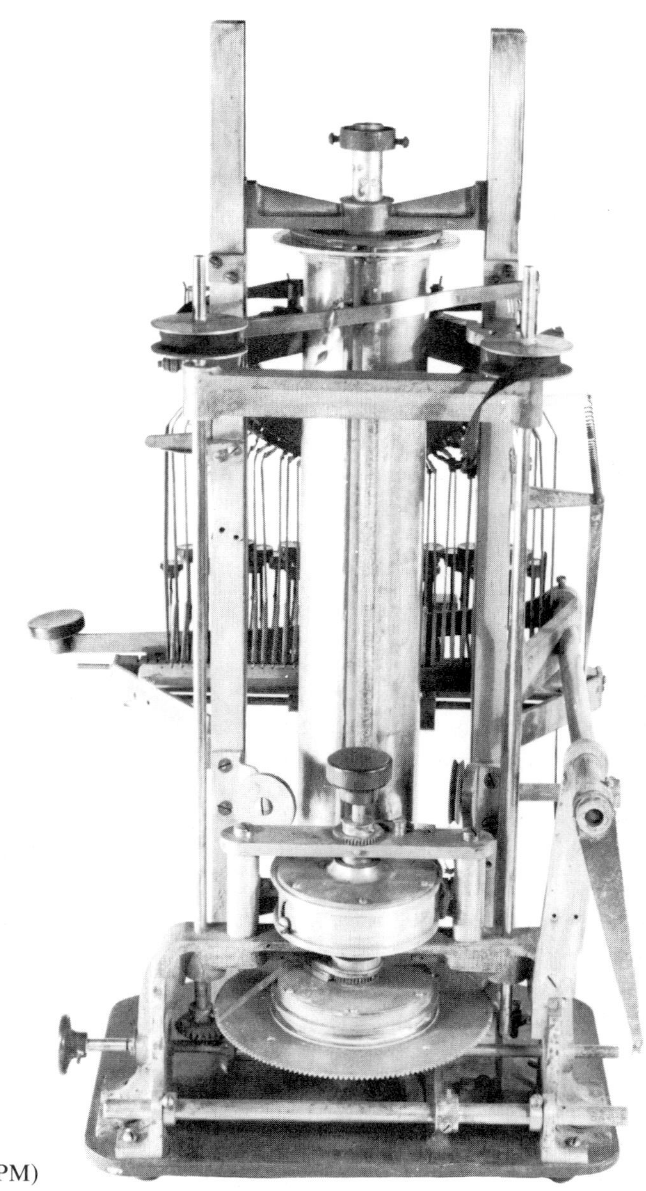

262. Hanson   (MPM)

products under the names *Autocrat*, *Reporters Special*, *Rex* and *Rex Visible*.
Various models were at different times distributed by Sears Roebuck, and American
Can Co. The manufacturers later produced the *National* and *Demountable*,

changing their name to Demountable Typewriter Co. in 1923.

Harrison Hill: See *Prendergast.*

Harris Visible: See *Harris.*

Harry A. Smith: See *Blickensderfer.*

Hartford: An American up-stroke machine with double keyboard, generally said to have been placed on the market in 1894; the patent is dated the following year. It was invented by John M. Fairfield and manufactured in the city after which it was named. Since it was non-visible, the platen had to be raised to reveal the typing and this was achieved by gentle depression of a lever to the side, which released the spring-loaded platen so that it raised itself automatically. It hardly seemed worth while, however, since it had to be lowered manually. A slightly improved Model Two was introduced in 1896 and, in 1905, the Model Three with four-row keyboard and shift was offered simultaneously with the double-keyboard model. Models Two and Three were also called *Cleveland* after the factory moved to that city (Fig. 193).

Haughton: Listed[42] but without details. *Houghton,* similarly listed[28], is perhaps the same man. Research has failed to locate either the inventors or their machines. The closest is a type-setting machine patented by a Thomas A. Houghton of Michigan in 1904, and an A. T. Houghton unsuccessfully applied for a British patent in 1902, but Webster Co. already lists it in 1898.

Hauptargel: A machine for the blind, permitting the embossing of up to eight copies at a time, was presented at a German congress of blind teachers by a man called Hauptargel of Liepzig.[6]

Hazen: A design similar in appearance to a *Hall Type-Writer* but using a swinging sector instead of a square index is described and illustrated[6] and attributed to a patent granted to Marshman Hazen of New York. An examination of patent office records, however, has failed to reveal any trace of such a patent being granted. The only one issued to Hazen is one jointly held with Richard Uhlig for a swinging-sector design with four-row keyboard, 1908. (Fig. 263)

Heady: See *Mignon.*

Heiss: A US patent granted to Frank Heiss of New York in 1892 was for a two-row electric swinging-sector machine.

Helios: Small German type-wheel machine with two-row keyboard and triple

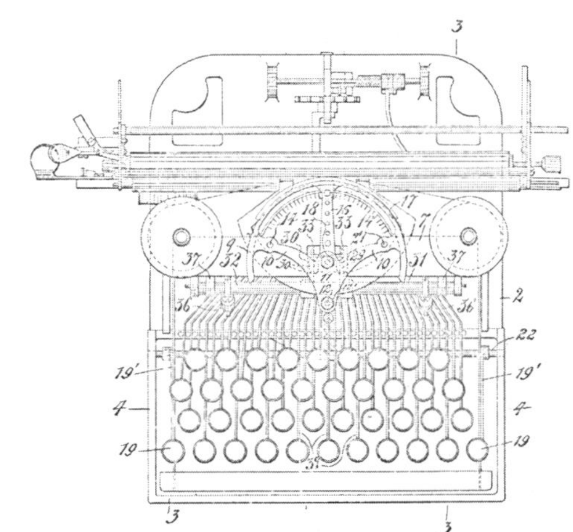

263. Hazen and Uhlig   (USP)

shift manufactured in 1908, first by Justin Bamberger, then by Deutsche Klein-maschinenwerke; in 1909 by Kanzler Schreibmaschinen A.G., which formed the Helios Schreibmaschinen GmbH to handle the machine, and from 1914 by the firm of A. Ney which marketed it as the *Helios Klimax*. The type-wheel had four rows of characters corresponding to the triple-shift positions, and printing was through a ribbon. A three-row keyboard model is recorded[28] but appears not to have been marketed. It was also known under the names *Bamberger*, *Ultima*, and a single known instance in AC labelled merely *Portable Extra* (Fig. 1).

Helios Klimax: See *Helios.*

Herald: See *American (b).*

Herrington: G. H. Herrington of Kansas was first granted a US patent in 1881 for a sophisticated plunger machine: turning a knob selected the character and depressing another brought plunger on to paper by means of a hammer. It was apparently not manufactured, but a second patent granted in 1884 to Herrington and D. G. Millison for a primitive vertical type-wheel device was sold in large numbers. It was manufactured by McClees, Millison and Co. of Chicago. It was similar to the *Dollar* with which it competed. (Fig. 264)

Hess and Stoughton: See *Century (c).*

Higgins: An untraceable electric machine reportedly patented in 1887. (A Higgins was connected with the *Waverley*, but this lead proved futile.)

293

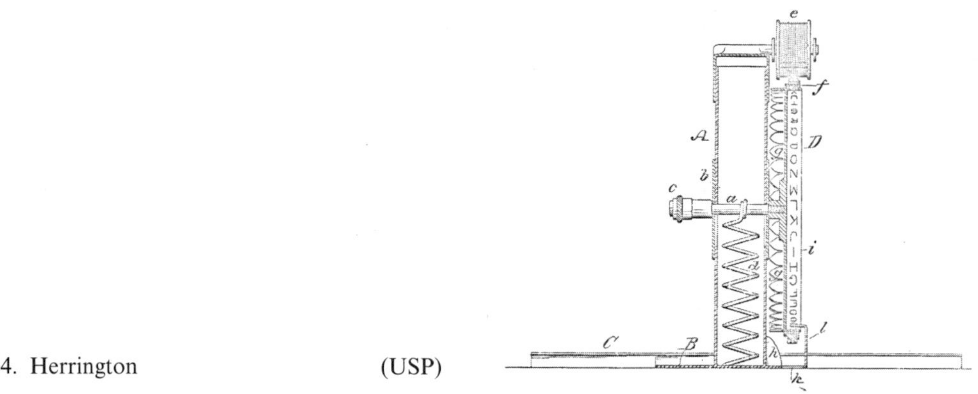

264. Herrington   (USP)

Hilgenberg: An up-stroke circumferential plunger machine was patented by C. Hilgenberg of Illinois in 1882. (Fig. 265)

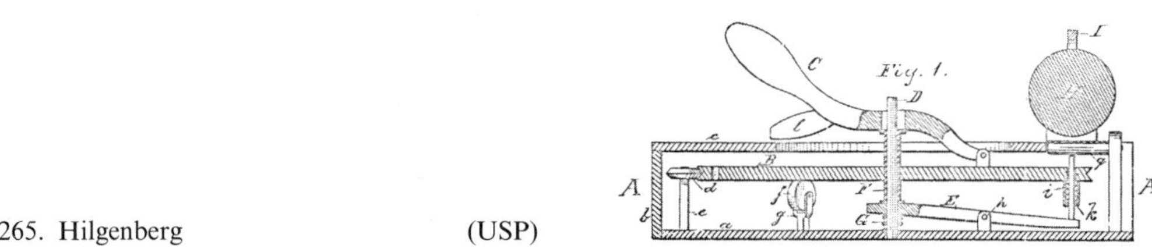

265. Hilgenberg   (USP)

Hitter: Jean A. Hitter Jr of Louisiana was granted an 1878 patent for a design in which type-plungers were carried on a horizontal wheel geared to a knob on the left. This was spun to select the character, whereupon another knob was depressed to bring plunger onto paper. Roller inking.

Hochfeld: A four-row machine with triple shift printing 180 characters was invented by Wilhelm Hochfeld of Hamburg in 1930. The idea was to be able to print in two different alphabets or types. A special key put the machine into register with one or the other position, whereupon the shift key was used in the conventional way. The type-bars were fitted with four characters each.

Hooven Automatic: See *McCall.*

Horton: The first oblique front-stroke machine, this instrument was patented in 1883 by E. E. Horton and manufactured in limited numbers by the Horton Typewriter Co. of Toronto. It had a double keyboard and three rows of inclined type-bars in an arc around the printing-point on the elevated platen. A vertically

294

moving ribbon was used, partially obscuring the typing, but otherwise it was the first type-bar machine to offer visible printing. (Fig. 176)

Houghton: See *Haughton.*

Hülsen: A syllable typewriter patented in 1896 by E. von Hülsen of Germany. Usually listed[6 etc.] among the 'unknown' machines.

Hurtu: See *Ford.*

Huth: A British patent for a full keyboard front-stroke machine was granted to A. H. Huth of London in 1893.

Hyndman: A machine of this name is listed[42] and also under the fuller title of *Hyndman's National*[27], but apart from a statement hinting that it belonged to the index class, no further details could be found. Webster Co.'s 1898 Ribbon Catalogue lists it.

Hyndman's National: See *Hyndman.*

Idea: See *Pacior.*

Ideal: German oblique front-stroke machine with four-row keyboard patented in 1897 by E. E. Barney of Groton and manufactured from 1900 by the bicycle and sewing-machine works of Seidel & Naumann A.G. A lever on the right of the keyboard operated the carriage return and line spacing. Later models were straight front-stroke machines. The company also manufactured the *Ideal-Polyglott* for simultaneous use of two or three western and/or oriental languages; also the *Ideal-Duplex* for two western alphabets, e.g. Roman and Russian, and *Ideal Oriental* for one western and one oriental. Combinations of the four-row keyboard with double shift offered additional capacity to meet these demands. The company also manufactured the *Erika.* (Fig. 266)

Imperial (a): Successors to the *Moya,* these down-stroke machines were manufactured in England by the Imperial Typewriter Co. Model A, 1908, had a curved three-row keyboard with double shift and interchangeable type-head. The type-bars were positioned in front of the platen. Model B, 1915, was essentially an identical machine, also marketed in France as the *Typo.* It was a fine, well constructed instrument. Model D, announced in 1919 and produced in 1921, substituted a straight three-row keyboard with double shift, retaining the down-stroke design and the feature of easy interchangeability. A smaller portable model was introduced in 1923. The Imperial went front-stroke from 1927, although the Model D continued to be marketed as the *Imperial Junior.* Also[22, 29] under

266. Ideal   (TMT)

the names *Ajax*, *New Imperial* and *Lloyd*. A German machine called *Faktotum* introduced in 1912 was virtually identical. (Fig. 194)

Imperial (b): A type-wheel machine with a three-row keyboard and double shift, manufactured in Germany by Karl Heinrich Kochendörfer in 1900. With inking by roller, it was almost the same as the Blickensderfer. Only a few machines were made. The company previously produced the *Eureka*.

Imperial (c): See *Triumph Visible*.

Imperial Junior: See *Imperial (a)*.

Imperial Visible: See *Triumph Visible*.

Improved Corona: See *Standard Folding*.

Improved Crandall: See *Crandall*.

Improved Type Writer: See *Sholes and Glidden Type Writer* and *Remington.*

Index Visible: This rarity in typewriters was introduced in 1901 by the Index Typewriter Co. of New York, and bore a more than passing similarity to the *New England* which made its appearance in London the previous year. The machine had a fixed 'keyboard' and a spring-loaded type-wheel to which a cord was attached. This cord was fitted to the index finger of the right hand which served as a pointer: by indicating letters on the 'keyboard' it exerted pull on the cord and turned the type-wheel. A key on the left brought the carriage forward for printing. Ribbon for the impression. (Cf. earlier patents of *Perce* and *Gilman & Kempster.*)

Ingersoll: One of the most primitive of all. Manufactured by Robert Ingersoll and Bros, this machine consisted of a series of blocks, one for each character, which rode along a rail passing over the line on which printing was being performed. The block with the desired character was brought over the printing-point by

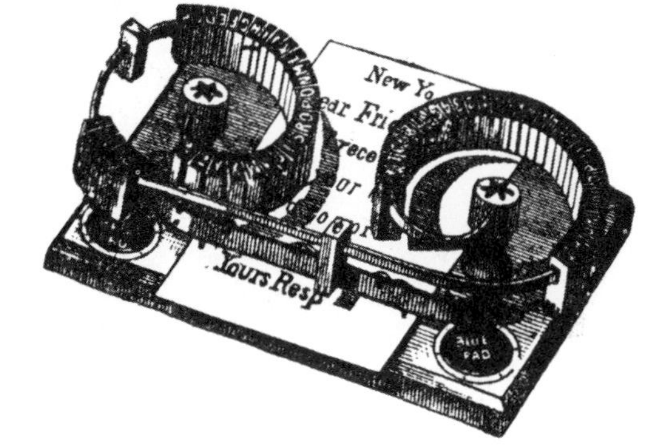

267. Ingersoll   (27)

pushing the other blocks out of the way, inking itself on a pad on the base. There was no guide for spacing. Also marketed as the *Nason.* Such a device would make a typist grateful indeed for a *Simplex!* (Fig. 267)

International (a): An up-stroke type-bar machine with a three-row keyboard invented by Lucien Stephen Crandall, who was already manufacturing the type-sleeve instrument bearing his name. The patent was filed in 1886 and granted in 1889 when it was manufactured in limited quantities in Parish, N.Y. in the same year. A four-row keyboard model appeared in 1893. The earlier model printed through a wide ribbon that passed from front to rear. It was a very heavy piece of machinery (approximately 36 lb.). A model with a double universal keyboard is in MPM. Manufacture ceased in about 1898. (Fig. 268)

268. International (a) (MPM)

International (b): One of the many linear index machines similar to the *Odell*. It was produced by the New American Manufacturing Co. and was also marketed as the *New American*.

Iskra: See *Pacior*.

Jackson: One of the few 'grasshopper' machines, this one was strongly reminiscent of the earlier Maskelyne. It was patented by Andrew Steiger (of Yost fame) in 1896 and introduced by the Jackson Typewriter Co. of Boston in 1898. The type-

269. Jackson (GC)

bars fanned out radially from the printing-point, resting on an inking pad; upon depression of a key, the bar first raised itself off the pad and then struck forward at the platen. Four-row keyboard, single shift. Few were made. (Fig. 269)

Janus: See *Meteor.*

Jeffreys and Edwards: A British patent granted to these two Londoners in 1894 was for a linear index design.

Jewett: A double keyboard up-stroke machine introduced in 1892 by Jewett Typewriter Co. of Des Moines, Iowa. It was named *American Standard* but the

270. Jewett   (SC)

word 'Standard' was contested by Remington which sued, and the machine was soon renamed after the president of the Company, George A. Jewett, who also financed the manufacture of the *Dennis Duplex* from which it developed. Numerous models were offered but none embodied major changes: Model Two: 1895; Model Four: 1897; Model Nine: 1898; Model Eleven: 1901. The first major departure was the *Jewett Visible* in 1904. A model with 122 keys was offered as a bilingual machine in 1902: it had one complete Roman plus another (e.g. Greek) alphabet. The Jewett was also assembled in Germany from 1898 and called *Germania* and *Germania-Jewett.* (Fig. 270)

Johnson: US patent for a Mechanical Typographer granted to Charles Johnson in 1869.

Johnson Book Typewriter: A company to market a machine of this name is listed[27] as having been registered in 1896. The product was a book typewriter for which J. W. Johnson was granted a US patent in 1899.

Jundt: A machine reportedly[42] invented by a Wisconsin watchmaker called Molle was to have been manufactured in 1908 in Dayton, Ohio and named Jundt. Nothing came of the project, but see *Molle.*

Junior (a): A very small type-wheel portable patented by Charles Bennett of New Jersey in 1901 and introduced in 1907 by the Junior Typewriter Co. of N.Y. It weighed about 4 lb. and measured approximately 11″ by 5″ by 2″. It had a diminutive three-row keyboard with double shift, squeezed up against a type-wheel which inked from rollers and struck forward at the printing-point with a locking pin for correct alignment. The mechanism was ingenious but operation was rather unpractical, not only because of its diminutive size but also because of a system whereby the keys were worked in columns—depression of a key on the top row caused the two lower keys to be depressed as well. The hairsprings which located the keys in the raised position were also inefficient and are commonly missing on machines that have survived till now. The same basic design, using a ribbon instead of ink rollers, was later marketed as the *Bennett* from 1910 by the Bennett Typewriter Co., being manufactured in the Elliot-Fisher factory. Both Junior and Bennett were supplied with neat rectangular leather covers which clipped over the machine for portability (Fig. 182).

Junior (b): A German toy typewriter with a tiny carriage, circular index, spacing and printing keys and change of case. The 'keyboard' is a dummy, selection of

271. Junior (b)　　　　　　　　　(AC)

the characters being made by rotating the knob on the index. Roller inking. Circa 1920. (Fig. 271)

Juventa: A conventional front-stroke portable with a three-row keyboard and double shift, manufactured from 1922 by S. A. Industria Dattilografica of Milan.

300

It was exported to many countries and was marketed under the names *Agar*, *Agar-Baby*, *Ardita*, *Diadema*, *Fidat*, *Merkur*, and, later, *Sabb*.

Kahn: A French shorthand machine invented in 1903.[15]

Kaley: Another of the numerous segmental type-disk designs with the type at the ends of the teeth, this compact device with pad inking was patented in 1885 by J. A. Kaley of Ohio.

Kanzler: This German thrust-action machine was invented in 1903 by Paul Grützmann with patents in the name of Deutsche Schreibmaschinen GmbH. Manufactured for a short time by A. G. für Schreibmaschinen-Industrie as the *Hansa*, the name of the company then changing to Kanzler Schreibmaschinen A.G. It had a curved four-row keyboard printing eighty-eight characters but using only eleven type-bars so that each bar printed eight characters. Depression of a key not only thrust the type-bar forward, as on the Wellington, but also located it in one of four vertical positions, so that the corresponding character was opposite the printing point. Depression of the shift key raised the level of the type-bars to bring the other four positions into register. Models One to Three were for different lengths of carriage; Model Four with minor changes appeared in 1912, in which year production ended. Also marketed as *Kanzler-Rapid* and *Chancellor*. (Fig. 188)

Kanzler-Rapid: See *Kanzler*.

272. Karli   (DTU)

Karli: The only example of this linear plunger device known to have survived is in DTU. The knob on the right is used to slide the type-carrier for character selection whereupon the central key is depressed for printing. The illustration is the rear view of the machine. (Fig. 272)

Kenbar: Small German portable reported[28] to have been sold in England. No date or details.

Kent: Electric type-wheel machine announced by the Kent Writing Machine Co. in 1892 but never manufactured. The type-wheel was mounted vertically in front of the platen and moved along its axis for typing, the paper remaining stationary. It was supposed to be fully electric, but the machine had keys and key-levers as on a manual, so that at least part of the printing operation was mechanical. It had a double keyboard.

Keystone: This machine of the swinging-sector class, with three-row universal keyboard and double shift, was patented in 1898 by the Keystone Typewriter Co. of Harrisburg, Pa. and introduced the following year. It looked like a Commercial Visible but worked like a Hammond: upon depression of a key, the type-sector was swung to the corresponding letter whereupon a hammer struck the paper against it from behind. Ribbon for the impression. It is said[40] to have been marketed in Germany as the *Grundstein*. (Fig. 183)

King: No details other than the name.[42] Webster Co. Catalogue (1898) lists it.

Kitzmiller: US patent, granted to George Kitzmiller of Virginia in 1904 and assigned to the Electric Typewriter Co. of the same state, was for a front-stroke type-bar design.

Klaczko: Max Klaczko, who was responsible for considerable developmental work on typewriters in Germany in the early part of the century, built a bilingual machine in 1913 with a three-row keyboard for typing in German and Russian. The type-bars had five characters each: two German, two Russian, and a figure. Not manufactured.

Kleidograph: Machine for the blind, introduced by N.Y. Institute for the Blind in 1894. It had twelve keys and cell-spacing bar. Paper was inserted from the rear and wound itself around a drum.

Klein-Adler: See *Adler.*

Kneist: Another of the numerous (Eureka, Graphic, Fortuna) copies of the *Hall Type-Writer*, this machine was 'invented' in Germany by O. Mayer and J. Funcke and introduced in 1893 by Wunder & Kneist. The machine was too similar to the *Hall* to warrant separate description. A model for the blind was introduced in 1901.

Knickerbocker: There is a sole reference[42] to a company called Knickerbocker Typewriter Co. whose assets, including a factory, changed hands in 1912. It was to be reorganized as the Defiance Typewriter Co. but no further details can be located.

Knoch: See *Ford.*

Kochendörfer: A battery-operated electric machine was patented in 1901 by the German sewing-machine manufacturer Karl Heinrich Kochendörfer, who had already produced the *Eureka* and *Imperial (b)* typewriters. The machine had a three-row keyboard and belonged to the thrust group—depression of the keys, as befits an electric machine, was a mere 2–3 mm. It was apparently never manufactured.

Kohl: A cipher machine, invented by Alexis Kohl of Denmark and a direct copy of the Writing Ball, is on display AMP and dates from circa 1885, according to correspondence in the Museum's files. (Fig. 210)

Kosmopolit: This handsome German machine by the manufacturers of the earlier *Hammonia* used a similar design for the base, with its four claw feet reminiscent

273. Kosmopolit   (TMV)

of sewing-machine bases of the period. Invented by J. C. Koch in 1888 and produced by the sewing-machine factory of Guhl & Harbeck, it was of the swing-sector class, with a pointer attached to a key serving to select the characters from a curved scale and to print them upon depression. Two semi-circular rows of rubber upper and lower case type were attached to the underside of the sector which rode along a rod at the rear of the machine and could be raised for paper insertion. The sector was also pivoted at the rear, the indicator projecting from the front above a curved comb which served to lock it for alignment during printing. Inking was from a pad on either side of the printing point which also served to mask out all but the character required. Platen and feed rollers were at the front of the machine and the paper exited towards the operator. The sector housing moved a space after each impression. Also sold as the *Cosmopolit* and *Cosmopolitan*. (Fig. 273)

Kratz-Boussac: There is a sole reference to a machine by this name[36]; no date is offered. It is described as a circular index machine consisting of a metal disk with the letters around its periphery and raised characters around its edge. Letter selection was by turning a handle, which was depressed for printing. Inking by pad.

Kromarograph: Musical typewriter invented by Lorenz Kromar in Vienna in

1904. The intention was the centuries-old dream of writing improvised music and the machine was thus connected to the keyboard of a piano and recorded the notes played, in a code of longer and shorter strokes.

Lambert: This unusual spherical index machine reportedly took seventeen years to perfect, its inventor, Frank Lambert of New York, finishing it in 1896. The earliest patent date is 1884. It was manufactured by the Lambert Typewriter Co. of N.Y. and later by the Gramophone Co. of England: in its honour, this organization changed its name to the Gramophone and Typewriter Co. and marketed the Lambert from 1900 till 1904 before giving up on it. The German subsidiary of the company also produced it, as did Sidney Hebert of France, who persisted with it longer than the others.

It consisted of a small circular disk of spherical section with a hole in the middle and the type in three staggered circular rows around it (the machine had double shift). This disk was fixed to an arm beneath a circular plate with the characters similarly reproduced on it, in such a way that the combined unit swivelled and rocked in any direction. If the plate was depressed at any point, the arm swung out in the opposite direction, presenting the corresponding character to the printing-point, a pin locking it in its slot. Further depression of the entire plate brought the type into contact with the platen. The top of the paper was clamped to the platen; the page curled around it as typing progressed and had to be uncoiled upon completion. It had a quaint warning bell arrangement: a small steel ball rolled in a tube, the end of which almost touched the edge of the bell. The tube was so inclined that the ball remained at rest at the other end; a small coil spring on the carriage passing the tube flicked the ball against the bell. (Fig. 6)

Lanari: A shorthand machine invented in 1905 by the Italian Enea Lanari. It had twenty-three keys in three rows.

Larson: A machine with a vertical type-wheel travelling on rods over a platen, with an index for character selection. Invented by S. P. Larson of Chicago, and

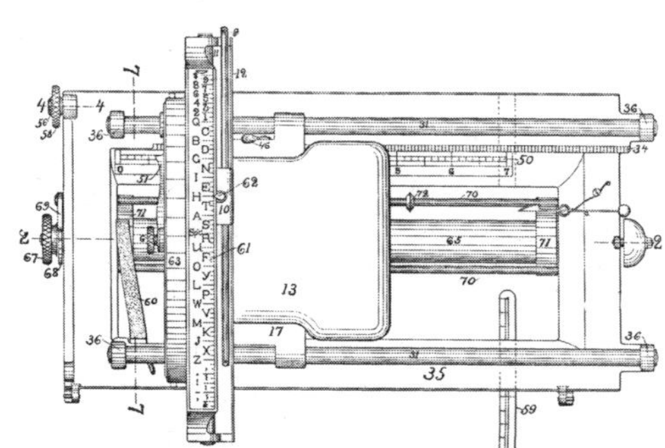

274. Larson   (USP)

covered by a patent filed in 1894, and granted the following year. (Fig. 274)

Lasar: A down-stroke offering visible typing was reported[42] to have been manufactured around 1898–1900 by the Lasar Typewriter Co. of St Louis, Mo. Apparently,[27] it infringed so many patents that it was soon withdrawn. The inventor, G. H. Lasar of Missouri, had protected himself with over twenty patents from 1889 onwards—but even that was obviously insufficient!

Leframa (a): See *Faktotum.*

Leframa (b): See *Sun.*

Legato: A musical typewriter with a type-sleeve and electric assistance was invented in Germany by Oskar Fischer in 1912 and was to have been manufactured, but nothing further was heard of it, possibly because of the intervention of the First World War.

Leggatt: A small portable by this name was introduced in 1922 but never produced. It was bought up by Rochester Industries whose *Rochester* portable was reportedly[42] similar.

Leming: A down-stroke machine with type-bars in a complete circle around the printing point, surmounted by a circle of key plungers. Patented by A. G. Leming of Arkansas in 1883. Not manufactured.

Leo Joseph: A machine with a single key;[6] no further details offered.

Levesque: Another machine with one key, listed[6] without further details.

Liberty: See *Molle.*

Lignose: A standard three-row front-stroke machine introduced in 1924 by A. G. Lignose of Berlin.

Liliput: Primitive circular index machine similar in design to the Simplex (a) but better made. The horizontal disc was spun by means of the knob on top and was then pressed down on to the paper. Inking by roller. Two paper feed rollers were provided on the carriage but these served merely to advance the paper, for printing was performed on a separate anvil. Three models of progressive sophistication were offered. Manufactured from 1907 by Justin Bamberger and Deutsche Kleinmaschinenwerke, where the Helios was briefly made. Also marketed as *Express* and *Gnom.* (Fig. 8)

Lindeteves: See *Wellington.*

Lindgren: A US patent was granted in 1881 to J. F. Lindgren of Illinois for a horizontal type-wheel machine with an indicator, depression of which caused

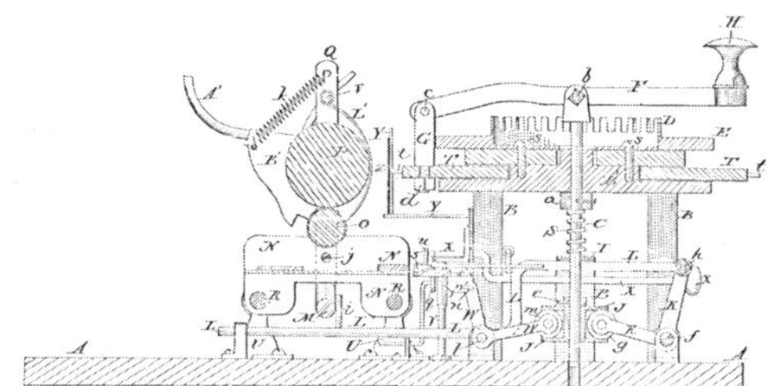

275. Lindgren   (USP)

platen to rock forward against the type. Roller inking. (Fig. 275)

Linowriter: A Smith Premier Model Ten, rebuilt in 1910 by the Empire Type Foundry of Buffalo N.Y. so that the letter order on the keyboard would conform to that on linotype machines. The idea was to offer a typewriter to linotype operators which would obviate the need to relearn the keyboard. For a description of the machine, see *Smith Premier.*

Little Giant: See *Simplex (a).*

Livock & Hermann: An oblique up-stroke machine, the only one of its kind. This remarkable American invention was described[6] as consisting of type-bars resting obliquely beneath a flat paper-carriage, the angle being such that the flat surface faced the operator. The type-bars, therefore, struck upwards from the rear. Upon depression of a key, the corresponding type-bar moved upwards against the paper at the same time as a hammer containing an ink pad dropped down from above it thereby, presumably, ensuring an impression visible to the operator on the upper side of the paper. No year was given and no details could be found in patent office records.

Lloyd (a): See *Imperial (a).*

Lloyd (b): See *Stoewer.*

Logomatografo: See *Clavigrafo.*

Logotype: A shorthand machine invented in 1923 by Edna Robenson of Atlanta Ga., and manufactured by the Atlanta Model Machine Co.
Consonant keys were in rows to the left and right, and formed grooves, one

for each finger except the index. It printed upper-case Roman characters on paper tape with orthographic fidelity but with erratic spacing.

Long: A miniature shorthand machine was patented in US by Eugene McLean Long in 1900 and apparently manufactured[28 etc.] for it was offered for sale in Paris in 1901. It was made of aluminium, small enough to be carried in the pocket and simple enough to be operated by one hand. It consisted of four keys which could be depressed in any of three positions to print a code of strokes on paper tape. It is sometimes called the *Long and Callaghan*, this presumably being the name of the manufactured item, for Callaghan does not appear in the patent. A *Mr McClean* is also sometimes[15] thrown in for good measure! (Fig. 276)

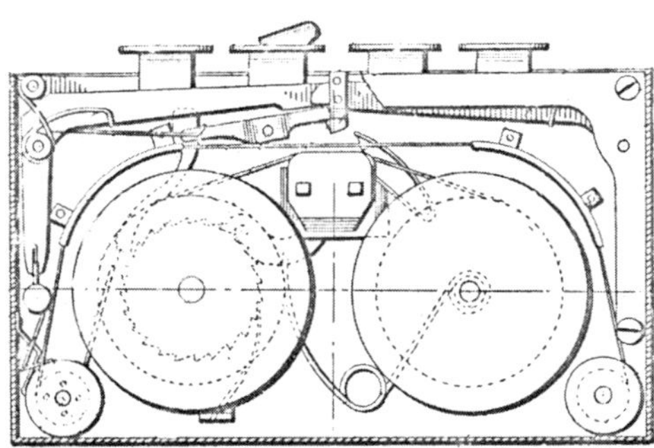

276. Long   (USP)

Long and Callaghan: See *Long.*

Longini: A primitive type-wheel device designed for printing signs in large upper-case letters, similar in purpose to the *Dart* but less sophisticated. The Longini was manufactured in Belgium in 1906 by H. E. Longini of Brussels and consisted of a large type-wheel mounted co-axially with a separate indicator wheel which was turned to select the character, whereupon a key on top of the machine was depressed. This assembly rode along a rod for alignment. Inking by roller.

Lucas: An interesting down-stroke machine consisting of a two-row circular keyboard above a circle of upright type-bars with platen beneath. US patent granted to G. L. Lucas in 1885.

Mackness: A shift-key machine announced in 1895 by C. F. Mackness of Scotland is mentioned[27] without additional details except, redundantly, that nothing came of it.

Magnetographe: French patent granted in 1877 to Eugène de Neufbourg for an electric machine.

Mahron: Type-wheel machine invented by a German, listed[6] but with no further details.

Manhattan: This was virtually identical to the successful Remington Model Two, and was introduced in 1898 by the Manhattan Typewriter Co. of New Jersey after most of the other machine's patents had expired. It had only moderate

277. Manhattan   (LC)

success before production ceased. Several further attempts were made to revive it—one in 1904 by the Imperial Typewriter Co. of Newark is mentioned,[28] as is another in 1905 by the Blake Typewriter Co. of the same city. (Fig. 277)

Manograph: A cheap index machine made in 1906 by Gebr. Heilbuth of Hamburg.

Marriott: An American book and billing typewriter patented in 1904 by J. H. W. Marriott of Maryland. It was of down-stroke design, with electric assistance, and was presumably equally useful for typing on sheets of paper as in bound books. It was never manufactured.

Martin: A primitive circular index design with flat paper-table, patented in US by J. Martin of Ohio in 1882.

Maskelyne: A superb design, invented in 1889 by John Nevil Maskelyne and his son and patented by them the following year. The machine had a grasshopper action and embodied, among other refinements, full differential spacing. On the first two models, the type-bars lay horizontally in front of the platen, resting on an inking pad and fanning out radially from the printing-point. This was somewhat similar in appearance to a Jackson or to half a Williams. Upon depression of a key, the corresponding type-bar left the pad and described the familiar hopping movement, coming down on to the platen in front. A three-row keyboard with double shift offered no less than ninety-six characters, and differential spacing was by the application of four (instead of the usual one) universal bars in order to advance the platen two, three or four spaces according to the width of the letter being depressed. As if this were not enough, a further sophistication was the possibility of using a single space to print diphthongs.

Model Three, also known as the *Maskelyne Victoria*, appeared in 1897. It retained the differential spacing of the earlier models but the action of the type-bars was changed to a unique arrangement whereby the ink pad faced *downwards* and the type rested against it face *upwards*, so that upon depression of a key the type-bar first lowered itself from the pad and then performed a somersault onto the paper. The mechanics of this design are beautiful to observe, as anyone privileged enough to have seen it will readily confirm.

The Maskelynes failed. The quality of the manufactured article failed to match the design and the ambitions of the inventors. It was too good, and it offered too much; consequently it would have been too expensive a proposition to build it as it deserved. In striving to make it more competitively priced, the manufacturers cut corners they could ill afford to cut, and their product proved too fragile to withstand the treatment of unloving hands. (Fig. 196)

Masspro: A small front-stroke machine with a three-row keyboard and double shift invented by George F. Rose of Standard Folding fame and manufactured by Mass Production Corporation of N.Y. in 1932.

Mauler: A machine for the blind invented by a man of this name was reported in 1887.[50] It consisted of a horizontal type-wheel with radial segments attached to its upper surface. On the outer edge of each segment was an embossed braille

278. Mauler   (50)

cell, and on the inner edge was the corresponding Roman character. The blind operator selected the character by feeling the braille cell and lowered a handle which brought a frame containing the paper down onto the Roman character. (Fig. 278)

McCall: One of a number of typewriters and attachments designed to type automatically from a roll of perforated paper similar to that used in a player piano. The use of perforated paper was also applied to automatic telegraph transmission. The present device was the invention of T. A. McCall of Ohio and was to have been produced by the McCall Automatic Typewriter Co. in 1906. The patent date is 1908. Other machines and attachments operating on similar lines were *Autist, Auto-Typist, Hooven Automatic,* etc.

McCool: An inexpensive type-wheel instrument invented by William A. McCool and covered by a patent filed in 1903 and granted in 1910. It was made by Acme-

279. McCool   (GC)

Keystone Manufacturing Co. of Beaver Falls, Pa. to whom the patent was assigned. The machine had a three-row universal keyboard and double shift; not many were produced. (Fig. 279)

McKinley: Carl McKinley of Washington patented an interesting device in 1880. A type-wheel mounted on an oblique shaft was geared to a horizontal indicator, depression of which caused a hammer to strike paper against type.

McKittrick: US patent granted in 1881 to G. McKittrick of N.Y. for a revolving pin barrel machine with two-row keyboard.

McLaughlin: This 'unknown'[6, 42] machine was an up-stroke electric, the subject of numerous patents granted to J. F. McLaughlin of Philadelphia from 1887.

McLoughlin: A primitive circular index machine of *Simplex* inspiration was reported[27] to have been marketed in 1884 by McLoughlin Bros., toy dealers of New York. A circular disk with rubber type around its edge was fixed to an

310

indicator plate, and this assembly rotated by means of a handle which selected the letter and locked in a corresponding tooth on a raised edge, printing being performed by depression of this handle. The index was hinged at the front of the base and paper was fixed beneath it. Pad inking.

Megagraph: A colossal typewriter designed for printing posters and billboard news bulletins. It measured 6′ by 5′ 10″ by 3′ 4″ and weighed over 400 lb. Patented in England in 1898 by Charles Augustus McCann, it printed upper and lower case and was provided with differential spacing. Inking by roller. It was not a best-seller!

Melbri: See *Gerda.*

Melographe: Musical typewriter reportedly[28] invented by a Brazilian priest called José Joachim Lucas in 1925.

Melotyp: See *Nototyp-Rundstatler.*

Menier: Listed[42] but without details.

Mentor: German oblique front-stroke portable with four-row keyboard, this machine was designed by Ing. Hanefeld, Heinrich Grützmann and Max Pfau, and manufactured by Metallindustrie A.G. in 1909. Also named *Thuringia* and *Monofix.*

Mercantile: See *American (b).*

Mercury: A small type-wheel machine with three-row keyboard and double shift, patented in England by F. Myers in 1887 and manufactured by the Mercury Typewriting Machine Co. of Liverpool. The design was interesting in that the type-wheel was off-set to the right, thereby making all printing visible to the operator. The keyboard served only to rotate the type-wheel to the desired character, the impression being made by means of a separate key which brought the platen up against the type-wheel. Roller inking. Myers was responsible for two other previous designs which were never manufactured: the first was a type-bar machine with a circular keyboard (see *Myers*), the second a type-wheel

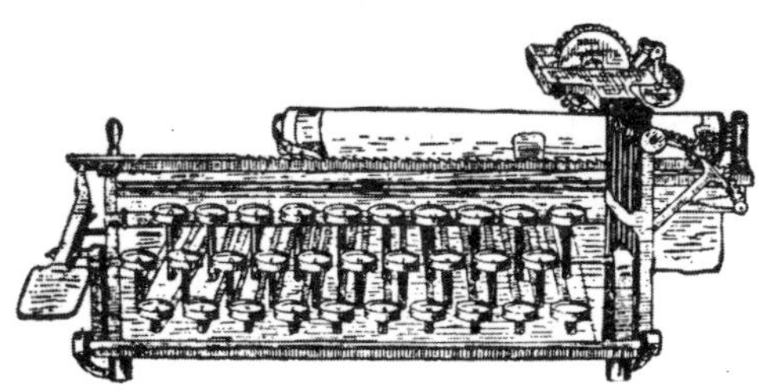

280. Mercury   (27)

device with four rows of keys. The third enjoyed only limited success. (Fig. 280)

Mergier: A French type-wheel patent granted in 1892 to Guillaume Emile Mergier for a design which permitted any number of keys forming the letters of a word to be depressed simultaneously, the machine printing all the characters selected so long as they followed each other in alphabetical order in the given word.

Merkur (a): See *Phönix.*

Merkur (b): See *Juventa.*

Merritt: One of the few linear plunger machines, this novel device was patented by M. G. Merritt of Mass. in 1890 and manufactured first by the Merritt Manufacturing Co. of Springfield, Mass. and later by the Lyon Manufacturing Co. of N.Y. The printing system of the instrument consisted of a carrier containing the type-plungers running longitudinally under the platen; an indicator attached to this carrier selected the characters along a letter index on the front of the machine. Depression of the indicator knob locked it in a slot and pressed the plunger up against the platen. Inking by roller. (Fig. 186)

Meteor: Front-stroke German portable of conventional three-row design with double shift. Introduced in 1911, it was produced first by Sächsische Strickmaschinenfabrik Meteor and, from 1922, by Vasanta Schreib– und Strickmaschinenfabrik A.G. Production ceased 1925. Also marketed under the names *Berolina, Doropa, Forte-Type, Janus, Pagina, Vasanta, Wilson.*

Meyer: See *Nouveau Siècle.*

Midget: See *Rochester.*

Mignon: An immensely popular type-sleeve machine, invented in 1903 by Dr Friedrich von Hefner-Alteneck and manufactured by the big firm of Allgemeinen Elektrizitäts-Gesellschaft of Berlin. The initials of this company, AEG, form one of the many names under which it appeared. A pointer suspended above a rectangular letter index selected the characters on the type-sleeve by providing a selector lever, geared to the type-sleeve shaft, with motion on two planes. This pointer was manipulated by the left hand; the right operated a space key and a printing key which brought the sleeve down onto the paper. Impression by ribbon.

On Model Four, a further key was provided for back-spacing. Type-sleeve and letter index were instantly interchangeable and alternatives of all kinds were offered.

This apparently primitive design continued to be made, under one name or other, until the 1940s. Model Two: 1905; Model Three: 1913. It was offered with braille type-sleeve and index in 1916, and an attempt at marketing it in the United States under the name *Yu Ess* (from the phonetic spelling of US) was shortlived. The operation was then transferred to France, where it appeared under the names *Eclipse, Heady* and *Stella.* A further attempt at re-introducing it into the United States in 1923 as the *Stallman* also failed. Model Four: 1923. In 1933 it was renamed *Olympia Plurotyp*, after the name of the manufacturer was changed to Olympia. An almost identical copy (the letter index was moved to the right) was independently manufactured in Czechoslovakia from 1936 to 1940 under the name *Tip-Tip.* A Mignon in TMT is labelled *Special.* (Fig. 20)

Mikro: An anachronistic front-stroke portable, printing only upper-case letters· from sixteen keys in two rows with double shift, manufactured by Paul Schütze GmbH of Dresden in 1925. Dimensions of carrying case: 6″ by 7″ by 3″, carriage separate.

Minerva: Braille machine made in Leipzig in 1928.

Miniature: See *Pocket (a).*

Minimax: There is a single reference to this machine[42] which was made in Germany in 1907 and looked like the *Ideal.*

Mitex: Three-row German portable of front-stroke design, with double shift, introduced in 1922 by Mitex Schreibmaschinen GmbH, which became the Tell Schreibmaschinen GmbH the following year when the machine was renamed *Tell.* It offered an early example of unit construction. Also produced in England as the *Bar Let.*

Model: One of Richard Uhlig's fifty typewriter designs. A company called the Model Typewriter Co. of Harrisburg was incorporated in 1909 but the product never reached the market.[42]

Modern: Another design by Richard Uhlig, this one dating from 1923.[42] It was never produced.

Molitor: An index machine listed[6] but without further details.

Molle: A front-stroke machine with three-row keyboard and double shift patented by John E. Molle of Oshkosh Wis. in 1913 and manufactured by the Molle Typewriter Co. of the same city in 1918. Molle was a watchmaker who made his first typewriter in 1906. A report that a machine of his invention was to be manufactured in 1908 and called *Jundt* obviously refers to that model. (See *Jundt.*) Nothing came of this effort, nor of a second machine which was ready in 1914. By the

281. Molle   (AC)

time manufacture was to begin in 1918, a third model was completed: the first machine to have actually reached the market, therefore, is labelled No. 3. Molle's aim in designing the machine was to simplify it as far as possible and make it easy to disassemble—he had suffered from the difficulties which some of the early designs presented to the repairman. But the machine was not a success and the company went bankrupt in 1922, whereupon it was resurrected as the Liberty Typewriter Co. of Chicago which briefly brought out the same product under the name *Liberty*. (Fig. 281)

Monarch Pioneer: Portable front-stroke machine with three-row keyboard built by John H. Barr in 1920 and manufactured by Remington Rand.

Monofix: See *Mentor*.

Moon Hopkins: A combined adding machine and typewriter of up-stroke design, invented by Hubert Hopkins of St Louis in 1902. The machine passed through several hands before being bought up by Burroughs Adding Machine Co. in 1921.

Moore: Charles Moore of West Virginia was granted a type-wheel patent in 1876, the most noteworthy feature of which is that a quarter interest was assigned to James O. Clephane, the 'demolition expert' (see page 144).

Morgan: George Morgan of Ohio was granted a number of patents in 1876 *et seq.* for complex type-wheel machines using a hammer for printing. Inking was by three-coloured roller with provision for selection of the desired colour. One of the patents was assigned to the Typographical Machine Printing Co. of Washington, but it did not reach production.

314

Morris: A small machine with a square index patented by Robert Morris of Kansas City in 1887 (filed 1886) and produced by the Hoggson Manufacturing Co. of New Haven, Conn. A carriage containing the index travelled along rods above the platen. The letters were selected by moving the index relative to a fixed pointer directly above the printing-point. A knob was provided for this purpose and depressing it brought the type into contact with the paper. Inking by pad. (Fig. 16)

Mossberg: Listed[42] but without details. The Mossberg and Granville Manufacturing Co. was responsible for producing the Granville Automatic, but there is no evidence that the machines are related.

Moya: A type-sleeve machine with a three-row keyboard invented in 1902 by Hidalgo Moya and manufactured in England in the following year. This first model was not commercially successful and was replaced in 1905 by an improved version on which, among other things, the ribbon spools were moved away from the direct line of vision as on the original model. The Moya was replaced, in 1908,

282. Moya, Model One   (ITC)

by a down-stroke machine called the *Imperial*. In France, it was offered as the *Baka 1*. (Fig. 282)

Müller: A rare attempt at crossing a type-wheel action with a syllable ambition was this German machine patented by F. J. Müller. It resembled the Hammond.[15]

Mulligan: A US patent was granted to B. T. Mulligan of New York in 1885 for a linear index machine with a knob that rode laterally to select the characters, and upon depression released a hammer striking from the rear.

Munson: Patented in 1889 by Samuel John Seifried who assigned two-thirds to Fred and Louis Munson of Chicago, it was introduced the following year by the Munson Typewriter Co. of the same city. It was of the type-sleeve class, with the peculiarity that the sleeve was mounted on a horizontal axis, parallel to the

315

platen. A wide ribbon, centrally mounted, gave this three-row machine its characteristic appearance. Depression of a key first selected the corresponding character by sliding the sleeve along its axis and rotating it, and then tripped a hammer which struck the paper against the character. The type-sleeves were easily interchangeable.

A slightly improved model was introduced in 1897 but, in the following year, the name of the machine was changed to *Chicago* and that of the manufacturers to Chicago Writing Machine Co. A further change of name and manufacturer occurred in 1912 when production was taken over by the Galesburg Writing Machine Co., the machine becoming the *Galesburg*. An aluminium model was introduced to mark the twenty-fifth anniversary of the Munson in 1915, and manufacture came to an end two years later. Also marketed as *Baltimore*, *Conover*, *Draper*, *Ohio*, and a single surviving machine, in LC, labelled *Yale*. (Fig. 11)

Murphy: A blind up-stroke machine resembling the Remington was the subject of this US patent granted in 1885 to J. R. Murphy of Pennsylvania.

Musigraphe: Musical typewriter patented in 1908 by the Belgian firm of Allman & Cie., but never produced.

MW: See *Gundka*.

Myers (a): A British patent granted to F. Myers in 1886 covered a down-stroke machine on which the type-bars were arranged in two semi-circles, radiating

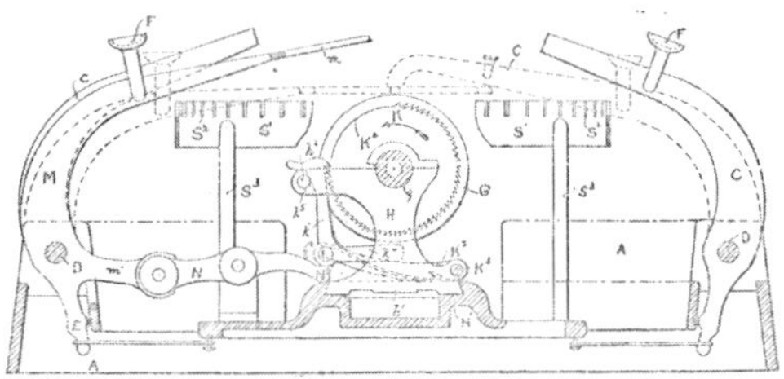

283. Myers (a)   (BP)

outwards, like fans, from the printing-point. The type-bars were set obliquely, raised towards the centre above the printing-point and were pivoted at their outer extremities. Keys were attached directly to the levers so that they formed a circular keyboard. There was no linkage—depression of the key brought the type-bar directly down on to the paper on a platen positioned centrally between the two semi-circles. It was never manufactured. See *Mercury*. (Fig. 283)

Myers (b): A book typewriter patented in the US by J. H. Myers in 1899. It was not produced.

316

Nason: See *Ingersoll.*

National (a): A front-stroke three-row portable with double shift manufactured by the Rex Typewriter Co. It was of lower profile than the *Rex*, but of similar mechanical design. It first appeared as the Model Two in 1916 or 1917, Model Three in 1918 and Model Five in 1920. It was for a time distributed by the American Can Co. Also marketed under the names *Crown, Express, National Portable* and *Portex* (a contraction of Portable Rex).

National (b): An up-stroke machine with a curved three-row keyboard and double shift, patented in 1889 by Henry Harmon Unz and manufactured by the National Typewriter Co. of Philadelphia. Ribbon inking. The inventor probably

284. National (b)   (LC)

took both spiritual and practical inspiration from an 1885 patent assigned to him by Franz Wagner, covering a not dissimilar up-stroke design. (Fig. 284)

National (c): See *Stenotype (b).*

National Portable: See *National (a).*

Natuatienne: A French machine listed but never produced.[28]

Neal and Eaton: Several patents for an electric swinging-sector machine with four-row keyboard were granted to these two residents of Massachusetts from 1892.

New American: See *International (b).*

New Century: See *Caligraph.*

New Century Caligraph: See *Caligraph.*

New England: This remarkable machine was introduced in England in 1900 and is virtually identical with the *Index Visible* which appeared in the United States the following year. For a description, see *Index Visible.* The manufacturer is unknown.

New Franklin: See *Franklin.*

New Imperial: See *Imperial (a).*

New Rapid: See *Rapid.*

New Sun: See *Sun.*

New Yost: See *Yost.*

Neya: A German front-stroke machine with three-row keyboard and double shift, invented by Arpad Krejniker and manufactured in 1925 by A. Ney of Berlin.

Niagara: A type-wheel machine with an indicator and circular index introduced by Blickensderfer Manufacturing Co. in 1902[28] or thereabouts. It used a number of Blick components, including the type-wheel, carriage and inking system. The

285. Niagara   (TMT)

type-wheel was mounted on a horizontal shaft at the outer end of which was a knob with which it was turned to select the character and depressed for printing, slots ensuring alignment. (Fig. 285)

318

Nickerson: This vertical platen machine is said[22] to have been first made by Walter Hanson, who died before it was completed, whereupon a Revd Lee had a hand in it and finally a Revd Charles S. Nickerson brought it to completion,

286. Nickerson   (MPM)

patenting it in 1909 and assigning it to his Nickerson Typewriter Co. of Wisconsin, formed in 1907 to manufacture the machine. He did not succeed in bringing it to the market, and the Automatic Telegraph and Telephone Co. later subsidized further development but to no avail. The only known example of this four-row keyboard, front-stroke machine is in MPM. Nickerson was previously granted an up-stroke patent in 1897. See *HANSON*. (Fig. 286)

Noco-Blick: See *Blickensderfer.*

Noiseless: A thrust-action machine developed conjointly by Wellington P. Kidder, inventor of the *Franklin* and the *Wellington*, and his Canadian collaborator on the latter machine, C. C. Colby. Experimental work on the machine continued from a first patent in 1896 until 1904, when the Parker Machine Co. was organized to bring the design to the manufacture stage. This in turn became the Silent Writing Machine Co. in 1908 and ultimately the Noiseless Typewriter Co. It was only in 1912 that the first machines were produced: these were the result of an invention eliminating cams altogether, bringing the type-bar almost to the paper and then using the momentum accumulated by an overthrow weight to make the actual contact.

Full-scale production of the design was undertaken in 1917 with the Model

Four—only relatively few machines had been sold in the two previous years. A portable model appeared in 1921, Model 5 in 1923 and the following year Remington bought out the company, and their Model 6 (1925) and later machines became known as the Remington Noiseless. Other makes produced noiseless machines using the same patents. Model 6 was the first with a four-row keyboard—previous models had three rows. The three-row Noiseless Portable (1921) became the Remington Noiseless Portable in 1924.

When word about the machine first leaked out, around the turn of the century, it was called *Silent*. (Fig. 215)

Nord: See *North's*.

Norica: Oblique front-stroke machine with four-row keyboard invented by Carl Fr. Kührt and produced in Germany in 1907 by Kührt and Riegelmann GmbH. Manufacturing rights were sold in 1909 to the Deutschen Triumphfahrradwerke A.G., who rebuilt it into a conventional front-stroke machine to which they gave their own name. The Norica went through several minor changes during its brief life, these consisting chiefly in changing the printing angle to make the machine less of a down-stroke and more of an oblique front-stroke design. It was *not*, however, a straight front-stroke, nor was the standard Triumph which developed out of it ever made as an oblique machine. The Triumph Perfect Visible (illustrated[22]) is the American, not the German *Triumph*.

North's: One of the relatively few down-stroke machines with the type-bars behind the platen, as in the Brooks and the Waverley, the North's was invented by Morgan Donne and George B. Cooper in 1890 and placed on the market by the North's Typewriter Manufacturing Co., London, in 1892. It was the successor to the *English*, with which Donne was also connected. The name was given the machine by Lord North who financed it—his untimely death brought the venture to an equally untimely close. Four-row keyboard with single shift and ribbon for the impression. The North's was made only in limited quantities. Also marketed in France as the *Nord*. (Fig. 12)

Nototyp-Rundstatler: A musical typewriter developed from the *Archo* in 1936. A second model was called *Melotyp*.

Nouveau Siècle: Primitive circular index machine virtually identical to the *Simplex (a)* and other similar devices. Manufactured in France by G. Meyer in 1912. Also marketed as the *Meyer*.

Nova: See *Sun*.

Nowak: Machine for the blind featuring interchangeable type-wheels for embossing and for printing, with a hammer striking paper against the character. This instrument was designed by a Viennese called Nowak and was apparently[6] manufactured by Sczepanski & Co. of the same city in 1902. US patent, 1903.

Oblique: Listed[42] but without details.

Odell: A linear index machine designed by L. J. Odell whose patent was filed in 1887 and granted in 1889. It was produced in four models offering progressive refinements: upper and lower case, instead of capitals only, appeared on Model Two in 1891. Inking by roller. It was marketed by the Odell Type Writer Co. of Chicago (and others) and was the most popular of the many machines (*Sun, International,* etc.) of similar design. Printing was effected by sliding the index till the desired character corresponded to a fixed pointer above the printing point, whereupon the index was depressed by means of the same finger-grip with which the character was selected. On models offering both cases, change of case was effected by rocking the index in its groove. A spring ensured a positive click action. (Fig. 25)

Official: This type-wheel machine with three-row keyboard and double shift was invented by C. E. Peterson of Minneapolis in 1898 and marketed in 1901. A

287. Official   (27)

hammer at the rear struck the paper against the type-wheel. Impression by ribbon. (Fig. 287)

Officine Galileo: An Italian machine by this name, with rectangular letter index

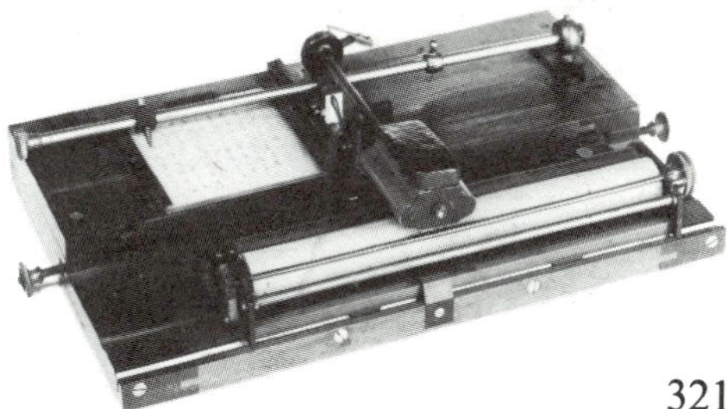

288. Officine Galileo   (STM)

and an indicator to one side, is in STM and dated 1904. (Fig. 288)

Ohio: See *Munson.*

Oliver: One of the great machines, it is probably the closest to an indestructible typewriter ever made. Of lateral down-stroke design, it had the inverted U-shaped type-bars banked to either side of the printing point, which gave the machine its unique appearance. There was a three-row keyboard, with double shift which moved the carriage in a front-to-rear direction. Ribbon for the impression.

289. Oliver   (AC)

Successive models incorporated improvements and sophistications, but the profile of the machine changed little.

It was invented by Revd Thomas Oliver from experiments originating in 1888, and the first patent was granted in 1891 for a segmental comb machine with a hammer for printing and roller inking. The lateral down-stroke design was patented next, in 1894; features of the first patent were, however, retained in the down-stroke instruments. The Oliver Typewriter Co. turned out its Model One in the same year, but full-scale production began two years later. Model Three: 1898, Model Five: 1906, Model Seven: 1914, Model Nine: 1916, Model Eleven: 1922. Production in the United States ceased in 1928, whereupon it was resumed in England with Model Fifteen labelled *British Oliver.* The production closed in 1931, after which a standard four-row portable using the name was manufactured in various European countries. The Oliver was marketed in Austria as the *Courier* and in Germany as *Fiver* and *Stolzenberg.* (Fig. 289)

Olympia Plurotyp: See *Mignon.*

Orientalis: See *Culema.*

Orpen: See *Dennis Duplex.*

Osborn: The machine listed[42] as unknown is, in fact, a type-wheel design patented in 1888 by A. A. Osborn of New York.

Otto: US patent granted in 1907 to Revd H. J. Otto for a pneumatic machine and a scheme to have all typewriters in a building piped to a compressor in the basement.

Pacior: Polish front-stroke with three-row keyboard and double shift introduced in 1921 by Pierwsza polska fabryka maszyn do pisania Wl. Paciorkiewicz . . . which merely means 'First Polish typewriter factory of Wl. P.' Also named *Idea* and *Iskra.*

Pagina: See *Meteor.*

Parisienne: Invented by the Frenchman Ernest Enjalbert in 1885, this circular index machine was marketed the following year. It consisted of a disk with rubber type, on the axis of which was an indicator which selected the characters from the letter index above. Depression of the indicator key made the impression on the paper to the back of the machine. Inking by pad.

Parker: An 'automatic' machine, with vertical platen, in which the element of automation consisted solely of means whereby the carriage could be controlled from the keyboard. It was the subject of a British patent granted to Roy Parker of N.Y. in 1908. Front-stroke design with double keyboard, it printed around the platen, which advanced vertically for line spacing.

Parks and Sheffield: These two New Yorkers were granted a US patent for an electric machine in 1883.

Patria: See *Germania (a).*

Peacock: E. E. Peacock of London was granted a British patent in 1884 for a design using radial type-plungers around the periphery of a drum above which was a circular index and selector. Roller inking.

Pearl: See *Peoples.*

Peerless: A non-visible machine of the up-stroke class with a double keyboard, patented in 1890 by Charles M. Clinton and James McNamara and manufactured in 1891 by the Peerless Typewriter Co. of Ithaca, N.Y. It was virtually identical to the Smith Premier, with which it fought a patent suit.

Peirce Accounting: A front-stroke machine with three-row keyboard and double shift, typing on a flat vertical paper-frame. Produced in 1912 by Peirce Accounting Machine Co. An example in MPM.

Pellaton: An electric type-wheel machine invented by a Swiss watchmaker called Georges Pellaton in 1932. A pointer was attached by a cord to the spring-loaded type-wheel shaft, so that locating the character with the pointer revolved the wheel to the corresponding position. The idea is taken directly from the *Index Visible*, in which the cord was attached to the index finger. On the Pellaton, however, printing was effected electrically by touching the corresponding contact with the pointer. Not manufactured.

Pember: US patent granted to Jay R. Pember of Vermont in 1873 for a syllable machine printing up to four letters at a time. The keys were arranged in four sets: one set for the thumb and another for the fingers, for each hand. A pedal brought paper and type together.

Peoples: A type-wheel machine patented in 1891 by C. J. A. Sjoberg of Brooklyn and assigned to the Garvin Machine Co. of N.Y. An indicator geared to the wheel selected the characters from a curved index, and upper and lower case were provided by means of a shift key which raised the level of the type-wheel. Roller inking, with a supplemental Sun-style reservoir. An optional model called *Pearl*

290. Peoples   (LC)

was also introduced: this was characterized by a novel arrangement whereby a piece of ribbon was looped around the type-wheel and moved a space at a time, independently of it. A further model using a conventional ribbon mechanism was patented in 1893 and named *Champion*. (Fig. 290)

Perce: W. R. Perce of Rhode Island was granted a US patent in 1883 for a primitive type-wheel device in which a selector, moving along a linear scale, exerted pull on a cord attached to a pulley on the spring-loaded type-wheel shaft. (Cf. later *Index Visible*.)

Perfect: See *Salter*.

Perfection: One of fifty inventions of Richard W. Uhlig, this 1906 machine was described[42] as having an interesting type-bar movement.

Perkeo: This copy of the Standard Folding and Corona machines was produced in Germany from 1912 by Clemens Müller A.G. in the plant previously used to manufacture the Albus in Vienna. It was marketed in France as the *Galiette*. See also *Albus*.

Perkins: A number of machines for the blind was developed by the Perkins Institute of New York over the years. A braille writer from 1900 had six keys and spacer, with conventional carriage, line-spacing levers etc. An improved model from 1905 appeared with a keyboard of what has become the conventional configuration for machines for the blind.

Perlita: See *Gundka*.

Perry: An interesting patent for a *Lambert*-type action was granted to Charles H. Perry of N.Y. in 1888. (Fig. 291)

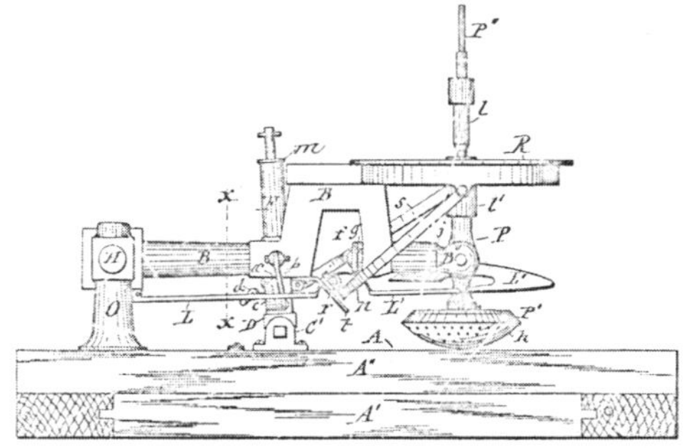

291. Perry   (USP)

Petit: Circular index machine with rubber type and indicator, patented by N. F. Petit of South Carolina in 1885.

Petty: See *Pettypet*.

325

Pettypet: A type-wheel machine using an index similar to the American Visible but with a pair of indicators and alignment teeth modelled after the Merritt, this anachronistic design was invented by the Viennese Podleci and others and was developed in the German Archo factory in 1930. It appears to have been the only one of several similar experimental designs to have been manufactured. It was also labelled *Petty*. Pettypet GmbH is reported[28] to have been founded to market the machine. A later version, called *Stylotype*, was developed by Alfred Unger of Berlin and eventually manufactured by the Stylotype Kleinschreibmaschinen GmbH. It, too, was a small portable but with a type-wheel controlled by a flat sliding indicator moved by an independent stylus. (Fig. 292)

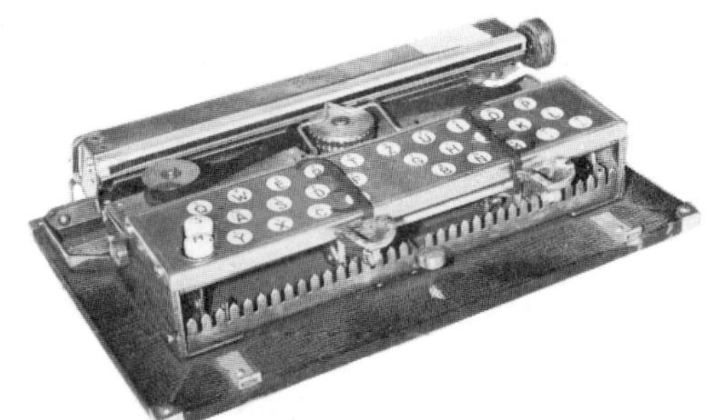

292. Petty   (BC)

Philadelphia: A type-wheel machine with a three-row keyboard, which was to have been manufactured by the Philadelphia Typewriter Co. It was patented by Byron Brooks in 1891.

Phipps: A US patent was granted to J. H. Phipps of Michigan for a conventional up-stroke machine. Dated 1880.

Phonautograph: For a machine by this name to have been developed by a man called A. C. Rumble sounds like a practical joke, but it wasn't! Invented in San Francisco in 1894, it used the vibrations of a phonograph membrane to produce voice prints which presumably a trained operator could read. According to the 1896 source[35] for this information, the machine was still in the process of development and the principle is still more or less that way today.

Phönix: German type-wheel machine with three-row keyboard and double shift, introduced in 1908 by Apparate- und Maschinenbau GmbH. The Phönix first appeared as the *Merkur*, while the name of the manufacturer was still Maschinenfabrik Merkur GmbH but the names were changed almost immediately. It was later changed again to *Sekretär*. The keys controlled the movements of the type-wheel by the rocker action of a plate beneath the keyboard.

Phonograph Typewriter: This marriage between a phonograph and a typewriter was intended to produce a machine which would operate electrically at the command of the human voice. According to the intentions of the inventor, Dr Frank Traver of Racine, Wis., the idea was that talking into the horn would move the type-bar corresponding to the letter enunciated.

Phonotyp: A shorthand machine invented in 1924 by an Englishman called E. Howard. It printed on paper tape from ninety-six keys by the chord method.

Phonotype Reporter: A shorthand machine patented in the United States by J. C. Zachos in 1876.

Pianograph: A musical typewriter of down-stroke design, with a piano keyboard, invented by a German called Hermann Wasem. It was not manufactured.

Piccola: See *Standard Folding.*

Picht: A famous German make of machines for the blind, both typewriters and shorthand, produced from 1899 by Oskar Picht. A number of different models was offered over the years, the two major ones being an instrument with six keys

293. Picht, type-wheel model   (TMT)

and space bar for embossing braille, and a type-wheel machine with roller inking but a braille letter index to permit a blind operator to print in visible characters. A sub-model embossed both braille and Roman letters for those who had not mastered the blind alphabet. The type-wheel machine was marketed from 1910 as a regular typewriter, by the substitution of a letter index with Roman characters for the braille index. The type-wheel and other elements were retained, double shift was offered, and printing was by a separate key. This model was called *Pionier.*

The instruments were first made by Wilhelm Ruppert of Berlin; after 1902 by Schreib- und Nähmaschinenfabrik Wernicke, Edelmann & Co.; from the following year by Alois Herde & Wilhelm Wendt; and from 1919 by Bruno Herde & Friedrich Wendt. A range of shorthand machines for the blind was also manufactured from 1909 on—these embossed paper tape instead of sheets. (Figs. 203 and 293)

Pionier: See *Picht.*

Pittsburg: See *Daugherty.*

Pneumatic: The first compressed air machine, patented by an Englishman called Marshall A. Wier in 1891. It had a three-row keyboard of small rubber balls

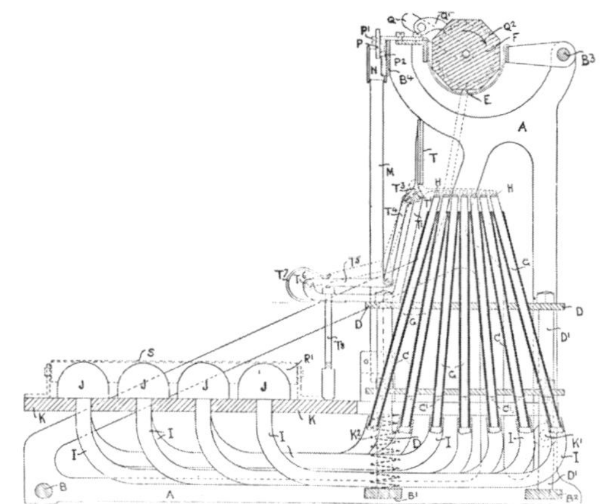

294. Pneumatic   (USP)

connected by tubes to type 'pistons', so that compression of the balls produced enough pressure in the lines to force the type into contact with the paper. It appears to have been manufactured. (Figs. 200 and 294)

Pocket (a): A primitive circular index machine with type fitted to the under-side of a disk which was twirled by a knob and depressed for printing. Inking by roller. Manufactured in England in 1887, and also known as *Miniature.*

295. Pocket (a)   (AC)

328

Dobson & Wynn and Miniature Pocket Typewriter Co. are both listed as manufacturers. (Fig. 295)

Pocket (b): A primitive circular index machine manufactured by the Pocket Typewriter Co. of Rockford. The index is horizontal and runs along a rack—

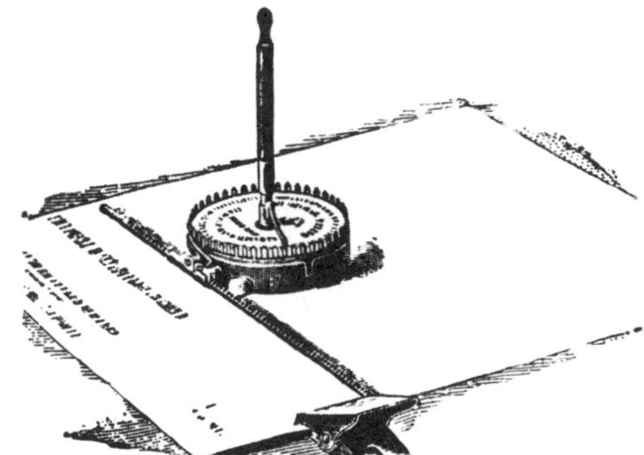

296. Pocket (b)   (LC)

there is no platen or carriage. First manufactured in 1894; several models all basically the same. (Fig. 296)

Polygraph: A down-stroke machine invented by Paul Riessner of Leipzig in 1903 and manufactured by the Polyphon Musikwerke A.G. The first model had a two-row curved keyboard, with square keys similar to those of the early Hammond Ideal which it clearly copied. In 1905, the same machine was offered with a straight

297. Polygraph   (27)

three-row universal keyboard. Ribbon inking. A third model of almost identical design followed in 1907 and production ceased in 1909. A later effort at re-introducing the machine in 1919 was short-lived. (Fig. 297)

Polytype: A type-wheel machine capable of printing up to four originals at the same time was invented in 1895 by Professor Felice Molinari of Milan and made in limited numbers by Prinetti and Stucchi. It consisted of a horizontal shaft on which two, three or four type-wheels were mounted above a flat paper-carrier designed for the corresponding number of sheets. Several models were made: two had keyboards, one a letter index. On the latter, a separate key was depressed to bring type-wheels into contact with the paper. This whole assembly was fixed in a frame for lateral and longitudinal movement. It was also offered as suitable for printing in bound books.

Populaire: French machine introduced in 1904—no other details.[28]

Portable Extra: See *Helios.*

Portefeuille: A European journalist who lived in America, Anton Daul, is re-ported[6] to have made an abortive effort at producing a typewriter the size of a wallet.

Portex: See *National (a).*

Postal: A type-wheel machine with three-row keyboard not dissimilar to the Blickensderfer, this double-shift instrument was invented by William P. Quentell and Franklin Judge and manufactured by the Postal Typewriter Co. of New York. It was covered by a patent filed in 1903 and granted the following year; Quentell himself, however, held many patents from 1896 on. Depression of a key first

298. Postal　　　　　　　　　　　　　(LC)

rotated the type-wheel and then brought it down onto the paper. Ribbon for the impression. It was a popular machine in its time, and several (similar) models were introduced, including Model Seven in 1908. The machine was widely exported throughout Europe. (Fig. 298)

Practical: See *Simplex (a).*

330

Prendergast: An electric swinging-sector machine using a hammer for printing was patented in 1903 by J. S. Harrison and H. Hill of Georgia and assigned to the Prendergast Electric Typewriter Co. of Maine. It got no further than that. Some sources[15] refer to the machine as *Harrison Hill.*

Presto: See *Senta.*

Promptografo: Vicente Alonso de Celada y Barona of Barcelona was granted a Spanish patent in 1881 for a stenographic machine of this name with fifteen keys printing Roman characters on paper tape advanced by clockwork. Several keys could be depressed simultaneously. Phonetic similarities were used to reduce the numbers of keys required. It printed 140 words a minute.

Proteus: See *Albus.*

Protos: One of the many chips off the old *Wellington* block (see also *Adler, Archo, Garbell,* etc.), this German three-row double-shift thrust-action machine was made in 1922 by Zimmer, Zinke & Co. and was almost an exact copy of the Adler 7. A shrunken *Protos Kleinschreibmaschine* which copied the *Klein-Adler* was introduced in 1924. The following year, manufacture came to an end.

Prouty and Hynes: First patent of a 'standard' front-stroke machine, filed in US by Enoch Prouty and Olive Hynes of Chicago on 5 January 1887 and granted 18 September 1888. The British patent is dated 15 November 1886. (Fig. 299)

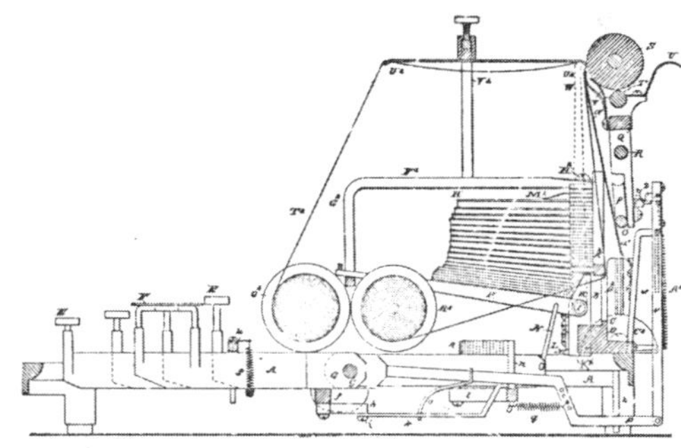

299. Prouty and Hynes   (USP)

Pullman: See *American (b).*

Punctograph: A machine for the blind similar to the improved *Daisy* except that the machine had a wooden frame with a line-spacing device, to which the paper was clamped. Built in 1885.

Raab: A pneumatic machine with up-stroke plungers actuated by depression of hollow rubber balls on the keyboard, this US patent granted to W. Raab of Iowa

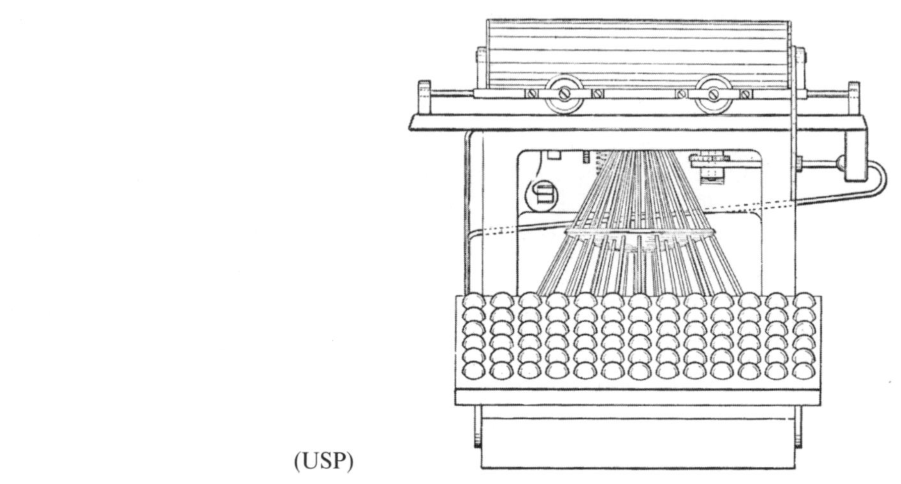

300. Raab   (USP)

in 1894 was virtually identical with the earlier *Pneumatic.* (Fig. 300)

Rapid: The first machine to use the thrust-action principle, later popularized by *Wellington, Noiseless* etc., was patented in 1888 by Bernard Granville of Chicago, later Dayton, Ohio. According to original literature in GC, it was manufactured first by Western Rapid Type-Writer Co. of Findlay, Ohio, and, from 1890 on, by the firm of Mead, Phillips and Granville of Dayton, Ohio. Instead of sliding on the same plane as later thrust-action machines, the type-bars of the Rapid radiated outwards from the printing-point in four horizontal

301. Rapid   (GC)

rows, the bars passing through square holes in two guide plates so that they all struck at the common centre. A four-row keyboard controlled the action, printing through ribbon. The machine was not a success and was soon withdrawn. It was also marketed under the name *New Rapid*. The inventor later designed another thrust machine to which he gave his name. (Fig. 301)

Rapide: See *Salter*.

Reed: US patent granted in 1892 to Charles J. Reed for an electric design.

Reichsfeld-Pad: There is a single reference[6] to this machine, exhibited by its Viennese manufacturer Alfred Reichsfeld in the Austrian capital in 1899. No further details.

Reliance: See *Daugherty*.

Reliance Premier: See *Daugherty*.

Reliance Visible: See *Daugherty*.

Rem-Blick: See *Blickensderfer* and *Remington*.

Remington: See references throughout the book. This famous machine developed from the *Sholes and Glidden Type Writer* which became the Improved Model One in 1878. It was of up-stroke design with four-row keyboard and it printed only upper case. Model Two introduced in 1878 printed both upper and lower case with single shift and was the prototype for all later Remington models (and many other makes) until 1908—these differed only in detail. Model Three was the same as Two but with wide carriage and extra characters; Model Four was upper case only, as on the first model. Model Five: 1888 was soon replaced by the more refined and popular Model Six (1894) and Model Seven (1896) which had additional keys. The first major departure in design was the introduction of the front-stroke visible Model Ten in 1908.

The machine was manufactured by E. Remington and Sons until 1886 when the firm of Wyckoff, Seamans and Benedict, till then sales agents, bought up the manufacturer's typewriter division. The machine was not labelled a Remington, however, until well into the 1880's; before then, it was simply called *Improved Type Writer*. The more specific title was applied after 'type writer' became a generic term. Towards the end of the century, it was the principal component of the Union Typewriter Co., a trust formed by some half-dozen manufacturers for the purpose of pooling their production and marketing resources and fixing prices. Of later models, few departed from the standard, front-stroke, four-row design. The *Remington Junior* introduced in 1914 was a three-row double shift

333

302. Remington Portable   (AC)

machine (see *Century*), and in 1920 a flat *Remington Portable* with four-row keyboard had type-bars which folded back for storage and were raised to an oblique
front-stroke position for printing. Thrust-action *Remington Noiseless* machines
were also marketed—see *Noiseless*. An abortive effort at remarketing the Blickensderfer Model Five was made in 1928. This product, identical with the original,
was labelled *Rem-Blick*. An alternative unrecorded name for this model is *Baby
Rem*, of which sole survivor appears to be a machine in AC. (Figs. 28, 150 and
302)

Remington-Fay-Sholes: See *Rem-Sho*.

Remington Noiseless: See *Noiseless*.

Remington-Sholes: See *Rem-Sho*.

Remmerz: A German machine invented by two brothers, H. & F. Remmerz,
using three type indices—one for upper, one for lower case, and one for figures.
It used an indicator for letter selection. Not manufactured. Date not given.[6]

Rem-Sho: The legal history of this machine is more momentous than its mechanical
specifications. Patented in 1894, it was manufactured in 1896 by the Remington-
Sholes Typewriter Co. which was founded by Zalmon Sholes and Franklin
Remington, sons of the typewriter pioneers. It was an up-stroke machine with
four-row keyboard and differed from its competitors mainly in a shift mechanism
which moved the entire type-basket rather than the carriage. The company
changed its name to the Fay-Sholes Typewriter Co. in 1901, after losing its court
case against Remington Standard who claimed exclusive right to the use of the
name Remington (even though the company no longer belonged in any way to
the Remington family). The machine was thereafter marketed as the *Fay-Sho*,
also *Fay-Sholes*, and the manufacturer became the Fay-Sholes Typewriter Co.
after its president, Charles Norman Fay.
    Soon afterwards (1904) it was offered as an adding typewriter, after the manu-

facturer had bought up a machine called Arithmograph; the name of the manufacturer is reported[42] to have been changed to Arithmograph Co. Appeals against the previous court ruling were pending, however, and were taken all the way to the Supreme Court which finally ruled in favour of Zalmon S. and Franklin R., and their machine was once again marketed under the previous name (*Rem-Sho*), and also *Remington-Sholes* and *Remington-Fay-Sholes*. A visible Model Ten was introduced in 1908, but the company was in financial trouble and went into receivership in 1909, and its assets were sold to the French firm Japy Frères.

The first models were beautifully cast in bronze; hence few of them are left because they were scrapped during the war for their metal content. (Fig. 157)

Reporters Special: See *Harris.*

Rex: See *Harris.*

Rex Demountable: See *Demountable.*

Rex Visible: See *Harris.*

Richardson: An indicator machine with a precocious golf-ball design was patented by H. B. Richardson of Mass. in 1884. (Fig. 2)

Riotor: A sole reference to this French down-stroke machine[36] described it as having type-bars resting on a vertical inking pad; upon depression of a key, the corresponding bar was projected till it was horizontally over the printing-point and then came down onto it. Date quoted was 1908.

Roberts: British patent for a shorthand machine granted in 1884. It had twenty keys, but a second model patented the following year cut the number to nine, using 'chord' operation.

Roberts Ninety: See *Blickensderfer.*

Rochester: A small portable invented by Wellington P. Kidder. A three-row double shift machine of the thrust group, it was introduced in 1923 by Rochester Industries Inc. with which Kidder was associated, the company having bought up the assets of the Leggatt Portable Typewriter Corporation in the previous year. A machine called a *Midget*, patented by Harry Bates of Chicago in 1924 (filed 1920) and perfected by Kidder, is also reported[28] to have been produced under the name *Rochester* in 1923.

Roe: A type-wheel design, patented by E. R. Roe of Illinois in 1885.

Rofa: A German down-stroke machine introduced in 1921 by Robert Fabig GmbH, the manufacturer of the *Faktotum* down-stroke before the First World War. The Rofa was introduced with a curved, three-row keyboard with double shift and roller inking; a later model produced in 1923 had a straight three-row keyboard. It was not a particularly good machine, but it enjoyed a large volume of sales owing to the great demand for typewriters following the war. Also marketed under the name *Correspondent* in Holland. (Fig. 172)

Royal Bar-Lock: See *Bar-Lock*.

Royal-Express: See *Salter*.

Sabb: See *Juventa*.

Saliger: Shorthand machine using type-bars and twelve keys patented in 1904 by Alois Saliger of N.Y.

Salter: A fine English down-stroke machine with the type-bars in front of the platen, this machine was patented in England by James Samuel Foley and John Henry Birch in 1892 and manufactured by the well-known scales-maker George Salter & Co. of West Bromwich. The type-bars were arranged in a semi-circle around the printing-point, much as on the Bar-Lock, except that the first Salter (Model Five) had a curved three-row keyboard and double shift. Inking on the

303. Salter No. 10   (AC)

very first few machines of this model was by roller, but this was replaced almost immediately by ribbon, and the machine was labelled Improved Model Five. The three machines of this model that the author knows to have survived all

336

have serial numbers around 2,600. A Model Six, introduced in 1900[14] had a straight three-row keyboard, as did Model Seven (1907) and Model Nine (1908). The first actual design departure was the introduction in 1913 of an oblique front-stroke model with four-row standard keyboard. Also marketed as *Rapide*, *Perfect*, and *Royal Express*. (Figs. 10 and 303)

Samenhof: A machine described[6] as similar to the *Hall* instruments (Eureka, Graphic, Kneist etc.) except that the type index was in the form of a comb with the characters on the ends of the teeth. It was invented by Dr L. Samenhof.

Sampo: A Swedish machine by this name, also called *Finnland*, is reported[28] to have been invented by Dr Rafael Herzberg of Helsingfors and manufactured by Husqvarna Vapenfabriks A.B. in 1894. 500 units were said to have been made, but none has apparently survived and details of the machine are unknown.

Samtico: A primitive machine marketed in 1899 by the Timings Burgess Co. of London. (The single source which lists this machine[6] is hopelessly confused.) It was one of three designs covered in a British patent granted in 1897 to Samuel Timings and John Burgess; the first was for an instrument with independent characters held in place by a rubber band; the other two were for *Simplex*-type machines. The second of these was novel: the characters were cut out of the disk as in metal stencils and an inked pad brought down over it left the impression on the paper beneath.

Sanger: An 1871 British patent for a shorthand machine.

304. Saturn   (MPM)

Saturn: A novel Swiss machine invented in 1897 by F. Meyer-Teuber and manufactured in limited quantities two years later by Feinmaschinenwerk E. Stauder.

The controls of this design consisted of a chart of seventy-two upper and lower case characters arranged in nine columns of eight each. Nine keys were placed one at the end of each column and, on the left of the chart, a small grip for the thumb and index finger was attached by a piece of cord to a lever on the right, so that the cord cut horizontally across the chart. Eight locating holes for this grip were provided. The desired character was selected, first by moving the grip (and, consequently, the cord) to the horizontal row containing the character, and then by depressing the key relating to the column in which the character appeared. These movements located the corresponding character on a plate with sliding rows of type; this plate was beneath the platen so that typing was non-visible. Depression of the column key brought the type first into position, then into contact with the paper, and a ribbon provided the impression.

The same machine was also described[6] under the name *Stauder* (the title actually used reads 'Stander' which is a typographical error) and the Saturn may have been marketed under that label, which was the name of the manufacturer. (Fig. 304)

Schade: A radial plunger machine copying the *Writing Ball*, this German invention by Rudolf Schade dates from 1896. The type-plungers were designed as on the Hansen machine except that lower-case plungers were towards the centre, followed by a circle or so of upper case and an outer ring of figures and signs, with the space key in the very centre. This unit travelled along rails for letter spacing, the paper remaining stationary. Printing through ribbon. The unit was designed to be clamped on to a table top.

The inventor, accused of plagiarizing the Writing Ball, insisted that he had struck upon this identical design without ever having seen the other, although by the time the invention appeared, the Hansen machine had enjoyed wide diffusion for many years. Very few Schades were made.

Schapiro: A type-wheel machine invented in 1894 by A. Schapiro of Berlin. In front of the machine was a linear letter-scale with an indicator attached to a rack. This rack meshed with a gear on the type-wheel shaft so that lateral movement of the indicator caused the type-wheel to turn correspondingly. A key depressed by the left hand brought the type-wheel down on to the platen. The manufacturer was apparently[42] famous for a duplicator of his invention called a Schapirograph and he soon abandoned the unprofitable typewriter to dedicate himself to the lucrative duplicator (Edison's typewriter met a similar fate).

Schenk: Listed[6] but without details.

Schiesari: An ingenious invention by the Italian Mario Schiesari in 1905, this syllable machine was offered both as a regular typewriter and as a shorthand machine because of the speed of which it was supposed to be capable. It was of down-stroke design with the type-bars behind the platen as on the *North's* and *Brooks*. Twenty keys with double shift did the printing and up to six keys could be depressed simultaneously, the type-bars having provision for eccentric lateral displacement to the right of $2\frac{1}{2}$ mm. for each key depressed, providing these followed each other in alphabetical order. In depressing the maximum of six keys, the last would therefore fall $12\frac{1}{2}$ mm. from the central printing-point. The carriage was made to advance the same number of spaces as keys depressed. A company called The Syllable Typewriter Co. was formed in N.Y. in 1914 to manufacture the machine but nothing came of it.

Schilling: See *Daugherty*.

Schmitz: A German patent from 1882 for an up-stroke machine with a transverse platen (as on early Sholes, Mitterhofer etc. models) and five keys printing a short-hand code composed of up to two vertical and three horizontal strokes: |≡|.

Schoch: An 1870 patent granted to Henry Schoch for a shorthand machine using type-bars fitted with pens which wrote a code of dots. A treadle or clockwork was suggested for paper advance.

Scripta: See *Gundka*.

Seifried: Several models of typewriters and shorthand machines for the blind. A Midget Writer was introduced in 1890, with three keys and a cell spacer embossing braille on paper inserted from the rear between two rollers. A braille shorthand version appeared in 1900.

Sekretär: See *Phönix*.

Senta: Front-stroke machine of conventional design with three-row keyboard and double shift manufactured in Germany by Frister and Rossmann A.G. from approximately 1913 to 1930, in which year both the three-row and the four-row model, introduced in 1926, were withdrawn. Also marketed as *Presto* and *Balkan*.

Service Blick: See *Blickensderfer*.

Sheehy: One of the many pin barrel patents, using a travelling type-wheel and stationary paper, was granted in 1884 to R. J. Sheehy of N.Y.

Sherman: Daniel H. Sherman of Ohio was granted a US patent in 1877 for a down-stroke machine with a semi-circular, four-row keyboard, and ribbon inking. An interesting fore-runner of later book typewriters.

Sherwood: A machine invented in 1900 by a man of this name living in Holland is described[6] as having a type-wheel striking from beneath the platen and keys that remained depressed until each printing operation was complete. Automatic carriage return. The 1901 British patent application, however, was not granted.

Shilling Brothers: See *Daugherty*.

Shimer: A design very similar to the Remington, this up-stroke machine was patented by Elmer S. Shimer of Milton, Pa. in 1891 and manufactured in small

305. Shimer   (QC)

numbers some years later. The only surviving example is in QC. (Fig. 305)

Sholes (F.): Fred Sholes was granted an 1880 patent for a design in which the type-bars struck downwards from behind the platen. This may well have resulted from one of his father's experiments. (Fig. 156)

Sholes (Z.): A three-row front-stroke machine in MPM is attributed to Zalmon Sholes, brother of Fred and son of Christopher Latham Sholes.

Sholes and Glidden Type Writer: The ancestor of the Remington, this machine printing only upper case was manufactured from 1874 until 1878, when it became known as *Improved Type Writer Model One.* Up-stroke, four-row keyboard, fully enclosed, it underwent several minor changes relating chiefly to carriage-return mechanism: the first series was offered with a sewing machine stand and treadle return (MPM), this being replaced by a large lever on the right (LC) which itself was later dropped (AC). See Chapters Six and Eight, also Figs. 148, 149, 192 and 306.

340

306. Sholes and Glidden   (AC)

Sholes and Miller: Fred Sholes and William Miller of N.Y. were granted a patent in 1879 for an awkward up-stroke design featuring type-bars which radiated horizontally outwards from the printing point. Depression of a key caused the bar to slide forward until the back of it dropped down a slope, throwing the front up against the platen. A spring then returned it to its original position.

Sholes Visible: A four-row keyboard machine with unique front-stroke action. The type-bars in vertical rows projected outwards from the centre of the machine towards the operator. Upon depression of a key, the corresponding bar moved from its rest position to a clearing between the rows and then straight through to the printing point. The procedure was reversed when the key was released. Printing through ribbon.

This is said to have been the last machine designed by Christopher Latham Sholes and is interesting in the light of the inventor's earlier aversion to type-bars: on this machine, they were made of one piece and were guided all the way from the rest position to the printing point, thereby presumably overcoming problems of alignment. The design made for slow operation, but printing was visible. The patent is dated 1891.

Louis Sholes, son of the inventor, undertook manufacture of the machine in 1901 through the Meiselbach Typewriter Co. and in 1909 with the C. Latham Sholes Typewriter Co. but failed on both occasions. It was reported to have been marketed under the name *Bonita Ball Bearing*[27] and *Meiselbach*[42] but others[22] deny that it was ever put out under that name—a dangerous assertion in the light of the many 'one-off' labels found on old machines. (Fig. 155)

Shortwriter: Stenographic machine produced in 1914 by the Shortwriter Co. of Chicago. Twenty-four keys with shift, so arranged that both hands printed

307. Shortwriter   (MPM)

virtually the entire phonetic alphabet. The vowel keys were centrally placed and printing was on paper tape. The machine appeared in several models and was widely distributed. (Fig. 307)

Silent: See *Noiseless.*

Silkman: A US patent was granted to E. J. Silkman in 1890 for an electric design.

Simplex (a): The most famous of the cheap primitive circular index machines, it consisted essentially of a disk cut into flexible segmental teeth with rubber type bonded beneath each tooth and a circular letter index above. The inventor was A. M. English of N.Y., whose patent was granted in 1892. The manufacturer was reported[28] to have been Robert Ingersoll and Bros of N.Y., the company already manufacturing a similar machine called *Dollar.* However, all machines which the author has ever seen were labelled as manufactured by the Simplex Typewriter Co. of N.Y., and the Ingersoll information is probably incorrect.

Countless models were produced until well into the 1920s and perhaps even later. The last model the author has been able to trace, the relatively sophisticated Model 300, retains the same essential principle of the original machines.

It was marketed in France from as early as 1896 as the *Eureka* and in England as the *Gladstone.* It was also known under a multitude of names including *Practical,*

342

Baby Practical, Baby Simplex, Little Giant, etc. Numerous further patents were granted in succeeding years, principally to Philip Becker and W. Thompson of N.Y., the original 1892 patent having been assigned to the former who was probably the prime mover of the manufacturing company. (Fig. 177)

Simplex (b): See *Smith Premier*.

Skrivekugle: The radial plunger machine invented in 1865 by Pastor Malling Hansen of Denmark and manufactured, from 1870, in various models. In English, the name has been translated as *Writing Ball* which also appears in the British patents. Full details and illustrations in Chapter Six, also Fig. 7.

Slocum: Several patents were granted to W. H. Slocum of N.Y. but none of his designs was manufactured. An 1879 patent covers a machine featuring a type-wheel on a vertical shaft, controlled by a two-row keyboard printing upper and lower case by means of a hammer. (Cf. *Hammond*.) This design incorporated differential spacing.

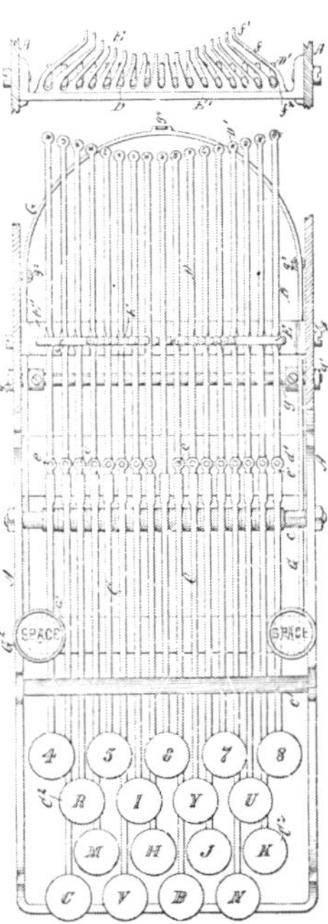

308. Slocum, 1886   (USP)

An 1886 patent was for an interesting up-stroke design in which the type-bars were pivoted halfway to allow them to pass obliquely upwards to the printing-point through a guide frame. This is a much neater arrangement than that on the similar but unrelated *American (b)*, which appeared some years later. (Fig. 308)

Smith (a): The 1900 British patent to Robert Henry Smith was for a shorthand machine which printed only two characters, separately or together and in different positions, providing a phonetic alphabet.

Smith (b): See *Emerson*.

Smith Premier: One of the big names in typewriter history, this machine was invented by Alexander Timothy Brown of N.Y. with patents filed from 1887 onwards (granted 1889) and manufactured from that year by the Smith Premier Typewriter Co. of Syracuse, N.Y. It was an up-stroke machine with double

309. Smith Premier   (AC)

keyboard and platen which swung up to reveal printing. A circular brush was provided for cleaning the type in the basket: this was wound up into position by means of a cranked handle, cleaning the type as it turned. Impression was by ribbon which zigzagged so that its entire width was utilized.

Several models of virtually identical specifications, principally Model Two (1895) and Model Four (1896), appeared until 1908, when the visible Model Ten was marketed. This was a front-stroke machine which retained the double keyboard. This model was sold in Germany as the *Simplex*. It was the last to appear with distinctive Smith Premier identity—the make, being part of the typewriter trust called Union Typewriter Co., later became a standard front-stroke four-row machine. (Fig. 309)

Soblik: A pneumatic machine invented by a German engineer living in Belgium

344

called Max Soblik. His US patent was filed in 1899 and granted the following year. The keyboard consisted of a number of rows of small holes, through which compressed air provided by two foot-pumps constantly passed. By placing a finger over one of the holes, the stream of air was diverted and was used instead to bring the corresponding character to the paper. These characters were on the ends of plungers set in a wheel which was kept revolving at 600 r.p.m. Since the rotation of the wheel was always in the same direction, any number of holes could be blocked on the keyboard simultaneously, and the machine automatically printed one character after another, on condition that they followed in alphabetical order. According to the inventor, this gave the machine a speed four times faster than a conventional typewriter. It did, however, possess the advantage that absence of levers, wheels, gears and hammers made it silent to operate.

Soennecken: A device by this name is illustrated[6] but neither the description nor the figure is clear enough to make the operation of the machine fully understandable. It is said to have been invented by F. Soennecken of Bonn (year not stated), and is described as being similar to the *Schapiro*, although this is difficult to see from the illustration. The similarities appear to include a linear letter scale with an indicator, but whether this machine also used a type-wheel is not shown.

Special: See *Mignon*.

Stainsby-Wayne: This long line of famous typewriters and shorthand machines for the blind started with a British patent granted in 1899 for a braille machine with a small keyboard of six keys and a cell spacer which rode on rails over a base-plate. Invented by Henry Stainsby and manufactured by Alfred Wayne and Co.

Stallman: See *Mignon*.

Standard Folding: The development of this highly successful design was started by Frank S. Rose of N.Y. in 1902 and patented in 1904; after his death in 1905, it was concluded by his son George, whose Rose Typewriter Co. marketed it in 1907. On this front-stroke portable with double shift and three-row keyboard, the entire carriage folded forward and came to rest above the keyboard, squaring the machine off for compactness. The manufacturing company changed its name in 1909 to the Standard Typewriter Co.; in 1912 the machine was called the *Corona* and, two years later, the company producing it became known as the Corona Typewriter Co. A number of minor improvements was made to the basic design over the years but by and large it remained unchanged until 1923, when a non-folding model with four-row keyboard was introduced. It sired a host of similar machines throughout the world, such as the *Erika*, manufactured

310. Corona (Standard Folding)   (AC)

in Germany, etc. Also marketed under the names *Atlas*, *Corona Piccola*, *Franconia*, *Improved Corona*, and *Piccola* (Fig. 310).

Star: See *Sun*.

Stauder: See *Saturn*.

Stella: See *Mignon*.

Steno: A 1922 patent for a shorthand machine with two pairs of type-wheels, the first pair covering opening consonants, vowels and syllables and the second for closing sounds etc. Not manufactured.

Sténo-Dactyle: A shorthand machine invented by Jules Lafaurie in 1899 and manufactured the following year. It was widely marketed throughout Europe, and received an enthusiastic welcome. Two sets of five keys each printed dots singly or in combinations on five parallel lines, giving the printing the appearance of musical chords. Later models printed numbers and dispensed with the lines.

Stenodattilografica: A shorthand machine invented in 1903 by the Italian Giulio Crespi. Two sets of six keys each printed numbers according to Lafaurie's (revised) Sténodactyle alphabet.

Sténoglyphe: A shorthand braille machine invented circa 1900 by a French Lieutenant Müller.

Stenograph: A shorthand machine patented in United States by Miles Bartholomew from 1879 on and manufactured by the United States Stenograph Co. of St Louis in 1889. It had four keys for each hand, plus central keys for the thumbs,

346

311. Stenograph   (GC)

and it printed a shorthand code of strokes in combinations of up to five, offering thirty-one distinct characters. A second model, for extra fast use, was supplied with clockwork paper advance. (Fig. 311)

Sténophile: A French shorthand machine which created quite a stir when first introduced. Invented by Charles Bivort, this 'veritable little mechanical marvel'[16] was introduced in 1906 after a number of years of development. It had two sets

312. Sténophile   (TMV)

each of ten keys, with an additional key between, and printing was by 'chord'. It was made in Germany briefly in 1913 and marketed as *Dictograph* but production ceased shortly afterwards both there and in France. (Fig. 312)

Sténo-Télégraphe: A telegraphic shorthand device to be used with any steno-graphic machine was demonstrated in France in 1886 *et seq.* by G. A. Cassagnes. It used perforated tape and was claimed to have a capacity of 12,000 to 24,000 words per hour.

Stenotiposillabica: Shorthand machines designed by the Italian Dario Mazzei. His patent dates from 1879. The machines had piano keyboards of forty-six keys and printed on paper tape by the chord principle. Each hand operated its own half of the keyboard. See *Ambrosetti* and *Tachigrafica.*

Sténotype (a): A shorthand typewriter which was a favourite in France for decades, invented in 1910 by Marc Grandjean. It printed twenty Roman characters on paper tape; the keyboard had two rows of ten characters each, separated by a space bar, and any number of keys could be depressed simultaneously. Up to 290 words per minute have been taken on it. Also known in England as *Stenowriter.* (Fig. 206)

Stenotype (b): A shorthand machine patented in 1911 by Ward S. Ireland of Dallas and manufactured by the Stenotype Co. of Indianapolis, later Owensboro,

313. Stenotype (b)   (AC)

Ky., and a succession of other companies. It used the fingers of the left hand to print opening consonants, and those of the right for closing consonants, with the thumbs printing the vowels. When it first appeared, it printed a code of long and short strokes but later models printed Roman characters with arbitrary combinations for economy (e.g. PB=N). The inventor designed another machine along roughly similar lines called *National.* (Fig. 313)

Stenotyper: A shorthand typewriter widely hailed after it was patented by John Franklin Hardy of Chicago in 1897, but enjoying only limited commercial success. It consisted of a keyboard of three keys per hand plus space key etc. which printed a shorthand code of dots and dashes on a sheet of paper the width of the machine. Any number of keys could be used at once. It was manufactured in Germany by makers of the Adler typewriter. (Fig. 314)

314. Stenotyper   (SML)

Steno-Typograph: A telegraphic shorthand machine invented by the German Angelo Beyerlen in 1884, while he was working in the United States. It had a piano keyboard of two sets of five keys each. The inventor also designed a machine for the blind which bore his name.

Stenowriter: See *Sténotype (a)*.

Sterling (a): This machine, variously listed as being manufactured in 1905,[28] 1910,[22] and 1911[42] is identical with the earlier *Eagle* and *Defi*, and undoubtedly

315. Sterling (a)   (MPM)

represents a later effort at manufacturing the same design. The inventor is listed as being the man responsible for the others of this design, and the manufacturer was the Sterling Typewriter Co. of N.Y. (Fig. 315)

Sterling (b): See *Bar-Lock.*

Stickney: A patent for an up-stroke design in which the carriage spaced prior to the impression, was granted in 1896 to Burnham Stickney of New Jersey.

Stoewer: A German front-stroke machine invented by Paul Grützmann in 1903 and manufactured by the sewing machine and bicycle company of Bernhardt Stoewer. The first model had a three-row keyboard with double shift; all later models were standard four-row machines. Models One and Two were also marketed as the *Cito* and *Lloyd.* In 1912, a three-row portable was introduced as the *Stoewer Elite,* also *Elite, DS,* and *De Esse.* It was replaced in 1926 by a four-row portable.

Stolzenberg: See *Oliver.*

Stüber: Syllable machine with sixty-six keys which printed on paper tape, up to two characters being printed simultaneously. Three models were made by the inventor, Dr Werner Stüber, from 1936 on, all of which were destroyed during the war.

Stylotype: See *Pettypet.*

Sun: A linear index machine in which the index rode at right angles to the platen was patented by Lee S. Burridge and Newman Marshman of N.Y. in 1885 and manufactured by the Sun Typewriter Co. of the same city. (See *Odell.*) Burridge then designed a moderately-priced front-stroke machine with a three-row keyboard and double shift which was patented in 1899 *et seq.* and introduced in 1901. Labelled the Sun No. 2, it enjoyed a long and successful career. Its distinguishing feature was that it inked from a roller against which the type rubbed on its way to the paper, and, in order to increase the effectiveness of the system, the roller itself replenished its ink-supply by striking against a larger roller that acted as a reservoir, this operation occurring each time a type-bar brushed past it. The system worked well. However, an unrecorded variation of the Model Two is in TMT: this uses a ribbon in place of the rollers, with vertical spools on both sides of the machine.

Larger and more sophisticated models of the Sun were introduced over the years —in all, nine were produced. A four-row model using a ribbon was introduced in 1907. The machine was extensively exported and appeared under the names *Carlem, Leframa, New Sun, Nova, Star.* (Fig. 190)

Surety: See *American (b)*.

Tacheografo: A small Italian shorthand machine invented in 1904 by Manlio Marzetti. It had eleven keys used singly or in chords, and printed Roman characters.

Tachigrafica: An improvement made in 1880 by an Italian mechanic called Bussadori of Mazzei's *Stenotiposillabica*, which met with no greater success than the original.

Tachigrafo Musicale: A musical typewriter invented by Angelo Tessaro of Italy in 1887. It was the first such machine to have been manufactured and sold.

Tangible: A machine for the blind was to have been produced by the Tangible Typewriter Co. in 1895 but this English venture did not succeed.[27]

Tarabout: A French shorthand machine invented in 1904 by Aristide Tarabout, which had a proliferation of keys and was designed to print entire syllables with the depression of a single key. Inking was by pad, the action of the type-bars being of the grasshopper type. Paper tape was used.

Taurus: A circular index machine, the size and shape of a large pocket-watch, this highly desirable object was actually manufactured in 1908 by Torrani & Co. of Milan. It printed on a narrow paper tape which passed beneath the characters and was pressed up against the index by depression of the pendant. Inking was by rollers. The characters were around the perimeter of the index which had a knurled outer rim for turning. The diameter of the machine was about $2\frac{1}{2}$ inches! (Fig. 189)

Taylor: An electric machine invented in 1910 by Joseph Taylor of Rochester was billed as the most perfect machine ever devised, but no manufacturer could be found for it.

Tell: See *Mitex*.

Thomas: A four-sided, square-section linear index with type on all sides was patented in US by H. J. Thomas of N.Y. in 1886. The index was set at right angles to the platen, and syllables and short words could be printed at one stroke.

Thuringia: See *Mentor*.

Tilden Jackson: Listed[42] but not described.

Tilton: C. E. Tilton of Mass. patented a linear plunger machine with ribbon inking in 1884. (Fig. 316)

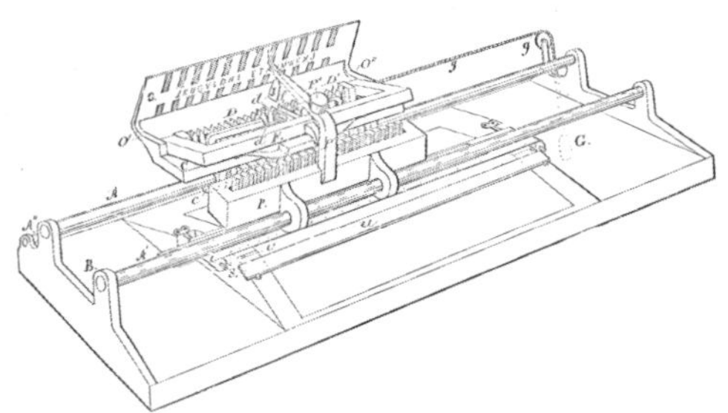

316. Tilton   (USP)

Tiphlotype: A machine for the blind, which embossed braille cells using six keys, invented by a Russian Count Kovaco and displayed at the Exhibition for the Blind in Amsterdam in 1885. Not produced.

Tipo-Stenografica: A shorthand machine invented in 1888 by Luigi Ranieri of Rome. It economized on keys by composing letters according to their component strokes (as on the first Foucauld machine).

Tip-Tip (a): See *Mignon.*

Tip-Tip (b): German patent from 1908, not described.[28]

Toepper: An experimental syllable machine using an index plate similar to the *Hall* was developed by Richard Toepper in 1895 but without success. A further syllable machine to be called *Blitz* was patented by him in Germany in 1896— this one, however, was of standard up-stroke design with three additional rows of keys for the common syllables and short words. It was apparently never manufactured.[6]

Toucheur Electrique: French machine listed[28] without details.

Travis: A type-wheel machine with four-row keyboard produced in 1905 by the Philadelphia Typewriter Co. after patents granted from 1896 to William H. Travis. Manufacture is said to have lasted for about six months.[22] (Fig. 317)

317. Travis   (GC)

Trebla: See *Darling.*

Triple Typewriter: Three Smith Premiers joined together and all operated from a single keyboard, thereby providing three simultaneous originals of anything typed. The model selected by the Cincinatti firm which is reported[27] to have performed this conversion is the double keyboard, up-stroke Model Two.

Triumph Perfect Visible: See *Triumph Visible.*

Triumph Visible: An oblique front-stroke machine with standard four-row keyboard, introduced in 1907 by Visible Typewriter Mfg. Co., Kenosha, Wisc. Also known as *Imperial, Imperial Visible* and *Triumph Perfect Visible.* It lasted only a few months, but the basic design, with minor changes to trim, was revived in

318. Imperial (Triumph Visible)   (MPM)   319. Burnett   (MPM)

1908 in the equally unsuccessful *Burnett*, produced by the Burnett Typewriter Co., Chicago. (Figs. 318 and 319)

Tucker: 'Tucker is a typewriter which is now no longer manufactured but I cannot establish whether it had a single key or a keyboard since the representative, Mr R. Stritter of Wiesbaden, absolutely refuses to favour me with any information on the matter.'[6] Those words, written in 1902, express Budan's rage at his inability to get information by mail. The present writer feels the same way about Smithsonian, Franklin Institute, and other American museums whose undeclared policy is quite clearly to ignore any specific requests for information coming from overseas.

Türk: A machine invented by a German called Türk, listed[6] but without details.

Typen: Perhaps the most primitive machine ever made, this type-wheel instrument was invented by Dr Henry Wetherill of Audubon, Pa. in 1924. The characters were around the edge of a wheel held vertically in a wire frame bent at the top to form a grip. A knob on the wheel turned it, inking by roller. That's all—yet the inventor claimed it took him fifteen years to evolve!

Type Writer: See *Sholes and Glidden Type Writer*.

Typo: See *Imperial (a)*.

Typograph (a): See *Addey's Typograph*.

Typograph (b): See *Cash*.

Typo-Sténographe: An 1877 patent for a shorthand machine with which vowels were represented as dots above and below the consonant: ba=$\dot{\text{B}}$, be=$\dot{\text{B}}$, etc. Invented by Smitter and Legros.

Uarda: German machine recorded[28] but without details.

Uhlig: Yet another of the fifty or so machines invented by Richard Uhlig, this one was to have been manufactured by Uhlig-Gunz Co. in 1910.

Ultima: See *Helios*.

Unda: Front-stroke machine with three-row keyboard and double shift invented in 1913 by Ing. Endemann. It was to have been manufactured in the following year, but this was postponed because of the outbreak of the war. It eventually appeared in 1921, produced by F. A. Bechmann GmbH, of Vienna.

Underwood: A conventional three-row portable model introduced in 1919. (Fig. 320)

320. Underwood   (TMT)

Union: A machine by this name is listed[42] among the 'unknowns' and research has revealed nothing conclusive. The only possibility is a three-row down-stroke patent granted to Lee Burridge in 1900 and assigned to the Union Type Writer Co. of New Jersey, which appears to have been unrelated to the Trust with the similar name.

Union Typewriter Company: Also referred to as 'the typewriter trust, about which so much has been said and so little known'.[27] It was formed in 1893 with a capital of $20,000,000. Remington was the principal member of the trust, which included Caligraph, Densmore, Smith Premier, Yost, Monarch, and Brooks. The trust fixed prices and pooled production and marketing resources but its members continued to give the appearance of competing against each other. A central board controlled all commercial and manufacturing activities of the member companies. A Yost representative was reported in the London *Express* to have admitted that the trust had 'kept prices at a high level for thirteen years'.[27] Anti-trust action apparently[28] broke up the trust in 1913 when Remington took over the entire operation.[31]

Universal: A syllable and word typewriter with a long history of woes, this machine never made it to the market place. A Dr Franz Joseph Müller founded the Universal-Silbenschreibmaschinen GmbH in Berlin in 1904 for the purpose of marketing a syllable machine from two patents granted independently in 1902 to Otto Rössler and Carl von Sudthausen. The first was a type-bar design, the second a swinging sector. A number of experimental models of the Universal were completed, using all manner of systems, but none was good enough to manufacture.

Universal Crandall: See *Crandall*.

U.S. Typewriter: The only record of this machine is a contemporary advertisement in LC in which it is illustrated. Manufacturer was C. Wing Machine Works,

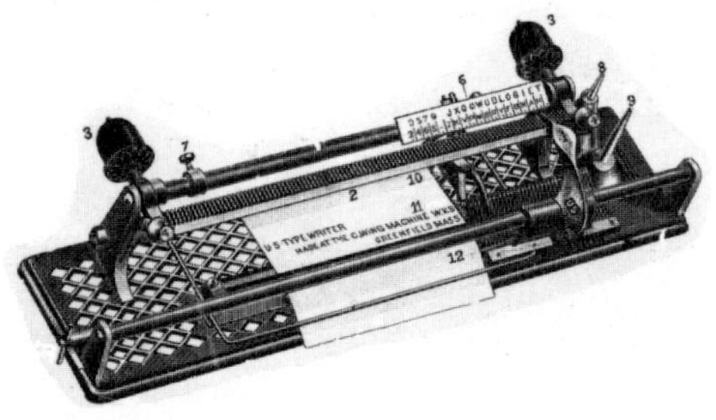

321. U.S.   (LC)

Greenfield, Mass., and the 1887 patent was granted to C. Wing of the same address. Type was carried on the ends of the teeth of a linear comb with character selection from the scale above. Roller inking. (Fig. 321)

Vari-Typer: See *Hammond.*

Vasanta: See *Meteor.*

Velograph: Circular index machine patented by Adolphe Prosper Eggis in 1886 and manufactured by Rymtowtt-Prince & Cie. of Geneva. It was the first Swiss typewriter. On the first model, the circular index, with rubber type underneath, was rotated by one of two knobs which served as indicators for upper and lower case respectively. Inking by pad. The disk was depressed to bring the type into contact with paper around a platen. A later model introduced in 1887 had certain improvements, including a central knob for rotation and depression. A further

322. Velograph   (CSM)

model, introduced in 1889 and also called *Commerciale,* was a linear index machine printing upper and lower case, similar in concept to the *Odell.* The Velograph was also offered as a cipher machine, printing a simple monalphabetic substitution

356

cipher by changing the position of the rubber type relative to the letter index on the upper disk. This model was also marketed as the *Eggis*. A further Swiss patent was granted to Czeslaw Rymtowtt-Prince in 1890. (Fig. 322)

Vélographe: A patent on a syllable machine permitting two keys to be depressed simultaneously was granted to the Frenchmen Riom and Carabasse in 1899. The machine had a keyboard of ninety keys with opening consonants and groups of consonants on the left, and vowels and final consonants on the right. The idea was that the typewriter would operate quickly enough to double for a shorthand machine. It was similar to the *Duplex* but was never manufactured.

Victor: A machine in which a vertical type-wheel was geared to a horizontal indicator, this device was patented in England by F. D. Taylor and J. A. White of Hartford in 1889 and was manufactured by the Tilton Manufacturing Co. of

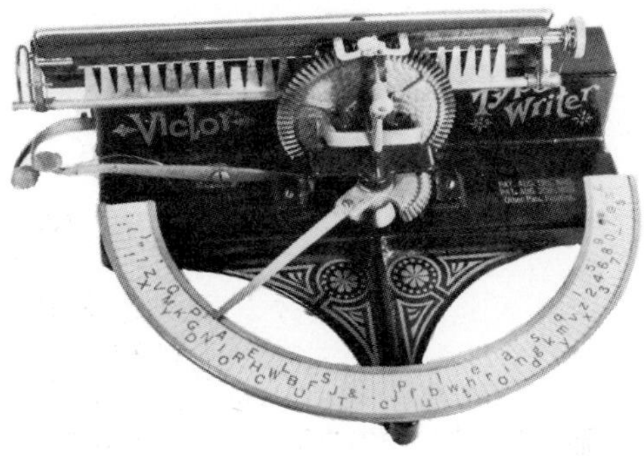

323. Victor   (LC)

Boston some years later. The gearing between indicator and type-wheel was such that half a revolution of the first produced a complete revolution of the second. Inking by pad. A hammer pressed the rubber type against the paper for printing. (Fig. 323)

Victoria (a): See *Gardner*.

Victoria (b): See *Famos (a)*.

Victorieuse: See *Gardner*.

Viktoria: Listed[42] but without details. It may be the same as *Victoria*.

Villey: A shorthand machine for the blind, invented in 1916 by Pierre Villey of France, himself blind from the age of three. It had twenty keys and printed on paper tape.

Virotyp: A miniature French machine built in 1914 for use primarily during the war. Invented by H. Viry, it consisted of a wheel with type-plungers around its perimeter beneath a circular letter index, and an indicator and knob with which to turn the wheel and select the characters. A strap on the left, for the left-hand little finger, and two hooks for the index and middle finger of the right made it possible to hold this machine firmly without resting it on a solid surface: the thumb and index finger of the left hand turned the knob while the thumb of the right depressed the printing key. Paper was gripped between two rollers beneath. It was also available with straps to attach it to the forearm for use on horseback and, later, was supplied with a base. The machine did not last on the market for long (Fig. 17).

Visbecq: An electric musical typewriter with a piano keyboard, invented in 1929 by Roger Visbecq.

Volks: See *Volksschreibmaschine.*

Volksschreibmaschine: A type-wheel machine manufactured in Germany by Friedr. Rehmann in 1898. An indicator selected the characters from a circular letter index mounted obliquely over a vertical type-wheel with rubber characters on its periphery, the indicator being geared to this wheel. Depression of a key on the left brought the type-wheel down on to the paper. Inking by roller. Also

324. Volksschreibmaschine   (DTU)

more briefly called *Volks.* An improved model, known as the *Diskret,* was introduced in 1899: this was also promoted as a cipher machine, the idea being that the letter index could be moved so that what was indicated did not correspond to the letter printed. The machine was also known as the *Discreet.* (Fig. 324)

Voyante: French typewriter, invented in 1898, with a five-row keyboard, each row controlling characters on a separate row on the type-wheel. It was never manufactured.

Wagner: Franz X. Wagner, the inventor of the Underwood typewriter, led an active and inventive life (see Chapter Eight) which began with developmental work on the Sholes and Glidden. His first patent, granted in 1880, was for a swinging sector machine with a four-row keyboard, assigned to Yost. An 1884 patent covered a swinging segmental comb design straddled by a four-row keyboard; the following year, he was granted a patent on an up-stroke design in which the type-bars rocked to select upper or lower case. This was assigned to Henry Harmon Unz (see *National (b)*) and Stephen Smith. An 1888 patent protected a type-wheel design with a linear index, after which he turned his attention to front-stroke inventions.

Wagner and Schneider: A Swiss firm of this name marketed several models of a machine for the blind in 1888. It used a horizontal circular index with the letters above and the characters to be printed or embossed below: it was offered either with Roman characters above for embossing braille cells (thus permitting the sighted to correspond with the blind) or vice-versa. Alternatively, a model with braille at both ends was available.

Waitt: A machine with radial flexible arms was patented by G. L. Waitt and assigned to the Edison Type Writer Co. of Philadelphia in 1887. (Fig. 325)

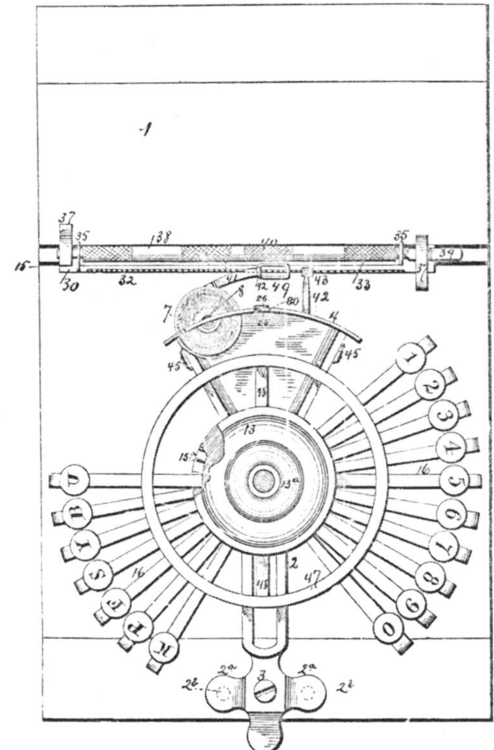

325. Waitt   (USP)

Walker: An invention described[42] as a noiseless machine which was also ribbon-less, basketless and with a motionless carriage was made by C. Wellington Walker of Stamford, Conn. in 1910. He is reported to have sold his rights to the invention in 1912, but nothing further was heard of this interesting machine.

Wallace: A three-row double shift machine invented by a professor of this name in 1912. Not manufactured.[42]

Wall Street Standard: See *Daugherty.*

Wanamaker: See *Wellington.*

Ward: Up-stroke machine with a circular two-row keyboard and centrally located platen, patented in 1878 and 1879 by Caleb T. Ward of N.Y. Roller inking. Not manufactured.

Waverley: A down-stroke machine patented in England in 1889 by Edward Smith Higgins and Henry Charles Jenkins and manufactured by the Waverley Typewriter Co. It was one of the few designs on which the type-bars stood in an

326. Waverley   (SML)

arc behind the platen (as on North's and Brooks). With a four-row standard keyboard, it inked first by roller but almost immediately after its introduction switched to ribbon. It incorporated differential spacing. Manufacture ceased in 1897. (Fig. 326)

Webster Miniature: An ambitious projected devised in 1898 by Joseph March Webster of Liverpool, who built a tiny visible portable machine with a universal keyboard plus a few special keys which printed selected phrases like 'Yours very

360

faithfully', 'Yours truly' etc. by a single depression. Dimensions, in leather case: 8″ by 6″ by 2″. It was not manufactured.

Wellington: Best and most popular of the thrust action machines, this invention of Wellington Parker Kidder was sold throughout the world, manufactured under licence in many countries, and extensively copied. Patented in 1892, it was not the first of the thrust machines nor did Kidder invent the action, but he was certainly responsible for its dissemination and popularity.

The machine was produced by the Williams Manufacturing Co. in Montreal as the *Empire* and in Plattsburg, N.Y. as the *Wellington.* In Germany, it was made under licence as the *Adler,* and was offered in various markets under the names *Davis, Wanamaker, British Empire,* etc. A machine in TMT is labelled *Lindeteves.*

A three-row keyboard with double shift operated the type-bars, which rested horizontally on a steel platform on which they slid as they were thrust against the paper. They were radially disposed in a segment of an imaginary circle with its centre at the printing-point, standing vertically on their ends under the upper cover of the machine. The design was noted for its relatively noiseless operation free from hammering. The shift key altered the level of the carriage. Printing was through a ribbon the width of the type-head. Later models included heavier office machines (1908/9) and lighter portables (1916), but the basic design remained unchanged. (Fig. 165)

Weltblick: See *Blickensderfer.*

Westphalia: An early German linear index machine patented in 1884 by Ernst Wilhelm Brackelsberg. The characters were on type plungers set in a metal bar beneath a letter index, with a handle with which the bar could be slid to select the desired character. This was similar to the arrangement on the later and better-

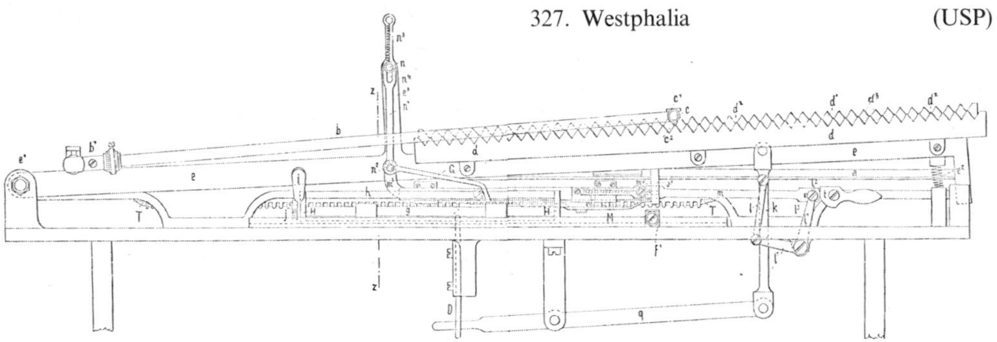

327. Westphalia   (USP)

known Odell etc. On the Westphalia, however, teeth provided alignment and the distance between them varied according to the width of the character to which they corresponded, thereby providing differential spacing. Double-sided carbon paper was used for printing—by placing a transparent sheet over the carbon paper, the printing was rendered visible. Only very few examples of this machine were made. (Fig. 327)

Whyte: A vertical platen machine, patented by J. G. Whyte in England in 1903. Its platen revolved for letter spacing and rode up a gentle spiral groove so that the printing was continuous, this being a vertical application of Wheatstone's (and others') idea.

Wilkins: Timothy Wilkins of Iowa placed a *Hall*-type machine in a frame with racks in both horizontal directions and made a book typewriter out of it. US patent granted in 1885.

Williams: This unusual machine, famous for its novel grasshopper movement, was first patented by John Newton Williams of N.Y. in 1890, after developmental work dating back to patents from 1875 onwards. It was marketed in 1892 when first the Domestic Sewing Machine Co. and then the Williams Typewriter Co. of Derby, Conn. made it.

  The machine was characterized by its two segments of horizontal type-bars which radiated in fan shape from the printing point on both sides of the platen. The type rested on a pad, and depression of a key caused the bar to hop from the pad to the paper and back again when the key was released. Since there was nowhere for the paper to go in such a design, two open-framed cylindrical baskets were provided beneath the platen on both sides, so that the paper had to be rolled into one, passed over the platen, and unrolled into the other as typing progressed.

  Model One had a curved three-row keyboard which was replaced within a year by a straight three-row. The only main design departure in successive models was Model Four with four-row standard keyboard, introduced in 1900. The company was liquidated in 1909. (Fig. 169)

Wilson: See *Meteor.*

Wilson & Torrey: Listed[6] but without details.

Wirt's Typewriter: An American machine designed in 1904 to print letters and musical notes to a total of 280 different characters, from a keyboard consisting of a mere twenty-eight keys. A sixteen-sided type-sleeve was used, with a hammer striking paper against the character. Ribbon inking. The sleeve was mounted on a horizontal axis so that its rest position was half-way along its total travel, for it was made to slide in either direction on its axis, and rotate as well. Shift keys governing its action were provided. The outcome of the invention is disputed: it was either never heard of after its initial announcement[28] or else it monopolized what there was of the musical typewriter market.[27]

World: An inexpensive swinging sector machine which enjoyed considerable success over a period of ten years or so was patented in 1886 by John Becker of Boston and was produced by the World Type Writer Co. of Maine, to whom the patent was assigned, later by the Pope Manufacturing Co. and the Typewriter Improvement Co. of Boston. The machine consisted of a semi-circular disk, to the curved edge of which was attached a strip of rubber type. The characters were reproduced on a semi-circular index and selected by means of a pointer fixed to the disk. Holes corresponding to each letter secured alignment, and printing was performed by depression of a bar which forced the rubber type on to the paper, inking by pad. Model One printed only capitals while Two, introduced in 1890, was for both upper and lower case. It was also marketed under the name *Boston*, and, apparently[6] *America*. (Fig. 23)

World-Blick: See *Blickensderfer*.

World Flash: An outfit called The World Flash Co. of Chicago is reported[27] to have taken out a patent on a machine similar to the Remington (which model is not specified, nor is the date) but electrically operated from a battery.

Worrall: A machine with pivoted horizontal type-levers with keys at one end and the type on the underside at the other was patented by T. D. Worrall of Washington in 1886 and assigned to Worrall Manufacturing Co. of Boston. This is the 'unknown' machine of this name sometimes[6 etc.] listed and illustrated.

Worth: A sole reference to this name[2] dates the machine 1892, but no other details.

Write Easy: See *Gundka*.

Writing Ball: See *Skrivekugle*.

Xcel: Second attempt by W. H. Bennington to market a syllable typewriter. Unlike the *Bennington* of 1903, the machine was now of front-stroke design, with a standard four-row keyboard plus an additional row of keys for common syllables and short words. The Xcel Typewriter Corporation was last heard of in 1922, the same year in which it was formed.

Yale: See *Munson*.

Yeremias: An electric machine invented by the Hungarian Arnold Yeremias in 1899. It is described[6] as a visible machine of great simplicity, with key depression of 1 mm. The first model had a simple keyboard, a second had double keyboard. Not produced.

Yost: Produced by George Washington Yost after he severed his relationship with the American Writing Machine Co., the instrument bearing his name first appeared in 1887. It was developed by Alex Davidson, Andrew Steiger and Jacob Felbel, the latter two of whom filed a patent that same year (granted 1889). The machine owed its novel up-stroke grasshopper action to a desire on the part of the inventors to produce a unique and patentable design.

The type-bars rested against a circular inking pad beneath the platen and hopped up against it upon depression of a key, falling back into place as it was released. A double keyboard was provided. Model Four: 1895, Model Ten: 1902 were virtually identical.

The first major design departure was Model Fifteen: 1908, which had a standard four-row keyboard and single shift replacing the double keyboard; it was of front-stroke visible design. However, the type-bars still rested on a curved inking pad and hopped up against the platen with the characteristic movement. Model Twenty: 1912 was almost identical. The Model One was also known as *New Yost* and the Model Fifteen as Model A.

The Yost was widely exported to Europe and elsewhere, where it enjoyed more success longer than in the United States. It was first manufactured in the Merritt factory in Springfield, Mass., later by the Yost Typewriter Co. of Bridgeport, Conn. (Figs. 152 and 191)

Young: J. L. Young of N.Y. patented a vertical type-wheel geared to a horizontal circular indicator in 1883.

Yu Ess: See *Mignon.*

Zerograph: Invented by Leo Kamm of London with patents from 1895. See page 225, Figs. 208 and 209.

Chronological List of all Inventions to 1867

(The first year in which a machine is on record is the only one quoted. Later models by the same inventor will be found described in the body of the text. At times, a machine was built years before it was eventually patented or otherwise made public—again, the earliest date is the one listed.

An asterisk denotes an entry that is either erroneously included in secondary sources or is otherwise to be considered of dubious authenticity. It is proposed that henceforth all those inventions now known to have been incorrectly included in past studies be dropped from future ones, and those that are merely speculative be laid aside until corroborating evidence can be located.)

Chapter Two		*Chapter Three*	
* ?	Le Roy	1808	Turri
1714	Mill	1808	Schönfeld
*1730	Sanderson	1822	Lechet
1745	Unger	1823	Conti
*1745	Cred	1827	Gonod
*1749	Carmien	1829	Burt
1753	von Knauss	1830	Galli
*1762	von Neipperg	1831	Drais
*1771	Hohlfeld	1833	Progin
1771	Payen	1835	Morse
1772	Jaquet-Droz	1836	Berry
*1779	Rochon	1837	Bidet
*1779	von Kempelen	1837	Vail
*1780	Pingeron	*1838	Knie
1782	Leschot	1838	Davy
1782	Maillardet	1838	Steinheil
1783	Neussner	1838	Dujardin
*1784	L'Hermina	1839	Perrot
*1784	Jenkins	1839	Foucauld
1790	Paris	*1840	Jaackson
1806	Bramah	1840	Bain

<table>
<tr><td>

1841   Wheatstone
1841   Baillet de Sondalo & Coré
*1843   Kohl
1843   Thurber
1844   Labrunie de Nerval
1844   Pape
*1844   Spencer
1844   Littledale
1845   Brett
1845   House, R. E.
1845   Saintard & Saint-Gilles
*1845   Leavitt
*1847   Prentice
*1847   Froment
*1847   Rohlfs & Schmidt
1847   Beach
1849   Bianchi
*1849   Sorensen
1850   Eddy
1850   Siemens
1850   Marchesi
1850   Wight
1850   Fairbank

Chapter Four
1850   Hughes, G. A.
*1851   Levitte
*1851   Larivière
*1851   Hirzel
*1851   Ehlwein
1851   Tollputt
1852   Jones
1854   Thomas
1854   Devincenzi
1855   Ravizza
1855   Clément
1855   Hughes, D. E.
1856   Cooper
1857   Francis
1857   Hood
*1857   Johnson
1858   Harger
1859   Guillemot

</td><td>

1860   Cox
1861   Codvelle
1861   de Avezedo
1862   Martin
1862   Michela
1863   Pratt
1863   De Mey
1863   Livermore
1863   Flamm
1864   Mitterhofer
1864   Halstead
1864   Peters
1864   Bryois
1864   Danel-Duplan
1865   House, G.
1865   Smith
1865   Hansen
1866   Gensoul
1866   Dilliès
1866   Peeler
1866   Hall
*1866   Sweet
1867   Sholes
1867   Lamonica
1867   Fontaine
*1867   Worral
*1867   Allen
*1867   Ferreira
*1860–70   Anonymous (Cookson)

</td></tr>
</table>

Scoreboard of Early Typewriter Inventions

First typewriter patent	Mill	Br.	1714
First machine known to have existed	Turri	It.	1808
First typewritten letter	Turri	It.	1808
First typewritten manuscript	Turri	It.	1808
First Italian machine	Turri		1808
First French machine	Gonod		1827
First American machine	Burt		1829
First German machine	Drais		1831
First British machine	Berry		1836
First type-bar design	Conti	It.	1823
First type-sector design	Burt	US	1829
First type-plunger design	Galli	It.	1830
First type-wheel design	Bidet	Fr.	1837
First machine for the blind	Turri	It.	1808
First shorthand machine	Gonod	Fr.	1827
First printing telegraph	Morse	US	1835
First keyboard	Conti	It.	1823
First feed rollers	Gonod	Fr.	1827
First upper and lower case	Burt	US	1829
First visible typing	Progin	Fr.	1833
First platen	Bidet	Fr.	1837
First ribbon	Bain	Br.	1840
First manufacture	Hughes Foucauld	Br. Fr.	1850s
First production figures	Jones	US	1852
First commercially successful	Hansen	Den.	1870

Transcription of Densmore Ledger Statement

I have not kept a positively accurate account of the whole cost, but in round numbers when I went to New York, in September of last year, under an arrangement whereby Mr. (or our) John Hoskin and John E. Waring were to try to sell the exclusive control of the whole thing, as nearly as I could calculate, there had been expended by Mr. McMoney (?), up to that time, $10,000. Of this, I had furnished $9,000 and Mr. Sholes $1,000; but this last $1,000 was only borrowed by me of Mr. Sholes. I am to run all the money risk, and am to pay him the $1,000, whether I ever get it out of the invention or not.

And last autumn, I told my brother Amos that it had cost that amount up to that time—that is, $10,000, or ten thousand dollars, and that he might have an interest in it at cost.

Since then, that is, some time last month, assigned to him and to my brother Emmett are undivided one-tenth interest each, in the whole (—?—) at cost up to that time; that is, at one tenth each of the whole cost of ten thousand dollars (up?) to first of October, eighteen hundred seventy one or one thousand dollars each at that time.

And it is because of this arrangement and the general propriety of it, that I have got these books to keep henceforth an accurate account hereafter of the enterprize.

I have named the invention the 'Type Writer' and the organization or company owning it, I (think?) better be called the 'Type Writer Company'; and accordingly, I have placed that name at the head of this statement, and shall herein keep the accounts of the enterprise in that name.

My thought is, ultimately, to have a (corporate?) organization under the law in that name for the invention; and to have another (corporate?) organization under the law, to make and sell (—?—) in the name of the 'Type Writer Manufacturing Company. [Quotation marks not closed.]

I have deemed this statement preliminary to the opening of these books, a proper one to be made, as a brief history of the condition of the invention and the enterprize at this time.

There have, up to this time, been three patents issued on the original invention —two to Sholes, Glidden and Soule, and one to Sholes. Besides these one has been applied for in the name of Sholes and Matthias Schwalbach—an ingenious (*sic*) German mechanic who has made every instrument thus far, and who has

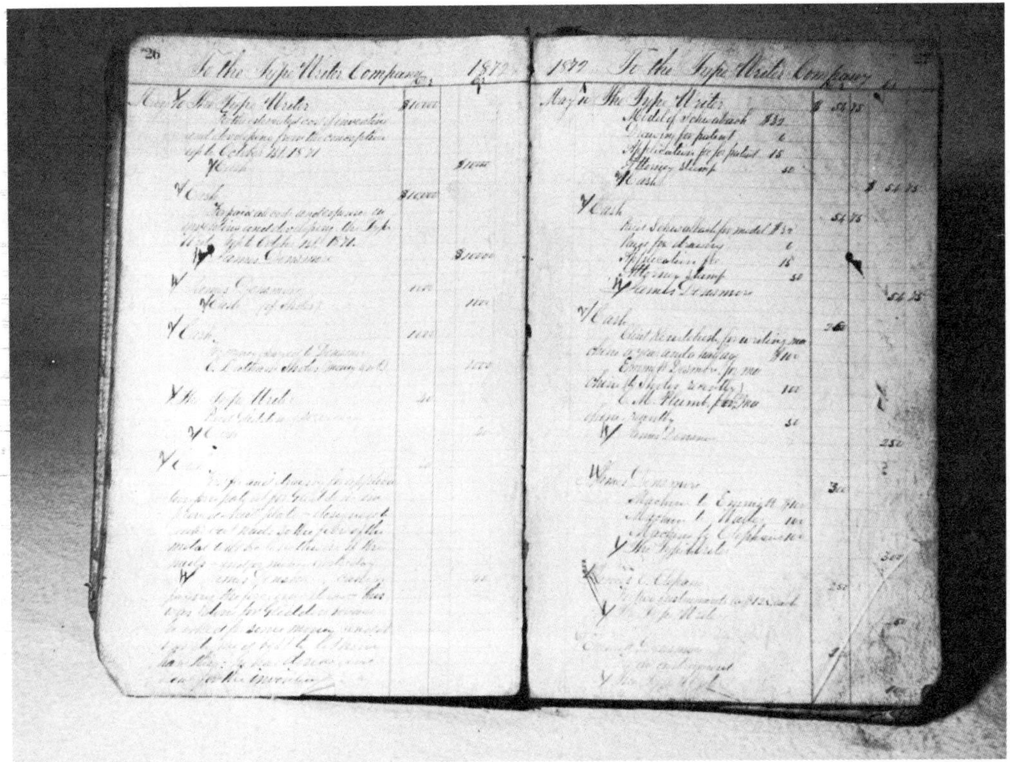

328. A precious document indeed is this ledger
opened by James Densmore in 1872 during the
final stages of development of the Type Writer.
It reads like a VIP roll call and virtually everyone
gets a mention: Glidden for services, Sholes for
a loan, Schwalbach for models, Clephane for
purchases, Walter Barron for history's first
recorded typewriter service call, and, of course,
the Densmore brothers themselves.   (GC)

contributed several good suggestions.

There are yet several other patents to be applied for—some in Sholes' name alone, and several in the name of Sholes and Glidden.

The interest assigned to Amos Densmore and Emmett Densmore are not absolute individual interests in the invention, ~~but~~ (*sic*) to be treated and used independently of the remaining interests; but is only to each one one-tenth interest in all income or profits growing or coming thereout.

James Densmore

(Several blank pages)

I have deemed this statement preliminary to the opening of these books, a proper one to be made as a brief history of the condition of the invention and enterprize at this time.

There are now three patents out, on the matter—two in the names of Sholes, Glidden & Soulé, and one in the name of C. Latham Sholes. There is another applied for in the name of C. Latham Sholes and Matthias Schwalbach. Mr. Schwalbach is the ingenius German mechanic who has made every instrument yet made and who has contributed several good ideas to the invention.

There are others yet to be applied for in Sholes' name, and also in the name of Sholes and Glidden.

(Statements of accounts receivable and payable follow.)

Abbreviations used in Text

AC Author's Collection
AMP Arts et Metiers (Musée de Techniques C.N.A.M.), Paris
BA Berliner Akademie, (East) Berlin
BC August Baggenstos Collection
BP British Patent
CSM Crown Copyright, Science Museum, London
DTM Danish Technical Museum, Helsingor
DTU Dresden Technical University, D.D.R.
FIP Franklin Institute, Philadelphia
FP French Patent
GC Dave Golden Collection
IP Italian Patent
ITC Imperial Typewriter Collection
LC Paul Lippman Collection
MHN Museum of History, Neuchatel
MPM Milwaukee Public Museum
PNY Perkins Museum, New York
QC Ed Quiring Collection
SAR State Archives, Reggio Emilia
SC Don Sutherland Collection
SIW Smithsonian Institution, Washington
SML Science Museum, London
STM Science and Technology Museum, Milan
TMT Typewriter Museum, Tilburg
TMV Technical Museum, Vienna
USP United States Patent
WC Robin Wyatt Collection

Index numbers throughout the text refer to the Bibliography.

Index

Numbers in italics refer to pages on which illustrations are located